A TRIBUTE TO
JESSE DOUGLAS & TIBOR RADÓ

THE PROBLEM OF PLATEAU

Published by

World Scientific Publishing Co. Pte. Ltd.

P O Box 128, Farrer Road, Singapore 9128

USA office: Suite 1B, 1060 Main Street, River Edge, NJ 07661

UK office: 73 Lynton Mead, Totteridge, London N20 8DH

Library of Congress Cataloging-in-Publication Data is available.

THE PROBLEM OF PLATEAU
— A Tribute to Jesse Douglas and Tibor Radó

ISBN 981-02-0556-2

Printed in Singapore by Stamford Press Pte Ltd

A TRIBUTE TO
JESSE DOUGLAS & TIBOR RADÓ

THE PROBLEM OF PLATEAU

Editor
Themistocles M. Rassias

World Scientific
Singapore • New Jersey • London • Hong Kong

Joseph Plateau (1801 – 1883)

PREFACE

This volume consists of articles written by eminent scientists from the international mathematical community, who present important research works concerning the problem of Plateau after its classical solution was given by Jesse Douglas and Tibor Radó some sixty years ago. These works provide insight and perspective on various problems associated with minimal surfaces and surfaces of prescribed mean curvature in which some exciting new results have been obtained in recent time.

The first section of the volume is devoted to some recollections written by several scientists, who have been colleagues or students of Douglas or Radó or have just communicated in some way with them. They describe their personalities, their friends and the Institutions where they worked in different countries, the environment and consequences of their extraordinary creativity.

The second section is devoted to articles which explore the relationship between Analysis and Geometry with main emphasis on concepts and methods that lead to a recent account of results in Global Variational Analysis.

Jesse Douglas was born in New York on 3 July, 1897 and died there on 7 October, 1965. His work covered several subjects in Function Theory, the Calculus of Variations, Differential Geometry, and Abstract Algebra.

The first important contribution of Douglas concerned the problem of Plateau. The problem was first posed by J.-L. Lagrange (1736–1813), and had aroused the interest of several mathematicians, including B. Riemann (1826–1866), K. Weierstrass (1815–1897), and H. A. Schwarz (1843–1921). The problem requires a proof of the existence of a surface of least area bounded by a given contour in space. The problem is named after Joseph Plateau (1801–1883) who performed experiments on certain shapes with soap bubbles in 1847. The problem of existence and uniqueness resisted solution until around 1930 when it was solved (independently) by J. Douglas and T. Radó. The paper "Solution of the problem of Plateau" by J. Douglas appeared in *Trans. Amer. Math. Soc.* **33** (1931), 263–321. For this work, Douglas received one of the first two Fields Medals ever awarded, at the International Congress of Mathematicians in Oslo in 1936. (The other medal was received by Lars V. Ahlfors, at the same Congress). The selection committee consisted of G. D. Birkhoff, Elie Cartan, C. Carathéodory, F. Severi, and T. Takagi. It was Carathéodory who presented the works of both medallists. In his laudatio, Carathéodory

[Bericht über die Verleihung der Fieldsmedaillen an L. V. Ahlfors und J. Douglas (Comptes Rendus du Congrès International des Mathématiciens, Oslo 1936, vol. 1, 308–314, Oslo, Brøggers (1937))] emphasized the novelty of Douglas' method and the fact that Douglas related Plateau's problem to the Dirichlet integral showing that it suffices to consider harmonic surfaces, and transformed the Dirichlet integral into a functional A(g) which was essential to his approach. Carathéodory compared this approach with the solution by Tibor Radó, saying: "Nach einer ziemlich langen Entwicklung, an der viele Mathematiker teilgenommen haben konnte Tibor Radó im Jahre 1930 das Randproblem der Minimalflächen und das damit zusammenhängende Plateausche Problem für alle geschlossenen räumlichen Jordanschen Kurven lösen, durch welche man eine Fläche von endlichem Flächeninhalt hindurchlegen kann" (After a rather long development, in which many mathematicians participated, in 1930 T. Radó was able to solve the boundary value problem of minimal surfaces and Plateau's problem connected with it for all closed Jordan curves in space which bound a surface of finite area.)

Douglas was again rewarded for his work on the problem of Plateau in 1943. He received the Bôcher Memorial Prize of the American Mathematical Society for his memoirs: "Green's Function and the Problem of Plateau" (*Amer. J. Math.* **61** (1939), 545–589), "The Most General Form of the Problem of Plateau" (*Amer. J. Math.* **61** (1939), 590–608), and "Solution of the Inverse Problem of the Calculus of Variations" (*Trans. Amer. Math. Soc.* **50** (1941), 71–128).

In addition to publishing several important papers on analysis and geometry, in 1951 Douglas contributed significant work in group theory, especially to the problem of determining all finite groups with two independent generators, a and b, such that every group element can be expressed in the form $a^r b^s$, where r and s are integers.

Tibor Radó was born in Budapest on 2 June, 1895, and died in New Smyrna Beach, Florida, on 12 December, 1965. Radó made fundamental contributions to several fields, notably to Complex Analysis (to the proof that a Riemann surface can be triangulated, etc.), Real Analysis and Measure Theory and the Calculus of Variations (continuous transformations, generalized Jacobians, surface area, etc.), Subharmonic Functions, Partial Differential Equations, Point Set and Algebraic Topology, and Logic. During the period 1928–1929, Radó was awarded an international research fellowship of the Rockefeller Foundation to study at Munich. In Munich, Radó met Carathéodory on

20 June, 1929 and later again on 12 June, 1930* Radó's most important work was concerned with Plateau's problem [See his paper "The Problem of the Least Area and the Problem of Plateau" (*Math. Z.* **32** (1930), 763–796) and his brilliant Springer Ergebnis-Bericht "On the Problem of Plateau" (Berlin, 1933)].

An earlier, rather complicated (and less general) solution of Plateau's problem by R. Garnier (1928) was simplified by Radó, and E. J. McShane, who published another solution in 1933, based on measure and integration, gave credit to Radó for introducing him to Lebesgue's ideas and techniques. New solutions were given later by R. Courant in 1936 and a generalization of the problem of Plateau on a Riemannian manifold by C. B. Morrey in 1948. Afterwards A. S. Besicovitch, as well as H. Federer, obtained a new existence proof for a surface of the type of the disc of smallest area bounded by an arbitrary closed curve, under the hypothesis that the closed curve bounds some surface of finite area. In 1956 W. H. Fleming was able to construct a Jordan curve which bounds no surface of smallest area and of finite topological type. In 1939 M. Morse and C. Tompkins, as well as M. Shiffman (independently), applied Morse theory to find a minimal surface with a prescribed boundary such that the surface gives neither a relative nor an absolute minimum to the surface area. Morse and Tompkins made use of the thesis by H. Lebesgue and the works of Douglas and Radó on the minimum problem, whereas Shiffman used Courant's papers. In 1964 R. Palais and S. Smale discovered a useful condition (a compactness condition on the function, called "Condition C" by them), which allowed them to apply the Morse theory to a wide variety of variational problems and deduce theorems related to the existence of solutions of problems in non-linear elliptic partial differential equations.

I would like to acknowledge special thanks to the help extended by the contributing authors in preparing this volume. On a personal level, the early inspiration and interest in the field engendered by Professor Stephen Smale of the University of California at Berkeley, who introduced me to the subject of minimal surfaces via a thesis assignment. In addition thanks are due to Professor Simon Donaldson for inviting me recently to visit Oxford University for discussing some Global Analysis Problems. I also would like to acknowledge the superb assistance that the staff of World Scientific Publishing Company has provided in the preparation of this publication. The continued support of the Director of European Programs Craig Sexson and Associate Dean Jeff

* as related by his daughter, Despina Carathéodory-Rodopoulou about Carathéodory's office sign-in book for visiting mathematicians..

Nonemaker, both of the University of La Verne in Athens, has made this volume possible. Their support and encouragement has been invaluable during the last ten years of my association with the University. Last but not least I wish to express my wholehearted thanks to my wife Ninetta and my children Matina and Michael who played such an important role with their patience, understanding and active support during the preparation of this volume.

Athens, Greece
May, 1992

Themistocles M. Rassias

TABLE OF CONTENTS

Preface v

Part I

Joseph Plateau and His works
Carl C. Grosjean and Themistocles M. Rassias 3

Remarks on the Mathematical Work of Tibor Radó
Erwin Kreyszig 18

The Yin and the Yang of My Relationship with T. Radó
Maxwell O. Reade 33

Vague Recollections of a Young C.C.N.Y. Poor-In-Mathematics Undergraduate Student of Jesse Douglas, an Old Famous Mathematics Professor
Edward Siegel 35

Jesse Douglas as Teacher and Colleague
Fritz Steinhardt 37

My Recollections of Jesse Douglas
Dirk J. Struik 41

Jesse Douglas, the Plateau Problem, and the Fields Medal: Some Personal Reflections
Henry S. Tropp 43

Part II

Classifying Pseudo-Riemannian Hypersurfaces by Means of Certain Characteristic Differential Equations
Luis J. Alias, Angel Ferrandez and Pascual Lucas 53

Affine Minimal Higher Order Parallel Affine Surfaces
Franki Dillen and Luc Vrancken 76

Cartan's Method and Plateau's Problem
Jerry Donato 87

Critical Point Theory and Multiple Periodic Solutions of Conservative Systems with Periodic Nonlinearity
Alessandro Fonda and Jean Mawhin 111

On the Uniqueness for Hypersurfaces with Constant Mean Curvature in R^{n+1} Bounded by a Round $(n-1)$-Sphere
Miyuki Koiso 129

On the Theory of Minimal Surfaces
Erwin Kreyszig 138

Area-Minimizing m-Tuples of k-Planes
Gary Lawlor 165

Morse Index of Complete Minimal Surfaces
Shin Nayatani 181

Removable Singularities of Stationary Fields
Thomas H. Otway 190

Second Variation Formulas for Willmore Surfaces
Bennett Palmer 221

A Jordan Arc in R^m with Positive m-Dimensional Lebesgue Measure
Harold R. Parks and Richard M. Schori 229

Some Problems and Remarks on the Eigenvalues of the Laplacian and Minimal Surfaces
Themistocles M. Rassias 237

Recent Developments on the Structure of Compact Surfaces with Planar Boundary
Ricardo Sa Earp 245

The Parametric Plateau Problem and Related Topics
Gerhard Ströhmer 258

The Role of Minimal and Rigid Surfaces in Theoretical Physics
Victor Tapia 285

On the Number of Rigid Minimal Immersions Between Spheres
Gabor Toth 327

PART I

The Problem of Plateau (pp. 3–17)
ed. Th. M. Rassias

JOSEPH PLATEAU AND HIS WORKS

CARL C. GROSJEAN
Laboratorium voor Numerieke Wiskunde en Informatica
Rijksuniversiteit–Gent
Krijgslaan 281, gebouw S9
B–9000 Gent, Belgium

and

THEMISTOCLES M. RASSIAS
Department of Mathematics
University of La Verne
P.O. Box 51105, Kifissia
Athens, Greece 14510

Plateau, Joseph–Antoine–Ferdinand, was born in Brussels on the 14th October 1801. Raised in an artistic environment (his father, a native of Tournai, had a remarkable talent for painting flowers), young Joseph was able to read at the age of six. He made rapid progress in primary school. One day, after having attended a session of "physique amusante", popular in the nineteenth century, he promised himself that sooner or later, he would try to unveil the secret of many mysterious facts which he had observed. He was sent to the Academy of fine arts by his father, evidently to continue the artistic tradition of his family. Unfortunately, he soon became an orphan, as he lost his mother when he was only thirteen, and his father at the age of fourteen. Together with his two sisters, he was taken in by his maternal uncle, master Thirion, a lawyer. After a serious illness, Joseph resumed the courses at the Academy. During the day, he attended with great success the lectures of an exquisite teacher named Van der Meulen, and in the evening, he diverted himself by carrying out experiments of entertaining physics. He constructed all his apparatus with his own hands and often surprised the audience, both by his amazing skill and the originality of his instruments.

At the age of sixteen, he started secondary school at the Athenæum of Brussels where he was fortunate enough to have teachers such as Vautier and Quetelet with whom he soon became on excellent terms. It is from these years onward that the affection of Adolphe Quetelet for his disciple Plateau became steadily more profound as time went on. It is also in that period that he made friends with young Verhulst who later became known as a mathematician and statistician, and with his class-mate Nerenburger who became an army general. Their scientific discussions contributed to their mutual education, for instance, it was the latter who conveyed to young Joseph the interest in astronomy. With Quetelet's permission, they often spent part of the night at the National Observatory.

After having brilliantly terminated his secondary studies in 1822, Joseph was morally obliged, at the suggestion of his tutor, to enter the University of Liège and to have himself registered as a student in the Faculty of literature and philosophy, despite his dislike for these disciplines and also for law. The next year, nonetheless, he acquired successfully his bachelor's degree in literature and entered the Faculty of Law. But, one day, having sneaked into the chemistry class-room out of mere curiosity, he started to feel an irresistable attraction toward the observational sciences come up in him. This gave him the necessary will-power to obtain his bachelor's degree in law, whereafter he got himself admitted as a candidate in physics and mathematics. Since then, he was able to devote himself plainly to these, his favourite disciplines.

Having become a tutor to his sister Joséphine, he settled definitively in Liège, and in order to provide for the subsistence of the family, Plateau accepted in 1827 a position as a teacher in mathematics at the Athenæum of this city. He managed to fulfil this task adequately while preparing at the same time his doctorate's thesis. On the 3rd June 1829, he acquired the doctor's degree in physical and mathematical sciences. His thesis was a remarkable study of the properties of the impressions which light can exercise on the eye. Unfortunately, the same year, his ardour for experimentation pushed him to carry out a very dangerous experiment consisting in looking directly into the bright sun during approximately twenty-five seconds. A long time after that imprudence, his eyes were still irritated. In 1830, his somewhat precarious health forced him to resign as a teacher at the Athenæum. He left Liège to return to Brussels, and there a new position was soon offered to him by a certain Mr. Gaggia, director of an institute of high standing. Plateau was engaged as a professor and before long, he succeeded in conquering the affection of his students. He also managed to occupy a distinguished rank within a pleiad of young scholars whom Quetelet liked to surround himself with, trying to direct them towards scientific careers. It was with the support of his influencial protector that in 1835, Plateau got nominated as a professor of physics at the University of Ghent where the vehicular language was French in those days. According to a declaration of one of his best students, Hubert Valerius, his teaching had a particular cachet : his way of talking was simple, straightforward and very clear. At the same time, he was a talented experimenter, each experiment having the same degree of rigour as his talking. Besides his lecturing, Plateau invested a good deal of efforts in promoting science. He spent much time to complement the apparatus of the physics laboratory and to adapt them to the most recently accomplished progresses.

On the 27th August 1840, Plateau married Miss Augustine–Thérèse–Fanny Clavareau, daughter of a director of the exchequer. From this marriage, three children were born : Félix who also became a professor at the University of Ghent and a member of the Royal Academy, Ernest who became an engineer at the Belgian Railroad Administration, and Alice who married G. Van der Mensbrugghe.

Around 1841, the brilliant professor underwent the first attacks of an inflammation of the choroid, undoubtedly a consequence of the fatal experiment of 1829. The disease gradually spread to both eyes, worsened during two years in spite of the most assiduous attendence and in the course of the year 1843, the blindness became total. But, as the illustrious Faraday declared later, "although bodily Plateau remained plunged in the darkness of a sad, profound night, the perspicacity of his mind, having become more intensive than ever, was to lead to the most brilliant discoveries and to conquer for Belgian science an immortal glory".

In 1828, in the days that he was still preparing his doctor's degree at the University of Liège, Plateau published an article about the impressions effectuated in the human eye by different colours. The same year, he studied the appearances resulting from two lines each rotating around a fixed point with a uniform angular velocity. He had invented in this manner a novel procedure to offer to the eye the spectacle of a variety of plane curves. In his outstanding thesis of 1829, he established the following two propositions : 1) an arbitrary impression of light on the retina takes an appreciable time for its entire formation as well as for its complete disappearance; 2) the total duration of perceiving an extremely short flash of light is approximately equal to one third of a second. These results enabled Plateau to explain a host of optical illusions.

In 1830, Plateau reconsidered the theoretical basis of the appearances generated by the simultaneous movement of two lines, and this led him to curious findings. For instance, he discovered that a perforated disk rotating at a suitable speed, enables one to make an object in very fast periodic motion appear immobile. In those days, that was a discovery of considerable importance because, ever since, stroboscopic effects have been constantly used in motion pictures and practically never is the name of the first inventor made known to the public.

In 1833, Plateau published a memoir entitled : *Essai d'une théorie générale des apparences visuelles.* It was the first part of an extensive work which was to comprise the study of the persistance of optical impressions on the retina, of accidental colours, of irradiation, of effects of the juxtaposition of coulours and coloured shadows. In particular, concerning the accidental colours, Plateau established that the pictures resulting from looking at a coloured object have the following characteristics : a very short persistance of the initial impression; appearance of the accidental image; usually, successive disappearances and reappearances of that image more or less frequently.

In order to explain all these phenomena, Plateau assumed that during the observation of a coloured object, the retina exerts a growing reaction against the action of the impinging light and tends to establish itself in some sort of opposite state. Consequently, after the disappearance of the object, it is spontaneously in that opposite state, whence the impression of an accidental tint. In this way an oscillatory state between the original and the accidental tint arises.

Plateau also applied himself to the study of irradiation, being the phenomenon in virtue of which a luminous object surrounded by a dark space appears more or less amplified. The clever physicist designed an apparatus which enabled him to formulate some remarkable laws. As for the other accidental appearances of simultaneity, Plateau expounded his general ideas in a *Supplement of the Treatise on Light* by Herschell, but the merciless infirmity which had struck him in 1843, did not allow him to treat the subject in detail. Nevertheless, he was able to make a very ingenious rapprochement, namely, that the phenomena of the simultaneous accidental colours are to *space* what the accidental colours by succession are with respect to *time*.

In regard to the theory of the above-mentioned phenomena, Plateau had many opponents, but he resisted them honorably, sometimes even with considerable success. Moreover, he never had a contradicter on matters of experimentally observed facts. As it has been stated by Prof. Delsaulx, "his experimental studies constitute a perfect model of research method, perspicacity of analysis, precision of the measurements and simplicity as well as clarity of exposition. There are no works in our (Belgian) great academic collections, capable of educating as surely and rapidly to the art of experimentation, a young person eager to contribute to the progress of science".

In 1840, Plateau made a fortuitous observation which provided him with the idea of an experiment which has remained famous ever since. His preparator had spilled some vegetable oil in a vase filled with a mixture of water and alcohol. The professor was surprised to see how the small oil drops adopted a spherical shape as if they were subtracted to the action of gravity. Then, he did on purpose what his assistant had done by coincidence : he introduced a large quantity of oil into a mixture of water and alcohol composed in proportions such that the densities of the oil and the mixture were equal. Then, the quantity of oil weighs exactly as much as the quantity of alcoholic liquid displaced and according to Archimedes's principle, it is as if the oil is in a state of weightlessness. It floats in the mixture, adopting the spherical shape. When a crank with at its free end a small circular disk is carefully inserted into the oil–mass and one transmits to the latter conveniently increasing rotational velocities, one succeeds in changing the shape of the oil–mass respectively into an oblate spheroid or a large ring with or without a small spherical mass at its centre, or finally a number of isolated spheres rotating around the axis constituted by the stem of the crank and at the same time rotating on their own. Notwithstanding several striking analogies between these phenomena and what is observed in the solar and planetary systems, Plateau warned in his time against overhastily drawing quantitative conclusions from his experiments in regard to astronomical facts. For example, in his case, when the ring breaks up, the isolated masses move on circular trajectories and not on elliptic ones.

After having published the above-mentioned results, Plateau had the intention to repeat his celebrated experiment using other basic materials, but his terrible eye-disease put an end to this project. After horrible sufferings, the invalid gradually regained forces, and soon he had the satisfaction to be freed of any financial cares for himself and his family. Indeed, on the 29th June 1844, he was nominated as a full professor and somewhat later, a royal decree which honours as much minister Ch. Rogier who signed it as the scholar who benefitted from it, secured his integral income.

Hardly recovered from illness, Plateau oriented once more his thoughts toward new research achievements, and to this end, he was generously aided by several friends and admirers. Once in possession of the means to subtract a liquid to the force of gravity, he asks himself what are the "shape-generating" forces which determine the equilibrium form taken by the liquid mass. He soon invokes a principle according to which a liquid exerts on itself a pressure equal to the product of the average curvature of the surface at any point considered, times a constant proportionality factor depending upon the nature of the liquid. He proves experimentally that this pression arises from an extremely thin superficial layer and he verifies this fundamental principle by a variety of procedures. Among the equilibrium surfaces which can enclose a quantity of weightless liquid, he finds besides the sphere, also the plane and the circular cylinder. He shows that such a cylinder becomes unstable when its length exceeds the perimeter of the (circular) section perpendicular to its axis of rotational symmetry. This proposition explains quite a number of natural phenomena. The three above-mentioned surfaces are of course not the only possible spatial figures with rotational symmetry. Plateau has examined three others which he called respectively the onduloid, the nodoid and the catenoid. Details concerning these geometrical figures must be omitted here, but it may be recalled from a statement of Professor Delsaulx, that Plateau mentally conceived almost without effort what analytic calculations only reveal painfully. His delicate reasonings often recall the acuteness of Foucault and his experimental research the method of Faraday.

A simple remark guided Plateau into a novel and magnificent research domain : since every convex equilibrium surface has its concave analogue, it follows that a very thin liquid membrane effectuated in the air must take on the same shape as a liquid mass in the state of weightlessness. In order to study such pellicular figures, he invented a special liquid which he called *glyceric liquid* enabling him to manufacture big bubbles persisting many hours. In addition to this, he succeeded not only in determining the pressure exerted by the air comprised in the laminar sphere, but also in calculating approximately the value of the action radius of the molecular attraction causing the surface tension.

Concerning the spatial equilibrium figures which have no rotational symmetry, Plateau formulated a general principle permitting to manufacture in the pellicular state every surface with zero average curvature about which one knows either the geometric origin or the equation in finite coordinates. He applied his principle to the realization of several equilibrium figures, one more complicated than the other, and the experimental observation always confirmed the theory.

In the study of the mutual combinations of liquid pellicles, he recalls that the superficial layers of each of them are in a continuous state of tension, and that the contractile force gives rise at each point to a pressure along the local normal equal to the double product of that tension by the average curvature of the surface. It is the coexistence of these two forces, tension and pressure along the normal, which has guided Plateau towards the study of the charming pellicular systems which he obtained by dipping into the glyceric liquid frames in iron wire depicting the sides of an arbitrary regular polyhedron. Here follow the laws discovered and verified by Plateau for every laminar system :
1) from each edge of the solid framework, a pellicle departs; 2) at one liquid edge no more than three pellicles can end up and then they form with one another angles of 120° ; 3) the liquid edges which end up at a same liquid point are always four in number and enclose angles of 109°28′. As for the theory of the generation of liquid pellicles, Plateau established a very peculiar principle, namely, that the surface layer of the liquid has a viscosity of its own, sometimes much higher or lower than the viscosity in the interior of the liquid mass.

Plateau's results in the above–mentioned domain have been expounded in a major work entitled : *Statique experimentale et thèorique des liquides soumis aux seules forces moléculaires.* This research has spread his scientific reknown throughout the two worlds; in Belgium as well as abroad, his colleague scientists have unanimously recognized the beauty of Plateau's experiments, their great importance for molecular physics and the unquestionable source of inspiration which they constitute.

The various research fields which were considered in the preceding paragraphs are those where Plateau made his main discoveries, but his scientific output is still more extensive. Indeed, he also contributed to pure and applied mathematics, especially to geometry.

Plateau's brilliant career was acknowledged at several occasions, not only by the scientific community, but also by the Belgian government. On the 15th April 1834, he was elected a corresponding member of the Royal Academy at Brussels and on the 15th December 1836, he became a full member. The majority of his papers and memoirs were submitted to and published by the Academy. In 1835, he was appointed extraordinary professor at the University of Ghent and this was followed by his nomination as a full professor on the 29th June 1844, although his blindness prevented him to fulfil his teaching duties. Twice, he received the quinquennial Prize for Physics and Mathematics, namely, in 1854 for the period 1849–1853 and in 1869 for the period 1864–1868. Named a Knight of the Order of Leopold on the 13th December 1841, he was promoted to the rank of officer on the 15th November 1859 and to the rank of commander on the 28th May 1872 on the occasion of the hundredth anniversary of the Royal Belgian Academy of Sciences, Literature and Fine Arts.

From 1841 onward, he was consecutively appointed corresponding member of the Société Philomatique de Paris, honorary member of the Physics and Natural History Society of Geneva, honorary or corresponding member of the Natural Science Society of the Canton of Vaux, the Royal Institution of London, the Batavian Physical Society of Rotterdam, the Physical Society of Frankfurt–am–Main, the Natural Sciences Society of Cherbourg, the Society of the Natural Sciences Fellows of Berlin and the Royal Physical Society of London. Plateau was also honoured with corresponding memberships of the Institut de France, the Academy of Sciences of Berlin and the Royal Dutch Academy of Sciences, a foreign membership of the Society of Sciences of Göttingen and the Royal Society of London. Finally, it should be mentioned that the University of Göttingen esteemed Plateau's work in such a measure that it added to one of its teaching schedules a special course on surfaces of zero average curvature and spatial equilibrium figures of liquids subtracted to the force of gravity.

To terminate, some particularities of the life of Joseph Plateau. He had a vivid and humorous character, his memory seldom let him down, and in contrast to most people, it became prodigious as he grew older. He was fond of visiting scholars and liked to welcome them at his home. In his conversations, he was devoid of any feelings of superiority, a characteristic of a great man. He respected the rights of priority of other scientists' inventions, as much as he appreciated being credited for his own. Plateau was a fervent Christian. He always felt distressed when he learned about a scholar taking advantage of the wonderful advances of science to promulgate materialistic or anti–religious doctrines. In his view, religion is a celestial balsam for all physical or moral sufferings, and he regarded it as a crime against humanity to try to deprive the wretched from it. As for him, the more he deepened into the secrets of nature, the more he bowed to the mysteries of the supernatural. Thanks to his sober life, he reached the old age, bodily and mentally in excellent state, exception made for his blindness.

Joseph Plateau died on the 15th September 1883. This sad event caused a profound emotion throughout the scientific world. Unanimous marks of regret were expressed in many academies and scientific societies which had honoured him with a membership or a medal. In his rectoral discourse on the occasion of the solemn opening of the academic year on the 16th October 1883, Prof. A. Callier pronounced a touching in memoriam, in which he recalled the noble life of the deceased, totally devoted to science and constantly pursuing Truth and intrinsic Beauty.

(inspired by G. Van der Mensbrugghe's homage to Prof. J. Plateau in the first Liber Memorialis of the University of Ghent)

The complete bibliography of Joseph A. F. Plateau

Memoirs of the Royal Academy of Belgium

Essai d'une théorie générale comprenant l'ensemble des apparences visuelles qui succèdent à la contemplation des objets colorés, et de celles qui accompagnent cette contemplation : c'est-à-dire la persistance des impressions sur la rétine, les couleurs accidentelles, l'irradiation, les effets de la juxtaposition des couleurs, les ombres colorées, etc., 1834. Mém. des membres, t. VIII.

Mémoire sur l'irradiation, 1839. Ibid., t. XI.

Mémoire sur les phénomènes que présente une masse liquide libre et soustraite à l'action de la pesanteur, 1re partie, 1843. Ibid., t. XVI.

Analyse des eaux minérales de Spa, faite sur les lieux, pendant l'été de l'année 1830, 1844. Ibid., t. XVII.

Recherches expérimentales et théoriques sur les figures d'équilibre d'une masse liquide sans pesanteur, 2e série, 1849. Ibid., t. XXIII.

Idem, 3e série, 1856. Ibid., t. XXX.

Idem, 4e série, 1858. Ibid., t. XXXI.

Idem, 5e et 6e série, 1861. Ibid., t. XXXIII.

Sur un problème curieux de magnétisme, 1864. Ibid., t. XXXIV.

Recherches expérimentales et théoriques sur les figures d'équilibre d'une masse liquide sans pesanteur, 7e série, 1866. Ibid., t. XXXVII.

Idem, 8e série, 1868. Ibid.

Idem, 9e série, 1868. Ibid.

Idem, 10e série, 1868. Ibid.

Idem, 11e série, 1868. Ibid.

Bibliographie analytique des principaux phénomènes subjectifs de la vision depuis les temps anciens jusqu'à la fin du XVIIIe siècle, suivie d'une bibliographie simple pour la partie écoulée du siècle actuel, 1877, t. XLII :

Première section : persistance des impressions sur la rétine.

Deuxième section : couleurs accidentelles ordinaires de succession.

Troisième section : images qui succèdent à la contemplation d'objets brillants.

Quatrième section : irradiation, 1878.

Cinquième section : phénomènes ordinaires de contraste.

Sixième section : ombres colorées, avec supplément.

Deuxième supplément à la bibliographie analytique pour 1878-1879. Ibid., t. XLIII.

Troisième supplément à la bibliographie analytique pour 1880-1881-1882. Ibid., t. XLV.

Communications of the Royal Academy of Belgium

First Series

Note sur un phénomène de vision. Bullet. de la séance du 6 décembre 1834; insérée dans le t. IX des Mémoires.

Sur un principe de photométrie, 1835 , t. II, p. 52.

Note sur un phénomène particulier qui se produit dans les yeux de l'auteur, 1835. Ibid., p. 84.

Notice sur l'anorthoscope, 1836, t. III, p. 7 : erratum à la page 65.

Note sur la figure de la nappe liquide qui s'écoule par une fente étroite, rectiligne et verticale, partant du fond d'un réservoir et s'élevant jusqu'au–dessus du niveau du liquide, 1836. Ibid., p. 145.

Addition à la note précédente, 1836, t. III, p. 222.

Sur un nouveau moyen de déterminer la vitesse et les particularités d'un mouvement périodique très rapide, tel que celui d'une corde sonore ou vibrante, etc., 1836. Ibid., p. 364.

Note sur l'irradiation, 1839, t. VI, 1re partie, p. 51.

Deuxième note sur l'irradiation, 1839, t. VI, 2e partie, p. 102.

Sur les phénomènes que présente une masse liquide libre et soustraite à l'action de la pesanteur, 1842, t. IX, 1re partie, p. 17.

Suite à ce travail, 1842. Ibid., ibid., p. 298.

Note sur une conséquence curieuse des lois de la réflexion de la lumière, 1842. Ibid., 2e partie, p. 10.

Sur un moyen de produire le vide à l'aide de la force centrifuge. Ibid., t. IX, août 1842.

Deuxième note sur le même sujet, 1843, t. X, 1re partie, p. 97.

Note sur des expériences d'optique, et sur un appareil pour vérifier certaines propriétés du centre de gravité, 1843. Ibid., 1re partie, p. 310.

Note sur une nouvelle application curieuse de la persistance des impressions sur la rétine, 1849, t. XVI, 1re partie, p. 424.

Deuxième note sur de nouvelles applications curieuses de la persistance des impressions sur la rétine. Ibid., ibid., p. 588.

Troisième note sur le même sujet, 1849. Ibid., 2e partie, p. 30.

Quatrième note sur le même sujet, 1849. Ibid., ibid., p. 245.

Sur les théories récentes de la constitution des veines liquides lancées par des orifices circulaires, 1856, t. XXIII, 1re partie, p. 757.

Second Series

Note sur une récréation arithmétique, 1863, t. XVI, p. 62.

Sur un phénomène de couleurs juxtaposées, 1863. Ibid., p. 139.

Une expérience relative à la vapeur vésiculaire, 1871, t. XXXII, p. 251.

Sur la mesure des sensations physiques, et sur la loi qui lie l'intensité de ces sensations à l'intensité de la cause excitante, 1872, t. XXXIII, p. 376.

Réponse aux objections de M. Marangoni contre le principe de la viscosité superficielle des liquides, 1872, t. XXXIV, p. 404.

Un mot au sujet du mémoire de M. Delbœuf sur la mesure des sensations. Ibid., p. 141.

Deuxième note sur une récréation arithmétique, t. XXXVIII, p. 469.

Notes sur les couleurs accidentelles ou subjectives, t. XXXIX, p. 100, t. XLII, pp. 535 et 684.

Sur des exemples curieux de discontinuité en analyse, t. XLIII, pp. 84 et 255.

Sur une loi de la persistance des impressions dans l'œil, 1878, t. XLVI, p. 334.

Un petit paradoxe, 1879, t. XLVII, p. 346.

Un mot sur l'irradiation, 1879, t. XLVIII, p. 37.

Sur la viscosité superficielle des liquides. Ibid., p. 106.

Une application des images accidentelles, 1880, t. XLIX, p. 316.

Third series

Quelques expériences sur les lames liquides minces, 1881, t. II, p. 8.

Une application des images accidentelles, 2e note, 1881. Ibid., p. 281.

Une petite illusion, 1882, t. III, p. 24.

Sur des sensations que l'auteur éprouve dans les yeux, 1882. Ibid., p. 241.

Sur l'observation des mouvements très rapides spécialement lorsqu'ils sont périodiques ; œuvre posthume, présentée dans la séance du 3 novembre 1883, t. VI, p. 484.

Quelques expériences sur les lames liquides minces ; 2e note, œuvre posthume. Ibid., p. 704.

Reports

Rapport sur une notice de M. Melsens intitulée : Sur la persistance des impressions de la rétine. Bulletin de l'Académie, 1re série, t. I, p. 477.

Rapport sur un mémoire de M. Duprez intitulé : Statistique des coups de foudre qui ont frappé des paratonnerres. Ibid., t. III, p. 463.

Rapport sur un mémoire de M. Bède intitulé : Recherches sur la capillarité. Ibid., t. VI, p. 405.

Rapport sur une note de M. Montigny relative à la vitesse du bruit du tonnerre. Ibid., t. IX, p. 6.

Rapport sur un mémoire de M. Bède relatif à la capillarité. Ibid., t. X, p. 47.

Rapport sur une note de M. Rousseau relative aux appareils servant à faciliter l'étude de la théorie des ondes lumineuses. Ibid., t. XI, p. 455.

Rapport sur un mémoire de M. Bède concernant la liaison entre les phénomènes de la capillarité et de l'endosmose. Ibid., t. XII, p. 111.

Rapports sur deux mémoires de M. Bède relatifs à l'équilibre d'une bulle d'air et à celui d'une goutte d'eau entre deux plans. Ibid., t. XIV, p. 442.

Rapport sur un mémoire de M. Valerius relatif à la distance focale des miroirs sphériques. Ibid., t. XV, p. 9.

Rapport sur un mémoire de M. Van der Mensbrugghe relatif à la théorie des courbes d'intersection de deux lignes droites tournant autour de deux points fixes. Ibid., t. XV, p. 613.

Rapport sur une note de M. Montigny relative à la scintillation des étoiles. Ibid., t. XVII, p. 435.

Rapport sur un mémoire de M. Lamarle relatif à la stabilité des systèmes liquides en lames minces. Ibid., p. 591.

Rapport sur deux notes de M. Brachet. Ibid., p. 435.

Rapport sur un travail de M. Van der Mensbrugghe relatif aux forces moléculaires des liquides. Ibid., t. XVIII, p. 124.

Rapport sur des notices de MM. Brachet et Vallée. Ibid., p. 314.

Rapport sur une note de M. Valerius relative à la constitution intérieure des corps. Ibid., t. XIX, p. 11.

Rapports sur deux notes de M. Delbœuf relatives à certaines illusions d'optique. Ibid., t. XIX, p. 154 et t. XX, p. 6.

Rapport sur un second mémoire de M. Lamarle relatif à la stabilité des systèmes laminaires. Ibid., t. XX, p. 220.

Rapport sur un travail de M. Delbœuf intitulé : Détermination rationnelle des nombres de la gamme chromatique. Ibid., t. XXI, p. 324.

Rapport sur un travail de M. Valerius sur un analyseur acoustique. Ibid., t. XXII, p. 203.

Rapports sur deux notes de M. Van der Mensbrugghe sur la tension des lames liquides. Ibid., t. XXII, p. 207 et t. XXIII, p. 440.

Rapports sur deux mémoires de M. Bède relatifs à la capillarité. Ibid., t. XXIII, pp. 4 et 440.

Rapport sur un travail de M. Montigny sur le pouvoir dispersif de l'air. Ibid., t. XXIV, p. 508.

Rapport sur un mémoire de M. Docq sur l'appareil auditif. Ibid., t. XXV, p. 79.

Rapports sur deux mémoires de M. Van der Mensbrugghe sur la tension superficielle des liquides. Ibid., t. XXVIII, p. 17 et t. XXXV, p. 460.

Rapport sur une note du même sur la viscosité superficielle de lames de solution de saponine. Ibid., t. XXIX, p. 345.

Rapport sur une note du même sur un principe de statique moléculaire avancé par M. Lüdtge. Ibid., t. XXX, p. 286.

Rapport sur une note du même sur un fait observé au contact de deux liquides. Ibid., t. XXXIII, p. 172.

Rapport sur un mémoire de M. Delbœuf sur des recherches relatives à la mesure des sensations physiques. Ibid., t. XXXIV, p. 250.

Rapport sur une note de M. Verstraete sur le phénomène de la vue. Ibid., t. XXXVI, p. 305.

Rapport sur un mémoire de M. Van der Mensbrugghe relatif à l'influence de l'électricité statique sur la tension d'un liquide. Ibid., t. XXXVIII, p. 17.

Rapport sur une note du même sur la théorie capillaire de Gauss. Ibid., t. XXXIX, p. 366.

Rapport sur une note du même sur la surface de contact d'un solide et d'un liquide. Ibid., t. XL, p. 272.

Rapport sur un mémoire du même sur le problème des liquides superposés dans un tube capillaire. Ibid., t. XL, p. 669.

Rapport sur un mémoire du même sur l'énergie potentielle des surfaces liquides. Ibid., t. XLV, p. 574.

Works not published by the Academy

Construire un triangle équilatéral qui ait ses sommets sur trois circonférences données. Correspondance math. et phys., 1827, t. III, p. 1.

Sur les sensations produites dans l'œil par les différentes couleurs. Ibid., 1828, t. IV, p. 51.

Sur les apparences que présentent deux lignes qui tournent autour d'un point avec un mouvement angulaire uniforme. Ibid., 1828, t. IV, p. 393.

De l'action qu'exerce sur une aiguille aimantée un barreau aimanté tournant dans un plan et parallèlement au dessous de l'aiguille. Ibid., 1830, t. VI, p. 70.

Lettre relative à différentes expériences d'optique. Ibid., 1830, t. VI, p. 121.

Sur quelques phénomènes de vision. Ibid., 1832, t. VII, p. 288.

Sur un nouveau genre d'illusions d'optique. Ibid., 1832, t. VII, p. 365.

Sur un phénomène de couleurs accidentelles. Ibid., 1834, t. VIII, p. 211.

Réponse aux objections publiées contre une théorie générale des apparences visuelles dues à la contemplation des objets colorés. Ibid., 1837, t. IX, p. 97.

Dissertation sur quelques propriétés des impressions produites par la lumière sur l'organe de la vue. Liège, 1829; in-4°.

Lettre sur une illusion d'optique. Ann. de chim. et de phys. de Paris, 1831, t. XLVIII, p. 281.

Des illusions d'optique sur lesquelles se fonde le petit appareil appelé récemment

Phénakistiscope. Ibid., 1833, t. LIII, p. 304.

Sur la persistance des impressions sur la rétine, traité de la lumière de Herschel, traduit par Verhulst et Quetelet, 1833, t. II, supplément, p. 471.

Sur les couleurs accidentelles. Ibid., p. 490.

Réponse à un article de M. Osann. Ann. de M. Poggendorff, 1836, t. XXXVIII, p. 626.

Réclamation relative à un instrument proposé par M. Doppler. Ibid., 1849, t. LXXVIII, p. 284.

Sur la limite de la stabilité d'un cylindre liquide. Ibid., 1850, t. LXXX, p. 566.

Physique, 1re partie (en collaboration avec M. Quetelet). Encyclopédie populaire belge. Bruxelles, 1851-1855.

Sur le passage de Lucrèce, où l'on a cru voir une description du Fantascope. Biblioth. Univers. de Genève, 1852, 4e série, t. XX de la partie scientif., p. 300.

Sur une production curieuse d'anneaux colorés. Journal le Cosmos, 2e année, 1853, 3e vol., p. 191.

Réclamation au sujet d'un passage du mémoire de M. Helmholtz Sur la théorie des couleurs composées *et rectification à un passage du mémoire de M. Unger* Sur la théorie de l'harmonie des couleurs. Ann. de M. Poggendorff, 1853, t. LXXXVIII, p. 172.

Réponse aux observations présentées par M. Chevreul. Comptes-rendus de l'Académie des sciences de Paris, 1863, t. LVII, p. 1029.

Statique expérimentale et théorique des liquides soumis aux seules forces moléculaires. Gand, 1873 ; deux vol. in-8°.

GENEALOGIE VAN DE FAMILIE PLATEAU

PLATEAU JEAN-BAPTISTE × WATTRIPONT CHRISTIANE-JOSÈPHE

PLATEAU JEAN-BAPTISTE-FRANÇOIS-JOSEPH × RIJS MARIE-ÉLISABETH-IGNACE-JOSÈPHE

PLATEAU ANTOINE-GHISLAIN-JOSEPH
Kunstschilder
ᛉ Tournai 19-I-1749
ᛦ Brussel 19-IV-1815
× 1792 DELIN ALBERTINE.
× 7-II-1801 THIRION MARIE-CATHERINE
ᛉ Fenal 13-II-1771
ᛦ Marche les-Dames 12-X-1814

PLATEAU JOSEPH-ANTOINE-FERDINAND
prof. Univ. Gent
ᛉ Brussel 14-X-1801
ᛦ Gent 17-IX-1883
× 27-VIII-1840 CLAVAREAU FANNY-AUGUSTA
ᛉ Luxemburg 1818
ᛦ Gent 14-II-1884

PLATEAU NATHALIE
ᛉ Brussel 1803
ᛦ St. Joos ten Noode 19-XII-1867
× DARDONVILLE arts te Spa

PLATEAU JOSEPHINE
ᛉ Brussel 1806
ᛦ Gent 1894

PLATEAU FELIX
prof. Univ. Gent
ᛉ Gent 1841
ᛦ Gent 1911
× LONGUEVILLE MARIA
ᛉ Tournai 1843
ᛦ Gent 1914

PLATEAU
1. FÉLICIE ᛉ Gent 1875
Epse A. BUISSERET 1898
2. JEANNE ᛉ Gent 1877
ᛦ Gent 1937
3. LUCIE ᛉ Gent 1881
ᛦ Gent 1944

PLATEAU ADOLPHE
ᛉ Gent 1840
ᛦ Gent 1842

PLATEAU ERNEST
ingenieur
ᛉ Gent 1848
ᛦ Gent 1924
× BRYANT JANE
ᛉ London 1851
ᛦ Haltert 1930

PLATEAU
1. ROBERT ᛉ Gent 1880
ingenieur
2. GABRIELLE
ᛉ Ledeberg 1886

PLATEAU ALICE
ᛉ Gent 1846
ᛦ Brussel 1919
× VANDER MENSBRUGGHE GUSTAVE
prof, Univ Gent
ᛉ Gent 1835
ᛦ Melle 1911

VANDER MENSBRUGGHE
1. MARCELLE ᛉ Gent 1874
Epse CH. DIEPEN
2. BERTHE ᛉ Gent 1876
3. ANDRÉ ᛉ Gent 1878
Dr juris
4. ROBERT ᛉ Gent 1881
ingenieur

The Problem of Plateau (pp. 18–32)
ed. Th. M. Rassias

REMARKS ON THE MATHEMATICAL WORK OF TIBOR RADÓ

ERWIN KREYSZIG
Department of Mathematics & Statistics
Carleton University, Ottawa, Canada K1S 5B6

Abstract

This paper begins with a sketch of Radó's life, largely based on the author's recollections. It then traces some main ideas in Radós's work on complex analysis, Plateau's problem, continuous transformations and surface area. A list of Radó's publications is included at the end.

0. Introduction

Although my paper "On the theory of minimal surfaces" in this Volume covers the main stages of the development of Radó's work on Plateau's problem, the title of the present article is intended to indicate that it would be impossible to survey Radó's other work in some detail, his deep and many-sided investigations on surface area being particularly difficult to explain sufficiently well within the space available. Accordingly, after a sketch of his life, based on my own recollections and help by Radó's former students, Professors Helsel and Reichelderfer, we shall survey the main areas of Radó's work in Secs. 1–5 and in a short concluding Sec. 6.

Tibor Radó (1895-1965) was born in Budapest, the capital of Hungary, then a city of about half a million inhabitants and the second largest cultural center in the Austria-Hungarian Monarchy after Vienna. Following the usual elementary and high school education, in 1913 Radó entered the Polytechnic Institute of Budapest, which at that time had a little under one thousand students of engineering. Two years later he joined the Austria-Hungarian Army. He fought on the Russian Front from 1915 until 1916, where he was captured by the Russians and taken prisoner of war. He was transported to a prison camp in Berezovka in Transbaikalia, a part of Eastern Siberia. There he met, as another prisoner, the Viennese Eduard Helly (1884-1943) a student and colleague of Hans Hahn (1879-1934) and wartime volunteer in the Austria-Hungarian Army. Helly had been wounded in September of 1915 and sent to that camp from the military hospitals of Kiew, Kursk and Woronesh. Helly became well known by his work in functional analysis during the early time of the creation of the theory of normed spaces. In 1912 and 1913 he had just published the first of the small number of his highly significant papers, containing "Helly's theorem" among other results and various novel ideas, when the War interrupted his successful mathematical activity, which he was able to resume only as late as 1920, two years after the end of the War when he was finally released and returned to Vienna.

Under incredibly hard conditions and physical and psychical stress and sufferings, Helly organized seminars and lectures in mathematics at that prison camp, in which Radó participated, together with Heinrich Elbogen, who was working on axiomatics,[1] and others. This was of great significance in Radó's life because it caused him to give up engineering and to become a mathematician. Thus, here we see a similarity to the career of his famous fellow countryman, the fifteen years older Friedrich Riesz (Riesz Frigyes, 1880-1956), who also started out as an engineering student (in 1897, in Zurich) and then changed over to mathematics.

Not only Helly, but also Radó was released as a prisoner of war as late as 1920 and returned to Hungary in October of that year. There he joined the University of Szeged in 1920 and in 1922 began to work under the guidance of F. Riesz.

Szeged had just become a new university town when Siebenbürgen had fallen to Romania, so that the University of Klausenburg (Kolozsvár), where F. Riesz and Alfred Haar (1885-1933) had been professors (both since 1912), had to be moved and all professors were forced to leave. Both Riesz and Haar spent a short transition period in Budapest and then founded the Mathematical Institute at Szeged, later called the *Bolyai Institute*, as well as the journal *Acta Scientiarum Mathematicarum.* The beginnings at Szeged, so close after the end of the War, were difficult, but Riesz and Haar created an inspiring atmosphere in which Radó began his professional career, contributing to the new journal already in the first volume, namely, a paper [2] on conformal mapping. In 1924 he married Ida Barabás de Albis. He remained at Szeged, first as an assistant and later as a Privatdozent, until 1929, being absent from town in 1928-29 as an "International Research Fellow" of the Rockefeller Foundation. That year brought him in close personal contact with Paul Koebe and Constantin Carathéodory, his mathematical contact being of earlier data, as can be seen from his publication [5] of 1923 on Riemann's mapping theorem, where he referred explicitly to well-known works of these two mathematicians. At that time, he also visited Leon Lichtenstein in Leipzig, whose work on partial differential equations and Plateau's problem had points of contact with Radó's interests, as we shall see.

In 1929 he was invited to Harvard University and Rice Institute for seminars and lectures. In 1930 he accepted an offer from The Ohio State University in Columbus, Ohio (in the sequel briefly called "Ohio State") as a professor; this marked the beginning of graduate work in mathematics at that University. That development was at first modest and slow because of lack of funds in the wake of the depression, so that no further outstanding mathematicians were hired until

1946. Radó successfully tried to make the best of the unfavorable situation, by creating a school consisting of his own Ph.D. students. The first of these was P.V. Reichelderfer, later professor at Ohio State and Ohio University in Athens, Ohio. Publication of joint papers [53, 54, 57, 58, 61, 72, 82, 84, 89] began in 1939 and stretched over twelve years, followed by a culmination in 1955, with the publication of a joint book [96] on continuous transformations, in the famous Springer Series.

The next of Radó's Ph.D. students to be mentioned here was R.G. Helsel, with whom Radó published two papers [65, 76] on surface area and the transformation of double integrals and whom we shall mention again below.

The third Ph.D. student was E.J. Mickle, with whom Radó published eight papers [77, 83, 88, 100-104] between 1949 and 1960, on upper continuous functions, surface area and measure theory. Mickle later became professor at Ohio State.

The situation of the Department at Ohio State improved considerably after the end of the War. The first outstanding mathematician hired was young Marshall Hall, Jr., (1910-90) coming from Yale; the hiring process began under Chairman W.R. Longley, but when Hall joined Ohio State in 1946, Radó had taken the chairmanship. Hall remembers:[2]

> "I was well received at Ohio State. Tibor Radó was then chairman. My name had been suggested to him by Saunders Mac Lane. I was highly favored and did not have to teach any of the sub-freshman or elementary courses. Within two years in 1948 I was promoted to be a Full Professor. Tibor Radó was primarily interested in Surface Area, but somehow had a side-interest in free groups. This interested me and we wrote some joint papers on the subject. [Mathematical Reviews mentions only [75] of our list.] ... At Ohio State it was a privilege to associate with Tibor Radó and others like Eugene [read: Erwin] Kleinfeld, but mostly with Herbert Ryser..."

What Hall expressed seems to be typical of Radó's attitude toward mathematics in general and research in particular; the author of this article could add similar experiences. Herbert Ryser (1923-85), whom Hall mentioned and who became known for his work in combinatorics, was also hired by Radó, and the third in that growing group of Algebra hired by Radó was Henry Mann (born 1905), known for his work in additive number theory and statistics (Mann-Whitney test). This was the beginning of the creation of the largest research group in the department, on Algebra. When Radó resigned from the chairmanship in 1948, Helsel became chairman and continued the work by hiring Erwin Kleinfeld and I.N. Herstein. Kleinfeld later went to the University of Iowa and Herstein to Chicago.

We now turn to Radó's work, beginning with complex analysis, which marks the first epoch.

1. Radó's Work in Complex Analysis, 1921-25

Most of Radó's papers show perfection and elegance reminiscent of those of his main teacher, the great F. Riesz, whose papers are worth reading to this day and whose proofs have hardly ever been improved or simplified. Paper [37], written short after Radó's beginnings in the United States, contains an interesting testimonial of that attitude similar in spirit to that of Riesz and in the last end expressing what Gauss called "*matura*" in his motto "*pauca sed matura*"[3].

Radó's first major field was complex analysis, in particular conformal mapping and Riemann surfaces; corresponding papers are [2–6, 8–11], published 1921-25, thus, when Radó's age was 26-30. That choice of field was certainly strongly influenced by Riesz's interests at that time. Riesz, then in his early forties, had just about finished his pathbreaking accomplishments in the early development of functional analysis, which he had obtained in an incredibly short time, beginning in 1907 with the Riesz-Fischer theorem, an unsuspected bridge between Hilbert's ℓ^2-spaces and Lebesgue's L^2-spaces, followed by the representation of bounded linear functionals on L^2 in 1907 and – a very famous result – on $C[a,b]$ in 1909, the creation of L^p-spaces and their duals, with compact operators on them, in 1910, the development of spectral theory of linear operators in 1913 and finally the creation of axiomatically defined normed spaces and compact operators and their spectral theory (an abstract Fredholm theory) in 1918 (submitted 1916, but delayed by the War).

Riesz's work of 1918-25 in complex analysis concerned many topics including boundary values of analytic functions (1923), inequalities (1921, 1923) and the theory of subharmonic functions (1924-26), his creation. However, whereas Radó's choice of complex analysis as the field of work was certainly influenced by Riesz, the topics chosen within the field were quite different from those on which Riesz was working; this is immediately seen by a comparison with Riesz's papers in his *Œuvres* [p. 22]. For reasons of space we can present here only some illustrative examples of Radó's accomplishments.

Radó's probably most outstanding contribution [11] to complex analysis concerns the proof of the triangulation of a Riemann surface, a result whose significance we are now going to characterize. The ingenious concept of a Riemann surface appeared first in Riemann's path-breaking thesis of 1851 and was of fundamental importance to the development of complex analysis as well as of topology. It was F. Klein (for closed surfaces, 1882) and P. Koebe (for open surfaces, 1908), who tried to sharpen and extend the concepts of Riemann introduced in his theory of algebraic functions, a phase of the development which was also affected by ideas of Poincaré. The breakthrough from Klein's intuitive-geometric ("anschaulichen") point of view to the modern abstract definition was accomplished by H. Weyl and Radó. In 1913, in his famous "Die Idee der Riemannschen Fläche", Weyl gave the first axiomatic definition of a Riemann surface.

In his definition, Weyl included as an assumption the validity of the second countability axiom (i.e., there is a countable basis for the topology). Now Radó

[11] was able to show that that requirement is superfluous, but follows from the conformal structure (defined by a family of homeomorphisms on the surface).

Thus, for a surface (a connected two-dimensional manifold), second countability is an essential hypothesis for triangulability (i.e., simplicial decomposability) because then one has available a countable open covering with desirable properties from which one gets the result. In contrast, for a Riemann surface one knows the existence of a covering with those desired properties from the start.

By definition, a two-dimensional manifold M has an atlas of charts (coordinate neighborhoods) that cover M, and the problem of whether it is countable is basic. In [11], following ideas of Prüfer, Radó showed that this need not be the case. The Prüfer-Radó construction involved a mapping of semicircular disks onto circular disks and identification processes. Radó emphasized that the point of the construction is the fact that the bounding diameter of a semidisk is mapped onto a single boundary point of a disk, and this phenomenon is impossible in the conformal mapping of the interior. Hence one cannot succeed in finding a similar example for the case of a Riemann surface.

Another essential point is the following. Whereas Weyl introduced Riemann surfaces by means of "analytic forms" ("analytische Gebilde", an extension of Weierstrass's "algebraische Gebilde"; see *Werke* IV, 13-45), on the basis of modern topology Radó was able to develop the topologic-metric properties of a Riemann surface without reference to analytic forms and thus giving them a final conceptual clarity that fits well into topology.

He also proved the existence of a nonconstant meromorphic function on any Riemann surface, which provided a holomorphic mapping of the surface into the Riemann sphere. Thus he proved that any Riemann surface can be regarded as a covering surface of the compactified complex plane.[4]

2. Radó's Work on Plateau's Problem, 1925-33

From the viewpoint of this Volume, this is the most important part of Radó's many-sided work, and its evolution is discussed in my article in this Volume, mentioned above, so that it will be sufficient to give a few additional observations.

Radó's interest in minimal surfaces and Plateau's problem is the second of his fields of work, after his beginnings in complex analysis just discussed, and it was inspired by A. Haar, Hilbert's Ph.D. student of 1909, who in 1927 himself gave a solution of the nonparametric Plateau problem by direct methods in the calculus of variations (Hilbert's creation!), showing analyticity by a method of Radó in [13], who helped Haar vividly ("lebhaft"). This is one incident of cooperation between Haar and Radó, as an example for many.

Radó took up work in the field only four years after the beginning of his mathematical publications, with the paper [13] on the analytic character of minimal surfaces in which he showed that any C^1-extremal function of the problem is analytic.

Of course, being in the neighborhood of Riesz, the greatest activist in Lebesgue theory after Lebesgue himself, Radó was aware of Lebesgue's contribution to Plateau's problem in his thesis of 1902, and he also had studied the deep and difficult work of S. Bernstein, resulting from Hilbert's problem on the analyticity of solutions of elliptic equations. Bernstein had solved the nonparametric Plateau problem under Lebesgue's conditions in 1910 and 1912. Radó simplified Bernstein's work and in 1927 supported Bernstein in defending his work against attacks by Ch. Müntz, in his article [16], which the editors of the *Mathematische Annalen* accepted against their usual policy of rejecting polemic papers, because they felt that [16] "may serve to clarify questions which often had not found sufficient attention." In [16], Radó also attacked Müntz for "missing geometric convergence theorems", resulting in a serious gap in Müntz's solution of Plateau's problem. He also helped to publicize Bernstein's basic work by translating a paper of 1914 [*Math. Z.* **26** (1927), 551-558] related to Bernstein's famous theorem that a minimal surface $z = z(x, y)$ defined for all (x, y) must be a plane, and by giving a simplified proof of that theorem in [23]. At that time, Radó was still at Szeged, but left soon for Munich and Leipzig, for Harvard and for getting settled in Columbus, Ohio. Thus, Radó's solution of Plateau's problem grew organically out from earlier work, but he was already in Columbus when in May of 1930 he submitted his solution [32] to the *Mathematische Zeitschrift*, where it appeared within a few weeks.[5] This was his second great success, after the triangulation of the Riemann surface.

In 1933, Radó summarized his experience in his well-known booklet [42] on Plateau's problem, recording the remarkable advances made since the appearance of Lebesgue's thesis. His excellent survey includes an analysis of several ways of stating the problem (four of them in the parametric case). In the nonparametric case, a minimal surface $z = z(x, y)$ with given boundary values satisfying the "three-point condition" is found by means of Haar's existence theorem and Radó's analyticity theorem. For the parametric case, Radó gives Garnier's results and then the solution of Douglas and of himself in all details, as well as McShane's solution, which he calls "the most direct method of solution." At the end he included a review of Douglas's solution for surfaces of the type of the Möbius strip. At the same time he added [33-36, 41] to his work, papers in which he made essential use of lower semicontinuity of certain integrals, a concept whose importance for the direct method in the calculus of variations had been emphasized by Tonelli and others. Paper [56] of 1941 on the generalization of McShane's lemma led Radó on to the extension of theorems on the lower semicontinuity of double integrals in [59], published in 1942, and on to work about generalized Jacobians, another field pursued, in part in cooperation with his Ph.D. student P.V. Reichelderfer, to be discussed in Sec. 4.

3. Radó's Work on Subharmonic Functions, 1925-37

As it often happens in mathematics, special cases and applications precede a general theory. This was the case for subharmonic functions, that is, upper semicontinous functions (not identically equal to $-\infty$) with the property that $u(x,y) \leq H(x,y)$ in any subdomain D', $\bar{D}' \subset D$, of the domain D of u, for any harmonic function H with $u \leq H$ on the boundary of D'. Indeed, special instances of such functions appeared in works by Hartogs and by R. Nevanlinna in complex analysis and by Poincaré, Remak and Perron in potential theory. The latter used them in 1923 in his famous method of existence proof. And it is interesting that Radó's work in the area began with an improvement of that method, in a joint paper [12] with F. Riesz in 1925.

It was F. Riesz who three years earlier had published his theory of subharmonic functions, created as a general instrument for unifying the various applications of the maximum principle and for systematically obtaining new such applications, various of which appeared in Radó's subsequent papers, relating to analytic and harmonic functions as well as to surfaces of nonpositive Gaussian curvature and to minimal surfaces. Indeed, Radó's next paper [23] of 1927 on subharmonic functions applied them in an elementary proof of Bernstein's theorem (cf. Sec. 2).

A year later, Radó extended Montel's theorem relating the subharmonicity of $\log u$ to that of $u \exp(ax+by)$ from smooth to continuous functions u, in a C.R. Paris note [28], whose method has since permitted speedy access to a number of fundamental problems.

Radó's subsequent work [39] of 1933, jointly with E.F. Beckenbach, on subharmonic functions and minimal surfaces is a new vista of Weierstrass's famous formulas from the standpoint of the maximum principle. It is shown that functions $x(u,v)$, $y(u,v)$, $z(u,v)$ are Cartesian coordinates of a minimal surface if and only if the logarithm of $[(x+a)^2+(y+b)^2+(z+c)^2]^{1/2}$ is subharmonic for every real constant triple a, b, c, entailing an extension of Schwarz's lemma and other theorems on the maximum principle from analytic functions to minimal surfaces. The two authors studied surfaces of nonpositive Gaussian curvature from a similar standpoint in [40], based on the fundamental observation that $K \leq 0$ if and only if $\log B$ is subharmonic, where $B = E = G$ and $F = 0$, i.e., the surface is referred to isothermic parameters. They showed that this has various consequences, for instance, that the isoperimetric inequality has an equivalent reformulation in terms of subharmonic functions, and that it also leads to the extension of several basic theorems from minimal surfaces to surfaces with $K \leq 0$. The characterization of functions with subharmonic logarithm in [40] suggested extensions to (the much simpler case of) convex functions of one variable in [43], leaving unsolved the general case.

In his fifty-page monograph [50] of 1937, that appeared four years after the monograph [42] on Plateau's problem, Radó lucidly summarized the development on subharmonic functions and their applications until 1936, emphasizing the matters just discussed, but fitting them into a much wider framework built through the activity of about twenty-five mathematicians during the first third of our century.

4. Radó's and Reichelderfer's Work on Continuous Transformations, 1927-55

Our discussion so far shows that the seemingly different fields of Radó's work are generically related, each growing systematically out of the previous one (just as in F. Riesz's many-sided work) and all started at an early time. The field to be considered now will confirm this impression. To that field, Radó devoted numerous papers (indicated below), some of them jointly with P.V. Reichelderfer (now residing in Athens, Ohio) as well as a portion of his third book [72] of 1948 on length and area and his entire fourth book [96] of 1955, co-authored by Reichelderfer. The motivation for all this work is quite interesting. It concerns the extension of the concepts of bounded variation and absolute continuity from one dimension to n dimensions. In 1925, Banach and Vitali had given definitions of those two concepts and of generalized Jacobians for continuous transformations on the closed unit square, leading to a theory that was very beautiful but without much applicability, particularly not to surface area in the calculus of variations. Hence new definitions were needed. Now a basic idea of a way out of the difficulties had appeared in papers by Z. de Geöcze as early as 1905-17, but in a most complicated form (and in part in the Hungarian language). It was Radó's merit to have overcome some of those difficulties in his papers [20, 25] of 1927-28, by developing a theory on the basis of suitable concepts (particularly those of the "essential part" and of the "kernel" of the image), but it took several years to master all the obstacles, part of them topological in nature. This was eventually accomplished, in part in collaboration with P.V. Reichelderfer and with E.J. Mickle, in the papers [47, 51-54, 58, 64, 71, 77, 81, 83, 89] from 1936 to 1951.

This work culminated in the joint book [96] of 1955 by Radó and Reichelderfer, which appeared in the Springer "*Grundlehren*" series, the title being "*Continuous Transformations in Analysis. With an Introduction to Algebraic Topology.*" In this book of roughly 450 pages, the authors treated the measure-theoretical and topological foundations of the theory of real functions of several real variables along the line of ideas initiated by Geöcze, Lebesgue, Tonelli and Vitali, and, of course, highlighted their own work. The book begins with a very careful introduction to so-called reduced cohomology theory of Radó [93] and continues with a topological study of continuous transformations in R^n, based on work by Federer; this is a detailed study of the topological index defined in terms of oriented frames. It is followed by a discussion of multiplicity functions and an outline of higher real variable theory.

The presentation in the book centers around bounded variation, absolute continuity and generalized Jacobians in R^n, as planned in the joint note [82] of 1949 and intended to be "comparable in utility and scope with the classical one-dimensional theory." The results on the transformation of multiple integrals obtained by the authors are very general and the range of their applicability is discussed in great detail; it depends on the scope of essentially absolutely continuous transformations, as is shown. The last part of the book concerns continuous transformations in R^2, consisting of a classical discussion of the topological index and related concepts in R^2 as well as an application of the results obtained in the earlier parts of the book.

The whole presentation excels by its clarity, care and great generality.

5. Radó's Work on Surface Area, 1926-60.

Related with Secs. 2 and 4 is Radó's work on surface area, which began with a *C.R. Paris* note [15] of 1926, culminated with his large book [73] in 1948 and was followed up in further papers for another twelve years. H.A. Schwarz's counterexample of 1880 (see Sec. 12 of my other paper) caused many definitions of area to be proposed, but by the time Radó entered the field, the situation had cleared up to some extent, and Radó's investigations were mainly based on two definitions, the Lebesgue area $L(S)$ of a continuous surface S (the lower limit of the elementary area of polyhedral surfaces (not necessarily inscribed S) approaching S in Fréchet's sense) as defined in 1902, and the surface area by Z. de Geöcze as developed in several papers from 1908 to 1917 (in part written in Hungarian), based on Lebesgue's lower continuity principle and on a modified projection principle discovered by Geöcze himself, who also recognized the fundamental role of topology in the parametric case. It is to a large extent Radó's merit that Geöcze's papers became better known. In 1927, Radó [19] proved for the nonparametric case that if $z = f(x, y)$ is continuous, then the Geöcze sum $G(D)$, D a subdivision of the unit square, approaches $L(S)$ as $\|D\| \to 0$. Simplifying a result by Geöcze, in 1928, Radó [25] proved for the parametric case that if the position vector of S is Lipschitzian, then $L(S)$ is given by the classical formula. Papers [63, 69, 80], the latter with his Ph.D. student E.J. Mickle, give an impression of the great effort that was necessary in developing the topological tools needed for an adequate theory, as presented in the book [73]. Actually, the paper [69] of 1945 just mentioned may be regarded as a program for [73], as Radó himself stated. That was a time of great activity in surface area, by C.B. Morrey, Jr., Y.W.T. Youngs, L. Cesari, E.J. McShane, A.S. Besicovitch and many others, with analytic topology independently developing valuable insights. In particular, in 1946 when Radó had his manuscript of [73] near completion, he became aware of parallel work by Cesari in Italy during the War, leading to close contact, as expressed by the joint presentation [91] at the International Congress of Mathematicians of 1950 in Cambridge, MA. It would be impossible to do justice to a comprehensive book like [73], and we are aware that many remarks should be added, on the "lower area" $a(S)$, which is shown to be equal to Geöcze's area and is equal to the Lebesgue area $L(S)$ (proved by Radó for $L(S) < \infty$ and $a(S) = 0$ and generally by Cesari in independent work). Radó's list of publications shows that a large number of his papers are directly or indirectly related to surface area and measure, up to the last of them, [104], jointly with Mickle published in 1960 and devoted to outer measures, with the theorems formulated for separable metric spaces, so that they can immediately be applied to surface area.

Three differential geometric papers [93, 94, 99], written during the period under discussion, aim at reducing differentiability assumptions.

6. The Final Period, 1960-65

Beginning around 1960, Radó, then in contact with Battelle Memorial Institute in Columbus, OH, in addition to Ohio State, developed interest in Turing machines and problems of computability. On these topic we had discussions almost daily, which I gratefully remember and which lasted until I left Ohio State. The publications [105-108] of 1962-65 belong to that short period, which came to an abrupt and completely unexpected end in 1965.

NOTES

[1] His paper of that time, entitled "Die axiomatische Methode in der Mathematik", was posthumously published in 1928.

[2] See P. Duren (ed.), *A Century of Mathematics in America.* Providence, R.I.: American Mathematical Society, 1988, Vol. 1, 369, 371.

[3] This has often been misunderstood, even by Schumacher, one of Gauss's closest friends, although Gauss commented on it quite clearly. For clarification and sources, see Sec. 7 of E. Kreyszig, On surface theory in E^3 and generalizations. In: G.M. Rassias (ed.) *The Mathematical Heritage of Carl Friedrich Gauss.* (Memorial Volume, to appear.) The second mathematician in the paper (whose name is not mentioned) is Fejér and the one "who was in Zurich in 1897" is F. Riesz.

[4] See L.V. Ahlfors and L. Sario, *Riemann Surfaces* (Princeton, NJ: Princeton University Press, 1960) for various aspects and relevant details.

[5] The publication of that extremely important paper had a history about which Radó was unhappy.

LIST OF PUBLICATIONS

1. On the roots of algebraic equations. *Math. és Phys. Lapok* **28** (1921), 32-37. In Hungarian.
2. Zur Theorie der mehrdeutigen konformen Abbildungen. *Acta Math. Szeged* **1** (1922), 55-64.
3. Bemerkung zu einem Unitätssatze der konformen Abbildung. *Acta Math. Szeged* **1** (1923), 101-103.
4. Sur la représentation conforme de domaines variables. *Acta Math. Szeged* **1** (1923), 180-186.
5. Über die Fundamentalabbildung schlichter Gebiete. *Acta Math. Szeged* **1** (1923), 240-251.
6. Bemerkung zur Arbeit des Herrn Bieberbach: Über die Einordnung des Hauptsatzes der Uniformisierung in die Weierstraszsche Funktionentheorie (Math. Ann. 78). *Math. Ann.* **90** (1923), 30-37.
7. Bemerkung zu einem Blaschkeschen Konvergenzsatze. *Jahresber. Deutsch. Math. Verein.* **32** (1924), 198-200. With K. Löwner.
8. Über die konforme Abbildung schlichter Gebiete. *Acta Math. Szeged* **2** (1924), 47-60.
9. Ein Beispiel zur Theorie der konformen Abbildung. *Tohoku Math. J.* **24** (1924), 164-167.
10. Über eine nicht fortsetzbare Riemannsche Mannigfaltigkeit. *Math. Z.* **20** (1924), 1-6.
11. Über den Begriff der Riemannschen Fläche. *Acta Math. Szeged* **2** (1925), 101-121.
12. Über die erste Randwertaufgabe für $\Delta u = 0$. *Math. Z.* **22** (1925), 42-44. With F. Riesz.
13. Über den analytischen Charakter der Minimalflächen. *Math. Z.* **24** (1925), 321-327.
14. Bemerkung über die Differentialgleichung zweidimensionaler Variationsprobleme. *Acta Math. Szeged* **2** (1925), 147-156.
15. Sur le calcul de l'aire des surfaces courbes. *C.R. Paris* **183** (1926), 588-590.
16. Bemerkungen zur Arbeit von Herrn Ch. H. Müntz über das Plateausche Problem (Math. Annalen 94, S. 53-96). *Math. Ann.* **96** (1926), 587-596.
17. Das Hilbertsche Problem über den analytischen Charakter der Lösungen der partiellen Differentialgleichungen zweiter Ordnung. *Math. Z.* **25** (1926), 514-589.
18. Geometrische Betrachtungen über zweidimensionale reguläre Variationsprobleme. *Acta Math. Szeged* **2** (1926), 228-253.
19. Sur le calcul de l'aire des surfaces courbes. *Fund. Math.* **10** (1927), 197-210.
20. Sur l'aire des surfaces courbes. *Acta Math. Szeged* **3** (1927), 131-169.
21. Sur l'aire des surfaces courbes. *C. R. Paris* **184** (1927), 63-65.
22. Bemerkung über das Doppelintegral $\iint (1 + p^2 + q^2)^{1/2} dx\, dy$. *Math. Z.* **26** (1927), 408-416.

23. Zu einem Satze von S. Bernstein über Minimalflächen im Grossen. *Math. Z.* **26** (1927), 559-565.
24. Sur un problème relatif à un théorème de Vitali. *Fund. Math.* **11** (1928), 228-229.
25. Über das Flächenmass rektifizierbarer Flächen. *Math. Ann.* **100** (1928), 228-229.
26. On surface area. Math. és Term.-tud. Ért. **45** (1928), 225-244. In Hungarian.
27. Sur l'aire des surfaces continues. *Atti Congr. Internaz. Mat. Bologna 3-10 Settembre 1928*, vol. **2**, 355-360.
28. Remarque sur les fonctions subharmoniques. *C.R. Paris* **186** (1928), 346-348.
29. Conformal representations of convex regions. Math. és Phys. Lapok **35** (1929), 1-9. In Hungarian.
30. Über zweidimensionale reguläre Variationsprobleme. *Math. Ann.* **101** (1929), 620-632.
31. Bemerkung über die konformen Abbildungen konvexer Gebiete. *Math. Ann.* **102** (1929), 428-429.
32. The problem of least area and the problem of Plateau. *Math. Z.* **32** (1930), 763-796.
33. Some remarks on the problem of Plateau. *Proc. Nat. Acad. Sci. U.S.A.* **16** (1930), 242-248.
34. On Plateau's problem. *Ann. of Math.* **32** (1930), 457-469.
35. On the functional of Mr. Douglas. *Ann. of Math.* **32** (1931), 785-803.
36. Contributions to the theory of minimal surfaces. *Acta Math. Szeged* **6** (1932), 1-20.
37. On mathematical life in Hungary. *Amer. Math. Monthly* **39** (1932), 85-90.
38. On surface area. Mathesis Polska **7** (1932), 1-18. In Polish.
39. Subharmonic functions and minimal surfaces. *Trans. Amer. Math. Soc.* **35** (1933), 648-661. With E.F. Beckenbach.
40. Subharmonic functions and surfaces of negative curvature. *Trans. Amer. Math. Soc.* **35** (1933), 662-674. With E.F. Beckenbach.
41. An iterative process in the problem of Plateau. *Trans. Amer. Math. Soc.* **35** (1933), 869-887.
42. *On the Problem of Plateau.* Ergebnisse der Mathematik und ihrer Grenzgebiete, Vol. **2**. Berlin: Springer, 1933. Reprinted New York: Chelsea, 1951.
43. On convex functions. *Trans. Amer. Math. Soc.* **37** (1935), 266-285.
44. The isoperimentric inequality on the sphere. *Amer. J. Math.* **57** (1935), 765-770.
45. On a converse of Kneser's transversality theorem. *Ann. of Math.* **36** (1935), 749-769.
46. A remark on the area of surfaces. *Amer. J. Math.* **58** (1936), 598-606.

47. On continuous transformations in the plane. *Fund. Math.* **27** (1936), 201-211.
48. A lemma on the topological index. *Fund. Math.* **27** (1936), 212-221.
49. Solution of a problem of F. Riesz on the harmonic majorants of subharmonic functions. *Duke Math. J.* **3** (1937), 123-132.
50. *Subharmonic Functions.* Ergebnisse der Mathematik und ihrer Grenzgebiete. Vol. **5**. Berlin: Springer, 1937.
51. On the derivative of the Lebesgue area of continuous surfaces. *Fund. Math.* **30** (1938), 34-39.
52. On absolutely continuous transformations in the plane. *Duke Math. J.* **4** (1938), 189-221.
53. On a stretching process for surfaces. *Amer. J. Math.* **61** (1939), 645-650. With P.V. Reichelderfer.
54. Cyclic transitivity. *Duke Math. J.* **6** (1940), 474-485. With P.V. Reichelderfer.
55. On upper semi-continuous collections. *Acta Math. Szeged* **9** (1940), 239-243. With J.W.T. Youngs.
56. On a lemma of McShane. *Ann. of Math.* **42** (1941), 73-83.
57. Note on an inequality of Steiner. *Bull. Amer. Math. Soc.* **47** (1941), 102-108. With P.V. Reichelderfer.
58. A theory of absolutely continuous transformations in the plane, *Trans. Amer. Math. Soc.* **49** (1941), 258-307. With P.V. Reichelderfer.
59. On the semi-continuity of double integrals in parametric form. *Trans. Amer. Math. Soc.* **51** (1942), 336-361.
60. On semi-continuity. *Amer. Math. Monthly* **49** (1942), 446-450.
61. Convergence in length and convergence in area. *Duke Math. J.* **9** (1942), 527-565. With P.V. Reichelderfer.
62. What is the area of a surface? *Amer. Math. Monthly* **50** (1943), 139-141.
63. On continuous path-surfaces of zero area. *Ann. of Math.* **44** (1943), 173-191.
64. On a problem of Geöcze. *Amer. J. Math.* **65** (1943), 361-381.
65. On the transformation of double integrals. *Trans. Amer. Math. Soc.* **54** (1943), 83-102. With R.G. Helsel.
66. Functions of rectangles. *Duke Math. J.* **11** (1944), 487-496.
67. Some remarks on the problem of Geöcze. *Duke Math. J.* **11** (1944), 497-506.
68. On surface area. *Proc. Nat. Acad. Sci. U.S.A.* **31** (1945), 102-106.
69. On continuous mappings of Peano spaces. *Trans. Amer. Math. Soc.* **58** (1945), 420-454.
70. The isoperimetric inequality and the Lebesgue definition of surface area. *Trans. Amer. Math. Soc.* **61** (1947), 530-555.
71. On two-dimensional concepts of bounded variation and absolute continuity. *Duke Math. J.* **14** (1947), 587-608.
72. On cyclic transitivity. *Fund. Math.* **34** (1947), 14-29. With P.V. Reichelderfer.

73. *Length and Area.* Amer. Math. Soc. Coll. Publ., Vol. **30**. New York: American Mathematical Society, 1948.
74. A note on convergence in length. *Bull. Amer. Math. Soc.* **54** (1948), 533-539. With M.C. Ayer.
75. On Schreier systems in free groups. *Trans. Amer. Math. Soc.* **64** (1948), 386-408. With Marshall Hall.
76. The Cauchy area of a Fréchet surface. *Duke Math. J.* **15** (1948), 159-167. With R.G. Helsel.
77. A new geometrical interpretation of the Lebesgue area of a surface. *Duke Math. J.* **15** (1948), 169-180. With E.J. Mickle.
78. On essentially absolutely continuous plane transformations. *Bull. Amer. Math. Soc.* **55** (1949), 629-632.
79. On the problem of Geöcze. *Ann. Scuola Norm. Sup. Pisa* (2) **14** (1948), 21-30.
80. On cyclic additivity theorems. *Trans. Amer. Math. Soc.* **66** (1949), 347-365. With E.J. Mickle.
81. Convergence in area. *Duke Math. J.* **16** (1949), 61-71.
82. On n-dimensional concepts of bounded variation, absolute continuity, and generalized Jacobian. *Proc. Nat. Acad. Sci. U.S.A.* **35** (1949), 678-681. With P.V. Reichelderfer.
83. On upper semi-continuous functions. *Proc. Amer. Math. Soc.* **1** (1950), 226-230. With E.J. Mickle.
84. On generalized Jacobians. *Trans. Amer. Math. Soc.* **68** (1950), 405-419. With P.V. Reichelderfer.
85. On identifications in singular homology theory. *Rivista Mat. Univ. Parma* **2** (1951), 3-18.
86. An approach to singular homology theory. *Pacific J. Math.* **1** (1951), 265-290.
87. A remark on chain-homotopy. *Proc. Amer. Math. Soc.* **2** (1951), 458-463.
88. On a theorem of Besicovitch in surface area theory. *Rivista Mat. Univ. Parma* **2** (1951), 19-45. With E.J. Mickle.
89. On generalized Lipschitzian transformations. *Rivista Mat. Univ. Parma* **2** (1951), 289-301. With P.V. Reichelderfer.
90. Sulla teoria delle omologie singolari. *Mem. Accad. Sci. Ist. Bologna. Cl. Sci. Fis.* (10) **8** (1950-51), 57-63.
91. Applications of area theory in analysis. *Proc. Int. Congr. of Math. Cambridge, MA, 1950*, Vol. **2**, 174-179. Providence, RI: American Mathematical Society, 1952. With L. Cesari.
92. On the infinitesimal rigidity of surfaces. *Osaka Math. J.* **4** (1952), 241-285. With T. Minagawa.
93. On general cohomology theory. *Proc. Amer. Math. Soc.* **4** (1953), 244-246.
94. On the infinitesimal rigidity of surfaces of revolution. *Math. Z.* **59** (1953), 151-163. With T. Minagawa.

95. On multiplicity functions associated with Lebesgue area. *Rend. Circ. Mat. Palermo* (2) **4** (1955), 219-236.
96. *Continuous Transformations in Analysis. With an Introduction to Algebraic Topology.* Die Grundlehren der mathematischen Wissenschaften in Einzeldarstellungen mit besonderer Berücksichtigung der Anwendungsgebiete, Vol. **75**. Berlin: Springer, 1955. With P.V. Reichelderfer.
97. On the space of oriented lines in Euclidean three-space. *Amer. Math. Monthly* **64** (1957), 79-89.
98. Lebesgue area and Hausdorff measure. *Fund. Math.* **44** (1957), 198-237.
99. On rigidity properties of developable surfaces. *J. Math. Mech.* **7** (1958), 419-432. With E. Kreyszig.
100. On covering theorems. *Fund. Math.* **45** (1958), 325-331. With E.J. Mickle.
101. Density theorems for outer measures in n-space. *Proc. Amer. Math. Soc.* **9** (1958), 433-439. With E.J. Mickle.
102. On reduced Carathéodory outer measures. *Rend. Circ. Math. Palermo* (2) **7** (1958), 5-33. With E.J. Mickle.
103. A uniqueness theorem for Haar measure. *Trans. Amer. Math. Soc.* **93** (1959), 492-508. With E.J. Mickle.
104. On density theorems for outer measures. *Rozprawy Mat.* **21** (1960), 48 pp. With E.J. Mickle.
105. On non-computable functions. *Bell System Tech. J.* **41** (1962), 877-884.
106. On a simple source for non-computable functions. *Proc. Sympos. Math. Theory of Automata (New York, 1962)*, 75-81. Brooklyn, NY: Polytechnic Press of Polytechnic Inst. of Brooklyn, 1963.
107. A generalization of Nelson's algorithm for obtaining prime implicants. *J. Symbolic Logic* **30** (1965), 8-12. With R.W. House.
108. Computer studies of Turing machine problems. *J. Assoc. Comput. Mach.* **12** (1965), 196-212. With S. Lin.

The Problem of Plateau (pp. 33–34)
ed. Th. M. Rassias

THE YIN AND THE YANG OF MY RELATIONSHIP WITH T. RADÓ

MAXWELL O. READE

Department of Mathematics
University of Michigan, Ann Arbor, MI 48109

During my three years at Rice University in Houston, Texas, I studied the theory of analytic functions and subharmonic functions as they related to the classical minimal surfaces. My advisor was Professor E.F. Beckenbach who had studied similar topics, as a National Research Fellow, under Professor T. Radó. When I received my Ph.D. in 1940, Beckenbach told me that he had arranged for me to do some post–doctoral work at Ohio State University where Radó was the "reigning" research mathematician. I assumed that I would continue my study of minimal surfaces and analytic functions of one complex variable.

It was late June 1940, when I drove into Columbus to introduce myself to Radó and to get a head start on future researches connecting minimal surfaces with subharmonic functions and the theory of functions. But I was surprised to find that Radó had changed his focus and had developed a school of students who were investigating "area functions", the extension of "functions of rectangles" to more general sets in the plane, and the "semi–continuity of certain multiple integrals". His advice to me was that I familiarize myself with the first four or five chapters of the book "Théorie de l'Intégral" by S. Saks. He showed no interest in discussing the possibility of extending certain function–theoretic ideas to minimal surfaces, as he and Beckenbach (and later, as Beckenbach alone), had done. That was a disappointment.

When classes began in September 1940, I was invited to attend Radó's seminar; "invited" is an euphemism — I was ordered to attend that seminar. The seminar included some fair mathematicians — Earl Mickle, John Milnor, Paul Reichelderfer, William Scott, Hugh Miser, among others. The main topic was the extension of certain functions of rectangles to additive and/or completely additive functions of Borel sets. However, I was interested and involved in several problems involving minimal surfaces. For example, did the symmetry of the boundary of a minimal surface imply the surface has the same symmetry ? How many surfaces did a simple closed Jordan curve in 3–space bound ? I wanted to read Seifert and Threlfall's little book, "Variationsrechnung" to further my knowledge of Morse Theory. But Radó was not interested; indeed,after the year spent studying functions of rectangles, the next year was spent studying the semi–continuity of integrals appearing in the theory of area, in particular, the integrals introduced by the Hungarian mathematician Geöcze. Even though I did attend most of the seminar meetings, I was a rather passive participant, although I did complete several papers derived from my doctoral dissertation with Beckenbach.

Radó was about five foot six, slender, and in very good physical condition. His manner at all times was didactic and direct — he did not brook much disagreement. For example, during my first year at OSU, I lived just north of Radó's house on Valhalla Drive. At Radó's suggestion we were to meet at the intersection of Valhalla Drive and High Street and continue together the walk of a mile or so to University Hall on campus. Radó delivered lectures on each of those walks — five times a week during the academic year 1940–41. When USSR and Germany signed their "mutual non–agression treaty", Radó claimed that that had been inevitable, the two countries were absolutely alike. When I disagreed and predicted the two countries would soon be at one another's throats, he said I was too young and uninformed to understand the situation. On those long walks he also extolled "hard work"; his contention was that a great deal of success in mathematical research was due to hard, long and constant application of the researcher's energy. He pointed to Paul Reichelderfer as an example of that precept (Radó and Reichelderfer eventually wrote "Continuous Transformations in Analysis..." in the "yellow peril" series).

Radó was a stern–appearing man, not given to light–hearted banter. I recall a reception at his house, a small gathering of his students, would–be students, and their wives. Radó almost insisted that the women, including Mrs. Radó, should converse in one room and the men in another. But he was thoroughly respected and everyone seemed to enjoy the gathering.

I did assume several of Radó's precepts. One was that you do not quote someone else's result unless you can prove that result yourself. Another was that when you write up your results for publication,there must be no "secrets", the motivation for the formal proof must be indicated. Radó must have picked up this idea from his teacher F. Riesz, whose papers are marvels of that principle.

Radó was a very patient and "giving" teacher to those who were working on problems of interest to him. However, he showed no interest in my wishes to continue my researches in an area which he and Beckenbach had opened up. That was the "Napoleon" in him. If you wished to have his attention, you had to work on problems and ideas that were of interest to him. Some fine mathematicians blossomed under his tutelage - Beckenbach, C.B. Morrey, J.W.T. Youngs, E.J. Mickle, to name a few.

I left OSU in June 1942, and saw Radó exactly once after that time. With his death, it seemed that an old–fashioned age had disappeared, the age when "Herr Professor" could and did dominate those who studied with him.

The Problem of Plateau (pp. 35– 36)
ed. Th. M. Rassias

VAGUE RECOLLECTIONS OF A YOUNG C.C.N.Y. POOR-IN-MATHEMATICS UNDERGRADUATE STUDENT OF JESSE DOUGLAS, AN OLD FAMOUS MATHEMATICS PROFESSOR

EDWARD SIEGEL
S. P. D.
International Center for Theoretical Physics
Strada Costiera, 11
P O Box 586
I-34100 Miramare
Trieste, Italy

I entered C.C.N.Y. in January, 1961, as a freshman student who was poor in mathematics and physics. In those days you were, as a student, forced to take a certain core group of mathematics courses, along with many others (140 credits needed to graduate with B.S.).

C.C.N.Y. is located about 1 kilometer from Columbia University, Jesse Douglas's alma mater in 1920. The west side I.R.T. subway (with its luminous blue lights) connects Columbia (downhill) to C.C.N.Y. (uphill) through Harlem, a dangerous neighborhood, and to Brooklyn College, 15–20 miles way, where Douglas once taught. Traditionally there has been much cross fertilization between Columbia University and C.C.N.Y.; many graduate students from Columbia taught part–time at C.C.N.Y.

C.C.N.Y. evolved from Townsend–Harris High School sometime in the early 20th century.

A professor Gill and, I believe, Jesse Douglas (B.S., C.C.N.Y., 1916) team–taught Calculus and Analytic Geometry in 1962, along with (and using the just written textbook of) Abraham Schwartz (who was department Chairperson) and Fritz Steinhardt a German professor with an Oxford accent who excited and inspired my interest in Differential Geometry.

I must say I found Calculus very boring (and still do), but old (by that time) professors Gill and Douglas did liven it up, with their witty anecdotes, especially about their nice Geometers (Analytic Geometry) versus how nasty analysts (Calculus) were.

Jesse Douglas would tell us that when he was a C.C.N.Y. student things were really tough, with the big war (World War I) going on, so that in peaceful and more affluent 1960–1965 we students should not complain. His course was given in 1962 and 1963, I believe.

He seemed to be intermittently in poor health and sometimes came late to class or did not show up at all. He came to C.C.N.Y. in 1955–56, 5 years before me, after having taught both at Columbia University and at Brooklyn College from 1942 to 1954.

I too lived in Brooklyn near Brooklyn College and made the 1–2 hours trip (each way daily on I.R.T. subway). One got the impression that he enjoyed students and talking with them in the more informal setting outside of the classroom versus in the rigid format of a formal class.

We students (and perhaps most faculty members) were somewhat in awe of him, but not once did he ever tell us about Plateau's problem or why he was so famous, indicating great modesty on his part.

He was a mysterious figure to us students. His untimely death in October 7, 1965 just after I graduated with a B.S. in June 1965 was not announced to even new alumni, and it was quite unknown to me until much later.

He was the only famous mathematician I had ever known, much less been taught by, and he did help me with my Calculus and Analytic Geometry, but not enough for me to become a mathematician, but only a physicist, which I did even more poorly in. This got me angry and so I became a physicist (and got my Ph.D. in Physics in 1970) to overcome my poor record in it.

The Problem of Plateau (pp. 37– 40)
ed. Th. M. Rassias

JESSE DOUGLAS AS TEACHER AND COLLEAGUE

FRITZ STEINHARDT

Department of Mathematics
City College of the City University of New York
New York City, N.Y., 10031, U.S.A.

It was my good fortune to have had Jesse Douglas as a teacher (summer 1941, Columbia University, in Differential Equations), to have shared an office with him at Columbia (General Studies Division) in the early 1950s, and to have been his colleague in the City College (CUNY) mathematics department where he taught for the last ten years of his life, 1955 - 1965. And my wife, when a student at City College (I did not know her then, as she avoided courses with reputed "tough" teachers) was Douglas' student in his 1961 Modern Algebra course. Thus while not particularly his personal or mathematical confidant, I can yet contribute a few first-hand and some second-hand observations and related items.

As a teacher Douglas ranks for me among the few truly impressive, charismatic expositors I heard, such as Heinz Hopf, George Polya, Emil Artin, Gilbert Baumslag, all but the last as their student. Douglas' charisma seemed to be related even more to his personality than to what he made of the material presented. His calm, quiet, completely lucid and cohesive style - he moved little, often sitting in his chair while he wrote on the board - gave his listener complete confidence in Douglas' mastery; what his unemotional but intense voice and perfect sentences projected was his bigger-than-life self-confidence.

One of the sources of this self-confidence was no doubt Douglas' oft-confessed gift of "total recall", even of events back in earliest childhood. (He told me, for instance, of very clear memories of walking with his mother shortly after learning to walk.) In other respects too his mind was acute way beyond the ordinary, not only as to his mathematical gifts and sharpness but also in his personal and professional sensitivities. Some of the over-sized sensitivity surfaced not infrequently in intense feelings articulated vigorously to colleagues, feelings of injustice and of less than fair appreciation of his accomplishments.

Indeed the volatility of this mix of great gifts and sharp emotions in Douglas' makeup worked to the disadvantage of the great American mathematician's external fortunes. By the middle 1930s he was acknowledged to be one of his country's half dozen most creative mathematicians; but the instability and occasional breakdowns of his nervous system did not allow him to find the permanent major position that would have been his due, certainly after his prize-winning memoirs and the 1936 Fields medal.

Thus Douglas held a number of temporary, minor lecturer positions (such as teaching Summer session and extension division courses at Columbia U.) and filled other posts for relatively short periods - at M.I.T., Brooklyn College, Yeshiva University - until he accepted a tenured professorship at City College (his Alma Mater) in 1955, at age 58. At Brooklyn College he served as Assistant(!) professor from 1942 to 1946, and I quote from a letter he wrote in 1945 while casting about for another position: "This semester I have a teaching load of 19 hours per week at Brooklyn College, including two advanced courses: complex variable and calculus of variations. I shall the more keenly appreciate a decent schedule of 6 teaching hours weekly, giving me the opportunity of continuing with my mathematical work..."

But earlier, in his late twenties and thirties, Douglas had spent time at the Princeton Institute for Advanced study and had traveled and lectured extensively, at Harvard, Princeton, Chicago, Paris, Goettingen, Rome.

Douglas' coming to City College was the result of untiring efforts by one of his former M.I.T. students, Abraham Schwartz, a member of the College's mathematics department. City College at the time had a first-rate student body and many gifted mathematics majors; the large mathematics faculty, by contrast, was largely undistinguished as far as research goes (exceptions included the great Emil Post, who died in 1954). City University, and a steep upgrading of criteria for faculty appointments,, were still about ten years away. In those circumstances many of the less creative mathematics department members felt threatened by the appointment, in the top rank of Professor to boot, of a luminary of Douglas' caliber. As a result, at the next following departmental elections the entire 5-man appointments committee was defeated, and the incumbent chairman was barely re-elected, after 21 ballotings in which the leader of the hacks came within inches of success! Things have vastly improved since that time, at least as far as the composition of faculty is concerned.

While Douglas, though always helpful and sympathetic, was perhaps a shade distant with students, he was quite approachable as a colleague and shared his opinions freely, often revealing his strongly-held attitudes about mathematical disciplines and fashions.

Thus for example, his attitude; in the years when it was all the rage, toward the "new math" (language of sets, together with related bits of logic, down to the secondary and even the elementary school level) was strongly negative. And in discussions on what should be the content of that grab-bag course "Advanced Calculus", Douglas indicated very little interest in foundational matters, including construction or even a postulational treatment of the number system. For him, mathematics only began on the other shore, as though - to paraphrase Kronecker - "God made the real numbers...".

A second example: To a young instructor, Douglas' recommendation for a textbook to use in the topology course, in 1965, was the 1934 (German) book by Seifert & Threlfall; a classic of course, but antediluvian in that fast-moving field by 1965 standards, with many new texts available. Surely being modern or "mainstream" meant little to Douglas.

Another of my recollections of Douglas also involves the choice of a textbook and again is revealing both of his indifference to "modernity" and of his acute sensitivity, already mentioned, to real or perceived slights to his achievements. He and I divided the over sixty registrants for the Differential Geometry course, at City College in 1963, into two sections. Ordinarily I would as a matter of course have deferred to him over the selection of a textbook to be used in both sections. But this turned out to be no ordinary problem; it led to amicable disagreement and the choice of two different books, as Douglas found nothing suitable for his section except for a somewhat antiquated and (to me) dull book from the 1930's.

His criterion? Most of the more recent Differential Geometry books had references such as "T. Rado' [1], [2] and J. Douglas [1] obtained the following general solution of the problem of Plateau..." or " This 'problem of Plateau'... was answered in the positive by J. Douglas (1931) and T. Rado' (1930)." Such offenses to Douglas' sense of fair or accurate attribution ruled out use of the books that contained them. It was not that he thought little of Rado's work; but whatever he thought of it, he did not consider that it contained the solution of Plateau's problem, as his own work certainly did; he could not tolerate the inaccuracy.

Since he was an excellent teacher and expositor, it is regrettable that Douglas left very little of his pedagogical creativity in written form. There is a 47-page "Survey of the Theory of Integration" (1941, Scripta Mathematica), and among the papers found after his death there was a hand-written chapter on "Interval Functions," most likely part of an Advanced Calculus text otherwise unwritten; just enough to make one wish for more.

Ill luck of sorts marked even the obituary that appeared in the New York Herald Tribune of October 8, 1965, written by someone who felt competent all by his incompetent self to enlighten his readers on the main claim to fame of Douglas (consistently spelled "Douglass" in the obituary). It begins "Squaring the circle has been an unsolved challenge to mathematicians since Euclid, but a lesser problem that had been troubling them for half a century was solved in 1936 by Dr. Jesse Douglass. It had to do with the plateau of graphs, and its solution won for Dr. Douglass the Fields Medal...". The writer goes on about "... that 'plateau' which had baffled mathematicians for 50 years...", and he later proceeds to quote a definition of "plateau" from Webster's Internationsl Dictionary, to wit "In a graph, a region of little or no change in the dependent variable;...". Douglas had a sense of humor and he might even have enjoyed a laugh at this earnest attempt. His students and colleagues were inspired by his gifts and enjoyed him as a person to be with.

December, 1990

The Problem of Plateau (pp. 41– 42)
ed. Th. M. Rassias

MY RECOLLECTIONS OF JESSE DOUGLAS

DIRK J. STRUIK
Department of Mathematics
Massachusetts Institute of Technology
Cambridge, Massachusetts 02139, U.S.A.

Jesse Douglas became a member of the mathematics department at MIT in 1930. He was, at 33 years of age, already a well–known scientist who had written an interesting doctor's dissertation in differential geometry under his teacher Edward Kasner at Columbia University in New York, had studied in Chicago, Paris and Göttingen, and, above all, had already been publishing on his solution of the problem of Plateau, subtle and highly original work for which he would receive the Fields Medal at the Oslo International Mathematical Congress in 1936.

We at MIT soon discovered that he was a remarkably sensitive mathematician with a wide field of creative power, equally versed in geometry as in analysis. I remember pleasant discussions with him, often on his research in the problem of Plateau, worrying about the work of his rival Tibor Radó, but also on colleagues past and present and faculty events.

He was a witty fellow, fond of anecdotes which he told spiced with his own little prejudices : my experience, he claimed, was that as a rule geometers were nice people, but analysts could be quite nasty. One day, at Göttingen, he told us, I was following a lecture of Professor Landau, when he mentioned the so–called Gibbs phenomenon in the theory of Fourier series. Said the professor : "Dieses phenomen ist von dem Englischen Mathematiker Gibbs (pronounced Jibbs) in Yale (pronounced jail) entdeckt". Only my respect for the great mathematician, Jesse said, withheld me from saying : "Herr Professor, what you say is quite correct. Only he was not English, but American, he was not a mathematician, but a physicist, he was not named Jibbs, but Gibbs, he was not in jail, but at Yale, and finally, he was not the first to discover it".

He loved to talk to some of his colleagues and was an affectionate soul, but basically he was a lonely man, unmarried, living in a Cambridge hotel, and often unwell. His health prevented him not unfrequently to come to his class on the regular schedule, so that Henry Philips, the head of our department, insisting on conscientious teaching, had to let him go, to my and others' regret. This was in 1936, the year he received the Fields Medal.

We heard from him from time to time, admiring his fine research, while he was living on fellowships. In 1942 he received a teaching engagement at Brooklyn College, New York, and from 1955 on he taught at City College, also in New York, his birthplace. I believe that he married, which may have lessened his loneliness. His mathematical research remained superb, in 1943 he received the Bôcher Prize of the American Mathematical Society. He died in 1965.

It is very pleasant to know that Professor Themistocles Rassias is editing this volume on the problem of Plateau, on which my old colleague Jesse Douglas has done such distinguished research. I am sure that Professor Rassias, who also did so much work on Plateau's problem, is eminently qualified to do this publication.

August 1, 1990

The Problem of Plateau (pp. 43– 49)
ed. Th. M. Rassias

JESSE DOUGLAS, THE PLATEAU PROBLEM, AND THE FIELDS MEDAL: SOME PERSONAL REFLECTIONS

HENRY S. TROPP
Department of Mathematics
Humboldt State University
Arcata, CA 95521-4959
USA

ABSTRACT

The exposition commences with a brief summary of Joseph Plateau's 19th century experiments which led to the well-known soap bubble demonstration of minimal surfaces on particular wire frames. This is followed by a brief account of the establishment of the Nobel Prize and the Fields Medal and the award of the latter in its first year of existence to Jesse Douglas for his mathematical proof of what was then the Minimal Surface Problem. The essay concludes with a personal response to the question: Why is there no Nobel Prize for mathematics?

Since this volume commemorates the sixtieth anniversary of Jesse Douglas' milestone publication, "Solution of the Problem of Plateau" [1] and the equally pioneering work of Tibor Radó [4;5;14], it is worthwhile to begin by briefly examining its experimental origins almost a century earlier.

Joseph Antoine Ferdinand Plateau (1801–1883) was a well-known Belgian scientist. His early work in physiological optics resulted in temporary blindness in 1829 during an experiment in which he stared at the sun for twenty-five seconds [3, p. 21]. Although his sight partially returned after a few days, the condition deteriorated slowly, and by 1843 he was totally blind. Despite this, he continued his experimental career [*ibid.*].

In the 1840s, Plateau began a series of experiments in which he studied molecular forces through the consideration of a weightless mass of liquid. Joseph Henry, the first Secretary of the Smithsonian Institution, was interested in this area of experimentation. The reason for Henry's interest appears in his *Introduction* to an English translation of Plateau's *Memoirs* [2(1863); p. 207]: "The experiments described in the first and second parts of the series have excited much interest, and have frequently been presented in popular lectures as precise illustrations of the mode of formation of Saturn's rings, and almost conclusive proofs of the truth of the hypothesis of La Place [sic] as to the genesis of the solar system."

Plateau's efforts in these experimental reports are centered on the study of molecular forces by making a liquid which effectively annuls the force of gravity. He accomplished this "by immersing a quantity of oil in a mixture of water and alcohol, the density of which was equal to that of oil" [2(1863), pp. 208–209; 3, p. 21]. After a variety of observations which included a system of small spheres resulting in a configuration similar to the rings of Saturn, Plateau began to vary his experiments by introducing metal wires so that, among other things, he could observe the forms of equilibrium that occurred.

It is in the Fifth Series of these *Memoirs*, subtitled "Theory and description of a new process for the realization of figures of equilibrium" [2(1865), pp. 411–435], that the experiments that mathematicians characterize as "Plateau's Problem" are described. In these studies of molecular forces, Plateau developed a special mixture of glycerine, soap, and water [*Ibid.*, p. 411] which had the property that, with suitable precautions, a bubble on film could last for up to eighteen hours. (On page 435 of this document, he gave his "recipe" for the preparation of this liquid, including the fact that he had his most success with five volumes of a glycerine prepared in London mixed with four volumes of soap and water, a mixture also described in detail). Throughout his report, he describes the shapes that occur when various frames are immersed in the soap solution, and states his conclusion that "the surfaces of all its films shall be a minimum" [2(1866), p. 289].

Many readers are familiar with viewing the various surfaces generated by immersing various geometric frames into a soapy solution. I first viewed these phenomena as a child on a visit to Chicago's Museum of Science and Industry. Despite the intuitive feel that the soap film, under the constraints of the wire contour and under its surface tension, did indeed produce "least" surface area, a satisfactory conclusion in experimental physics, it is obviously not going to satisfy mathematicians. Hence the search by a string of mathematicians ranging from Riemann, Weierstrass, Schwarz, Hilbert, Haar, Courant, etc., that eventually led to general solutions by both Tibor Radó [4,5,13] and Jesse Douglas [1]. According to Douglas, the pioneering mathematical memoir on the subject of minimal surfaces is that of Lagrange, 1760–61, on the calculus of variations.

The Fields Medal is a more recent development. It was established at the International Mathematical Congress in 1932 and the first two awards, presented at the 1936 Congress, were to Jesse Douglas and Lars Ahlfors [13]. The award was viewed almost from its inception as the Nobel Prize for mathematics. As the only international award in mathematics sanctioned by the international mathematical community, it clearly represented the same distinguished status for mathematicians accorded Nobel Laureates in physics, chemistry, medicine and literature. (A Nobel Prize for economics was added later and the peace prize is treated slightly differently from the others.) The two awards share common ground in that both Nobel and Fields wanted to avoid nationalistic links. Nobel, in his will, stated: "... in awarding the prizes no consideration whatever shall be given to the nationality of the candidates, so that the most worthy shall receive the prize whether he be a Scandinavian or not" [8, pp. x–xii]. Fields was even more emphatic: "... the medals should be of a character as purely international and impersonal as possible. There should not be attached to them in any way the name of any country, institution or person" [13, p. 174]. The latter request was ignored immediately when the Medal was named after Fields, who was no longer alive to protest.

Despite these similarities, the two awards differ sharply in at least two ways: criteria and renown.

Nobel's will, on the criteria for the award, clearly states: "The ... fund ... shall be annually distributed in the form of prizes to those who, during the preceding year, shall have conferred the greatest benefit on mankind" [8, pp. x–xi].

Fields' criteria were much less explicit. The award was to be made as much for promise as for accomplishment. In his words: "In coming to its decision the hands of the International Committee should be left as free as possible. It would be understood, however, that in making the awards while it was in recognition of work already done it was at the same time intended to be an encouragement for further achievement on the part of the recipients and a stimulus to renewed effort on the part of others" [13, p. 174].

In the matter of relative prestige, the difference is enormous. The Nobel Prize is annually recognized by many millions of people worldwide as the ultimate in its specified designations. The Fields Medal receives brief notoriety every four years at the International Mathematical Congress and its identity is familiar to only a portion of the mathematical community. I only learned of it when, in 1968, I was asked to write a notice on Fields for the Dictionary of Scientific Biography. Subsequently, I spent the academic years 1969–1971 at the University of Toronto with the late Kenneth May. There was a plaque on the fourth floor of Sidney Smith Hall which had a reproduction of the Fields Medal and a listing of recipients through 1966. In conversation with members of the mathematics department, I discovered that many were unaware of even the existence of the award. This was reinforced even more dramatically a few years later.

In 1972, while I had a brief association with the Smithsonian Institution, I went to the University of Maryland to hear a lecture by Jean Dieudonné. The lecture hall was full. The audience was mostly members of the mathematics faculty and graduate students. It was to the latter group that Dieudonné focused his remarks. He created a chart of various branches of mathematics that in his opinion contained the most potential for future mathematical research. The top of the chart contained the branches which, in his view, were the richest in unsolved problems and open-ended conjectures for research. The bottom of the chart were those that he viewed as being less rich in potentially interesting intellectual challenge. I wish now that I had recorded the talk and photographed the chalkboard. I didn't, nor did anyone else. All I remember is the bottom branch on his research potential chart: set theory. At the conclusion of the lecture, the chairman opened the floor to questions. A young man rose and asked: "Professor Dieudonné, what advice can you give about finding good research topics?" Dieudonné's response: "Look at the works of the Fields Medalists." Silence. Then, from somewhere in the hall came the question: "What is the Fields Medal?" Again, silence. I don't remember Dieudonné's response, but I do remember that my impression as I left the room was that few in that gathering could have responded to the last question. In fact, it was this experience that led me to write a paper on the subject [13].

As I worked on the paper and looked at the mathematical publications of some of the medalists, I realized that Dieudonné's response was in the spirit of Field's bequest. He really meant *all* of their works, not just that for which Prize recognition was achieved. As the years passed, I realized that the various Fields Medal Prize Committees had worked hard at the difficult task of trying to judge the promise of the individual as well as the significance of their accomplishments. Douglas' work represents this post-medal potential as well as already-reached distinction. After his milestone 1931 publication, he "... went on to consider surfaces bounded by any finite number of contours and to consider surfaces of higher topological structure ..." [9, p. 174]. His work after 1931 was not confined to the Plateau Problem. "In 1941 he published a complete solution of the inverse problem of the calculus of variations for three-dimensional space — a problem unsolved until then, although in 1894 Darboux had stated and solved the problem for the two-dimensional case" [9, p. 174]. Schaumberger also points out that in addition to some fifty papers on geometry and analysis, he made notable contributions to group theory, in particular, finite groups. Truly in the spirit of Fields' criteria [13].

Jesse Douglas also made an unpredictable contribution to the criteria for the Fields Medal: the age of the recipient. In 1936, he was thirty-nine years old. In 1950, the Committee took this as a "definition" of Fields' desire to award the prize to "young" mathematicians. For this purpose, "young" continues to be defined as not yet having reached the age of forty. This has been rigidly followed to the present day. The effect this has had on the Committee ignoring truly youthful mathematicians in favor of those in danger of reaching the age of forty before the next Congress is a matter of speculation.

One cannot leave the joint Prizes of Nobel and Fields without commenting on the question: Why didn't Nobel establish a prize for mathematics? There is not, nor will there ever be, a definitive answer. Synge did mention that it was not due to negative feelings about mathematics. In fact, Synge said, Nobel viewed mathematics as the Queen of the sciences [10]. The general mythology in the mathematical community is that the omission is due to the hostility between Nobel and the noted Swedish mathematician, Mittag-Leffler [11]. Nobel's hostility, goes the story, was so strong that he omitted mathematics in order to insure that Mittag-Leffler would not be awarded a prize. The source of the hostility, according to the mythology: Mittag-Leffler's attraction to Nobel's wife. Hogwash! Although this might be a rational reason, Nobel never married. Nor is there evidence that he had any such romantic inclination. So there is not even a reason for romantic jealousy. This does not, however, eliminate the possibility of personal animosity. Again, there is no evidence.

My personal conjecture involves the esoteric nature of mathematics. In his will, Nobel is very specific. This is not possible for mathematics. As I pointed out earlier, the Nobel awards are for accomplishments in the preceding year which *benefit mankind*. Mathematical events such as Douglas and Radó's milestone papers on the Problem of Plateau, Gödel's Theorem, Turing's 1936 paper, the solution of Hilbert's Tenth Problem, or the computer solution of the Four Color Problem are clearly significant mathematical milestones, but how does one measure their benefit to mankind? Fields recognized this difficulty when he decided to establish a prize in mathematics. He wrote of the difficulty, given the multiplicity of branches of mathematics, to make only one award. Instead, he recommended *at least* two awards (emphasis mine) and further gave the Committee freedom to designate specific branches for consideration at a given Congress [13, pp. 173–1744]. In some years there have been as many as four awards. Nobel must have recognized the difficulty assessing mathematical accomplishments and realized that he would have had to set up very different kinds of awards with a different basis than the five designated areas he had specified in his will. Also, according to Synge, Nobel was not a man to seek or take the advice of others [10]. Since mathematics did not fit into the philosophical framework underlying his establishment of Prizes, he omitted it. At least, that is my view.

Perhaps, as is so often true in mathematics, we are asking the wrong question. Why doesn't the Nobel Committee now establish a prize in mathematics? After all, they have added economics. Perhaps they view economics as another branch in which mathematics is used for the benefit of mankind.

The contents of this volume, and as further examples, the publications of the Editor, Professor Themistocles Rassias, are clear evidence that the award to Jesse Douglas for promise as well as accomplishment was precisely what J.C. Fields wanted [6,7]. (Professor Rassias' Doctoral Dissertation was written under the supervision of Fields Medalist (1966) Stephen Smale. Rassias' 1986 book contains not only some of the most significant results which have been achieved since Douglas and Radó, but he also poses several research problems for further work on the topic). Fields wanted a mechanism to encourage and reward mathematical research. I think he accomplished this beyond his wildest dreams. In the 1920s and 1930s when he formulated this award, mathematical research was almost non-existent in his native Canada and in the U.S. It was his dream to establish a research environment for mathematics in U.S. and Canadian universities. He had been exposed to that research world in the mathematical environment of Paris and Berlin between 1892 and 1902 [12]. He could not have foreseen the impact that Nazi Germany and World War II would have in moving the center of mathematical research from Europe to the U.S. He would, however, have viewed the establishment of mathematical research as an important role in the university as the realization of a life-long dream. In that sense, it is fitting that the "Nobel Prize for Mathematics" has his name attached to it. Symbolically, the 1936 Award to Jesse Douglas and Lars Ahlfors signals the emergence of North America as a new force in the world of mathematics.

Bibliography

1. Jesse Douglas, *Solution of a problem of Plateau*, Trans. Amer. Math. Soc. **33** (1931), 263–321.
2. Joseph Henry, English Translation and Introduction to the first six *Mem-*

oirs of Plateau's "Experimental and Theoretical Researches on the Figures of Equilibrium of a Liquid Mass Withdrawn from the Action of Gravity, & . . .," Annual Report to the Board of Regents of the Smithsonian Institution (1863), 207–285; (1864), 285–369; (1865), 411–435; (1866), 255–289.

3. Elaine Koppelman, *Joseph Antoine Ferdinand Plateau*, Dictionary of Scientific Biography XI, pp. 20–22, Charles Scribner and Sons, 1975.
4. Tibor Radó, *Contributions to the theory of minimal surfaces*, Acta Litt. Scient. Univ. Szaged **6** (1932), 1–20.
5. Tibor Radó, *On the Problem of Plateau*, Ergeb. Math. Grenzgeb. **V**, Leipzig (1933); reprinted New York, 1949.
6. Themistocles M. Rassias, *Global Analysis: Morse Theory and the Problem of Plateau* (Ph. D. Dissertation) U.C., Berkeley, 1976.
7. Themistocles M. Rassias, *Foundations of Global Nonlinear Analysis*, Teubner-Texte Math., Leibzig, Band **86**, 1986.
8. H. Schuck et al., *Nobel: The Man and His Prizes*, Elsevier, Amsterdam, 1972.
9. Norman Shaumberger, *Jesse Douglas*, Dictionary of Scientific Biography IV, pp. 173–174, Charles Scribner and Sons, 1971.
10. J.L.Synge, correspondence with author, 7 February 1972.
11. Edward Teller, *Why Einstein should have won the Nobel prize*, *Selected Studies: Physics-Astrophysics, Mathematics, History of Science*, ed. Th. Rassias and G. Rassias, North-Holland, Amsterdam, 1982, pp. 379–381.
12. Henry S. Tropp, *John Charles Fields*, Dictionary of Scientific Biography IV, p. 617, Charles Scribner and Sons, 1971.
13. Henry S. Tropp, *The origins and history of the Fields medal*, Historia Math. **3** (1976), 167–181.
14. Henry S. Tropp, *Tibor Radó*, Dictionary of Scientific Biography XI, pp. 259–260, Charles Scribner's and Sons, 1975.

PART II

The Problem of Plateau (pp. 53– 75)
ed. Th. M. Rassias

CLASSIFYING PSEUDO-RIEMANNIAN HYPERSURFACES BY MEANS OF CERTAIN CHARACTERISTIC DIFFERENTIAL EQUATIONS

LUIS J. ALIAS*, ANGEL FERRANDEZ** and PASCUAL LUCAS**

Departamento de Matemáticas, Universidad de Murcia,
Campus de Espinardo, 30100 Espinardo, Murcia, Spain

1. Introduction

The study of minimal surfaces has a long and rich history, and the connection between them and soap films motivated the celebrated Plateau's problem, which has remain completely unsolved for any non-planar contour until the last third of the nineteenth century. However, in the late twenties of this century, Jesse Douglas ([**18**], [**20**] and [**19**]) and Tibor Radó ([**41**], [**43**] and [**42**]) have, quite independently of each other, been successful in developing new methods for solving Plateau's problem. Douglas's work was important both for the simplicity of the method of proof, using calculus of variations of a certain functional, and for the result itself, since the desired minimal surface is nothing but that where the minimun of the above functional is achieved.

It is well known that minimal immersions of a differentiable manifold M in the Euclidean sphere S^n are just those immersions whose coordinate functions in the ambient Euclidean space are eigenfunctions of the Laplacian operator in the induced metric with eigenvalue $\lambda = -\dim(M)$. Moreover, Takahashi's result, [**45**], is particularly useful in studing isometric minimal immersions of Riemannian symmetric spaces into spheres, since it shows that such immersions correspond precisely to the isometric immersions into $\mathbb{R}^n$ which can be achieved by eigenfunctions of the Laplacian operator with the same non-zero eigenvalue. This will be the starting viewpoint of our study in order to obtain further natural extensions, all of them showing minimal immersions as trivial solutions.

The quoted theorem of Takahashi gives a characterization of minimally immersed submanifolds in nonnegatively curved space forms. That is given in terms of the

*Supported by a FPI Grant, Program PG, Ministerio de Educación y Ciencia
**Partially supported by a DGICYT Grant No. PB90-0014-C03-02

coordinate eigenfunctions of the isometric immersion $x : M^n \longrightarrow \mathbb{R}^m$. Actually, Takahashi's result is dealing with the eigenvalue equation

$$(*) \qquad \Delta x = \lambda x,$$

being Δ the Laplacian on M coming from the induced metric and λ a real constant. Then either $\lambda = 0$ and M is minimal or $\lambda > 0$ and M is minimal in $S^{m-1}(r) \subset \mathbb{R}^m$, where $r = \sqrt{n/\lambda}$.

Takahashi's theorem can be seen as a result of classifying submanifolds satisfying a certain differential equation in the Laplacian of the immersion. Then the following general problem comes out in a natural way:

> *Classify submanifolds by means of some Laplacian differential equation involving the isometric immersion.*

On the other hand, the equation (*) says that minimal submanifolds in nonnegatively curved space forms are the only ones whose immersion is associated to exactly one eigenvalue of its Laplacian.

Then, from this viewpoint and considering a first extension of Takahashi's theorem, B.Y. Chen, based on the equation (*), built up and developed a fruitful and interesting technique, the so-called *finite type submanifolds* (see [**11**]), chiefly directed to characterize certain families of Euclidean submanifolds. For instance, a Chen-type question states as follows:

> *Could you characterize Euclidean submanifolds whose isometric immersion is associated to two distinct eigenvalues of its Laplacian?*

In particular, if M^n is a compact hypersurface of the sphere S^{n+1} in $\mathbb{R}^{n+2}$ having constant mean curvature α and constant scalar curvature τ, then either M is a small hypersphere in S^{n+1} constructed in $\mathbb{R}^{n+2}$ by using eigenfunctions associated to only one eigenvalue of its Laplacian or M lies in $\mathbb{R}^{n+2}$ by means of eigenfunctions associated to exactly two distinct eigenvalues, which in addition completely determine the geometric quantities α and τ of M. Therefore, the family of Euclidean submanifolds which can be built by using only two eigenvalues of the Laplacian is large enough to pay attention on it, since it contains, among others, those spherical hypersurfaces with constant principal curvatures.

A second extension of Takahashi's theorem can be viewed as follows. For any isometric immersion $x : M^n \longrightarrow \mathbb{R}^m$ it is well known the formula $\Delta x = -nH$, that along with (*) yields to

$$(**) \qquad \Delta H = \lambda H,$$

where H states for the mean curvature vector field of the immersion. Let us denote by C_λ the family of submanifolds satisfying equation (**). It is not dificult to see that cylinders are in $\mathcal{C}_\lambda$ but they do not satisfy (*), so that $\mathcal{C}_\lambda$ contains Takahashi's family as a proper one. However, if M is compact, both equations define the same family. Then it seems natural to ask for the following geometric question:

Which is the size of $\mathcal{C}_\lambda$?

One hopes to find in $\mathcal{C}_\lambda$ other submanifolds apart from cylinders and those of Takahashi's family. Notice that this problem is closely related to those of Chen, because an immersion satisfying (**) is (i) either minimal or of infinite type, if $\lambda = 0$, or (ii) either of 1-type or of null 2-type, provided λ does not vanish.

Furthermore, it is worth exploring the existence of non-minimal submanifolds having harmonic mean curvature vector field.

As a third attempt to generalize Takahashi's condition, O.J. Garay, **[25]**, pointed out that if you extend the Laplacian in a natural way to $\mathbb{R}^m$-valued functions on M^n, then equation (*) characterizes those submanifolds whose coordinate functions in $\mathbb{R}^m$ restricted to M^n are eigenfunctions of its Laplacian, all of them associated to the same eigenvalue. There he deals with Euclidean hypersurfaces whose coordinate functions are eigenfunctions of its Laplacian but not necessarily for the same eigenvalue, expecting for enough examples apart from those given by Takahashi. Garay's condition can be written as a Laplacian coordinate equation as follows

$$\Delta x_i = \lambda_i x_i, \qquad i = 1, \ldots, m,$$

where $x = (x_1, \ldots, x_m)$, being x_i the coordinate functions; or even, as a matricial equation

$$\Delta x = Ax,$$

where $A = \mathrm{diag}(\lambda_1, \ldots, \lambda_m)$. Nevertheless, Dillen, Pas and Verstraelen, **[16]**, pointed out that Garay's condition is not coordinate invariant as a circular cylinder in $\mathbb{R}^3$ shows. Then they study and classify those surfaces in $\mathbb{R}^3$ satisfying the new following equation

$$(***) \qquad \Delta x = Ax + B,$$

where A is an endomorphism of $\mathbb{R}^3$ and B is a constant vector in $\mathbb{R}^3$.

One immediately asks for the geometric meaning of equation (***). Before giving an answer we first notice that both equations (**) and (***) are equivalent for surfaces in $\mathbb{R}^3$ (see **[16]** and **[21]**), but that situation is quite different for surfaces in the 3-dimensional Lorentz-Minkowski space. In this ambient space those equations show its power, bringing out the B-scrolls as an interesting and own family which cannot be given in the definite case and in his turn satifies (**) but not (***). Furthermore, it is worthwhile to set off that both equations work as constant mean curvature conditions, so that we are on the track of an isoparametric problem which allows us to reach the asked classification.

In the past few years we have made some contributions to each one of the problems above stated, so that this paper will be a sort of survey in the following sense. We will revisit our recent results concerning to the quoted extensions of Takahashi's theorem rather emphasizing on those we have got about hypersurfaces in the realm of Lorentzian geometry and that revision will include not only published or accepted papers for publication, but also unpublished and others in preparation results in order to make a self-contained article. Finally, we should like to take this opportunity to propose some open problems.

2. Spherical 2-type hypersurfaces

This section is devoted to get a nice characterization of those submanifolds that can be constructed by using exactly two eigenvalues of its Laplacian in terms of its mean and scalar curvatures, which in his turn allows us to solve a series of problems stated by B.Y. Chen in **[9]**. As an interesting consequence, provided the number of principal curvatures is bounded above, a classification of spherical Dupin hypersurfaces constructed in S^n by means of two eigenvalues is given.

A connected (not necessarily compact) submanifold M^n of a pseudo-Euclidean m-space Q^m is called *of finite type* if its position vector field x can be written as a finite sum of eigenfunctions of its Laplacian; more precisely, M^n is said to be *of finite k-type* if its position vector field x admits the following spectral decomposition

$$x = x_0 + \sum_{t=1}^{k} x_t,$$

where $\Delta x_t = \lambda_t x_t$, $t = 1, \ldots, k$, $\lambda_1 < \cdots < \lambda_k$, x_0 is a constant vector in Q^m and x_t $(t = 1, \ldots, k)$ are non-constant Q^m-valued maps on M^n. Otherwise, M^n is said to be *of infinite type*. In particular, if one of the eigenvalues λ_t vanishes, then M^n is said to be *of null k-type* (see **[11]**).

Let M be a hypersurface of the unit hypersphere S^{n+1} in $\mathbb{R}^{n+2}$ which we will assume (without loss of generality) centred at the origin of $\mathbb{R}^{n+2}$. Denote by x the position vector of M in $\mathbb{R}^{n+2}$ and by ∇ and D the Levi-Civita connection of M and the normal connection of M in $\mathbb{R}^{n+2}$, respectively. We also denote by σ, S and H (H') the second fundamental form of M in $\mathbb{R}^{n+2}$, the shape operator of M in S^{n+1} and the mean curvature vector field of M in $\mathbb{R}^{n+2}$ (S^{n+1}, respectively). If Δ denotes the Laplacian of M, then the following formula for ΔH was computed in **[12]**:

$$\Delta H = \frac{n}{2}\nabla\alpha^2 + 2\,\mathrm{tr}S_{DH'} + (\Delta\alpha + \alpha|\sigma|^2)N - (n\alpha^2 + n)x, \tag{2.1}$$

where $H' = \alpha N$, being N the unit normal vector field of M in S^{n+1}. Here $\nabla\alpha^2$ denotes the gradient of α^2 and $\mathrm{tr}S_{DH'} = \sum_{i=1}^{n} S_{D_{E_i}H'}E_i$, where $\{E_1, \ldots, E_n\}$ is a local orthonormal frame tangent to M.

Now, assume that M is of 2-type. Then its position vector in $\mathbb{R}^{n+2}$ can be written as

$$x = x_0 + x_1 + x_2, \qquad \text{with} \qquad \Delta x_1 = \lambda_1 x_1 \text{ and } \Delta x_2 = \lambda_2 x_2, \tag{2.2}$$

where x_0 is a constant vector in $\mathbb{R}^{n+2}$ and x_1, x_2 are $\mathbb{R}^{n+2}$–valued non-constant differentiable functions on M.

From (2.2) and the well known fact $\Delta x = -nH$, we have

$$\Delta H = bH + c(x - x_0), \tag{2.3}$$

where $b = \lambda_1 + \lambda_2$ and $c = \frac{1}{n}\lambda_1\lambda_2$.

REMARK 2.1. Through this section, we can assume that $c \neq 0$, otherwise last two authors have proved in [**23**] the non-existence of such hypersurfaces. Of course, if M is compact then $c \neq 0$.

From (2.1) and (2.3) one gets the following formulae:

$$n\alpha^2 + n = b - c + c < x, x_0 > \tag{2.4}$$

and

$$< \Delta H, X >= -c < x_0, X >, \tag{2.5}$$

for any vector field X tangent to M.

By using (2.4) and (2.5) a nice expression for the tangential component of ΔH is found:

$$(\Delta H)^T = -n\nabla\alpha^2. \tag{2.6}$$

On the other hand, from (2.1) one has

$$(\Delta H)^T = \frac{n}{2}\nabla\alpha^2 + 2\,\mathrm{tr}S_{DH'}. \tag{2.7}$$

Finally, an easy computation involving (2.6), (2.7) and Codazzi equation gives

$$S(\nabla\alpha) = \mathrm{tr}S_{DH'} = -\frac{3n}{4}\nabla\alpha^2. \tag{2.8}$$

Therefore, the following lemma is proved.

LEMMA 2.2. [**12**] *Let M be a 2-type hypersurface of S^{n+1}. Then $\nabla\alpha^2$ is a principal direction with principal curvature $-\frac{3n}{2}\alpha$ on the open set $\mathcal{U} = \{p \in M : \nabla\alpha^2(p) \neq 0\}$.*

Next lemma, which can also be found in [**12**], allows us to get a good information about the above quoted open set $\mathcal{U}$.

LEMMA 2.3. *Let M be a 2-type hypersurface of S^{n+1}. Then either M has constant mean curvature or $\mathcal{U}$ is dense in M.*

2.1. Main results

Before going any further, some computations are needed. For short, we write $h = (b - |\sigma|^2)\alpha - \Delta\alpha$ and $g = n\alpha^2 + n + c - b$, and use (2.1), (2.3) and (2.6) to get

$$cx_0 = n\nabla\alpha^2 + hN + gx. \tag{2.9}$$

Now, working on $\mathcal{U}$, choose a local orthonormal frame of principal directions $\{E_1, \ldots, E_n\}$ with associated principal curvatures $\{\mu_1, \ldots, \mu_n\}$, being E_1 in the direction of $\nabla\alpha^2$, so that $\mu_1 = -\frac{3n}{2}\alpha$. By using (2.9) we find the following auxiliar result.

LEMMA 2.4. *Let M be a 2-type hypersurface of S^{n+1}. Then the following formulae hold on $\mathcal{U}$:*

$$E_1(h) = \frac{3n^2}{2}\alpha E_1(\alpha^2), \tag{2.10}$$

$$E_j(h) = 0, \quad j = 2, \ldots, n, \tag{2.11}$$

$$nE_1E_1(\alpha^2) + \frac{3n}{2}\alpha h + g = 0. \tag{2.12}$$

Finally, an easy computation from (2.10), (2.11) and Lemma 2.3 gives

$$h = n^2\alpha^3 + k, \tag{2.13}$$

for a constant k, holding anywhere on M.

We are going to compute $\Delta\alpha^2$ in two different ways. First, by using (2.4) we find

$$n\Delta\alpha^2 = \Delta < cx_0, x >= -n < cx_0, H' > +n < cx_0, x >$$

and then, from (2.9), we get

$$\Delta\alpha^2 = -\alpha h + g. \tag{2.14}$$

On the other hand,

$$\begin{aligned} \Delta\alpha^2 &= 2\alpha\Delta\alpha - 2|\nabla\alpha|^2 \\ &= 2(b - |\sigma|^2)\alpha^2 - 2\alpha h - 2|\nabla\alpha|^2. \end{aligned} \tag{2.15}$$

Now, a straightforward computation (see [4]) yields to

PROPOSITION 2.5. *Let M be a 2-type hypersurface of S^{n+1}. Then the mean curvature α does not vanish anywhere on M.*

Next, we are going to prove one of the chief results of this section, which gives an affirmative answer to an open problem stated by B.Y. Chen [**9**, I.6].

THEOREM 2.6. *Let M be a 2-type hypersurface of* S^{n+1}. *Then M has constant mean curvature if and only if M has constant scalar curvature.*

Proof. If α is a constant, then h so is because (2.13) and then $|\sigma|^2$ is also a constant. As a consequence, we use the Gauss equation

$$|\sigma|^2 = n^2\alpha^2 - n(n-1)\tau + n \tag{2.16}$$

to get M has constant scalar curvature.

Conversely, suppose now M has constant scalar curvature. From (2.9) we find

$$|\nabla\alpha|^2 = \frac{1}{4n^2\alpha^2}\{c^2|x_0|^2 - h^2 - g^2\}, \tag{2.17}$$

that jointly with (2.14) and (2.15) leads to

$$4n^2(b - |\sigma|^2)\alpha^4 + (h - 2n^2\alpha^3)h + (g - 2n^2\alpha^2)g - c^2|x_0|^2 = 0. \tag{2.18}$$

Finally, from here, (2.13) and Gauss equation α must be a root of a polynomial with constant coefficients and therefore α is a constant. ■

Let M be a 2-type hypersurface of $S^{n+1} \subset \mathbb{R}^{n+2}$. Consider again the open set $\mathcal{U}$ which is dense in M unless it was empty and so M has constant mean curvature (see Lemma 2.3). Let p be any point of $\mathcal{U}$ and denote by $\gamma(t)$ the integral curve of $\nabla\alpha^2$ through the point $p \in \mathcal{U}$. Now, (2.13) allows us to rewrite (2.12) along $\gamma(t)$ as follows:

$$\frac{d^2}{dt^2}(\alpha^2) + \frac{3}{2}n^2\alpha^4 + \alpha^2 + \frac{3}{2}k\alpha + \frac{1}{n}(n + c - b) = 0. \tag{2.19}$$

Let $\beta = \left(\dfrac{d\alpha}{dt}\right)^2$. Then it is easy to see that equation (2.19) can be reduced to the following first order differential equation:

$$\alpha\frac{d\beta}{d\alpha} + 2\beta = -\frac{3}{2}n^2\alpha^4 - \alpha^2 - \frac{3}{2}k\alpha - \frac{1}{n}(n + c - b). \tag{2.20}$$

From this equation we obtain the following solution:

$$\begin{aligned} 4n^2\alpha^2\beta &= -\frac{3}{2}n^4\alpha^4 - 2n^2\alpha^2 - 6kn^2\alpha \\ &\quad - 4n(n + c - b)\ln(\alpha) + C_1, \end{aligned} \tag{2.21}$$

where C_1 is some constant.

On the other hand, from (2.9) one has

$$4n^2\alpha^2\beta = c^2|x_0|^2 - (n^2\alpha^3 + k)^2 - (n\alpha^2 + n + c - b)^2. \tag{2.22}$$

Therefore, (2.21) and (2.22) prove the following

THEOREM 2.7. *Let M be a 2-type hypersurface of $\mathbb{S}^{n+1} \subset \mathbb{R}^{n+2}$. Then M has constant mean curvature.*

The following result gives a nice characterization of compact 2-type hypersurfaces in the hypersphere $\mathbb{S}^{n+1} \subset \mathbb{R}^{n+2}$ and partially solves an open problem stated by B.Y. Chen [**9**, I.4].

COROLLARY 2.8. *Let M be a compact hypersurface of $\mathbb{S}^{n+1} \subset \mathbb{R}^{n+2}$ which is not a small hypersphere of $\mathbb{S}^{n+1}$. Then M is of 2-type if and only if M has non-zero constant mean curvature α and constant scalar curvature τ. Moreover, if M is of 2-type, α and τ are completely determined for the eigenvalues $\{\lambda_1, \lambda_2\}$ involved in the 2-type condition.*

Proof. The necessary condition follows automatically from Theorems 2.7 and 2.6. Now, if α and τ are constant, then $|S|^2$ is also constant and so (2.1) allows us to write

$$\Delta H = (|S|^2 + n)H + (|S|^2 - n\alpha^2)x, \tag{2.23}$$

where we have used $H = H' - x$. As a consequence there exist two constants, say r and s, such that $\Delta H = rH + sx$, with $s \neq 0$ because M is not a small hypersphere of $\mathbb{S}^{n+1}$. Therefore, we use Theorem 2.2 of [**11**, p. 257] to get that M is of 2-type. Last claim of the statement follows from Theorem 4.2 of [**11**, p. 276]. ∎

Next result gives a partial answer to another open problem stated by B.Y. Chen [**9**, I.1].

COROLLARY 2.9. *Let M be a compact 2-type hypersurface of $\mathbb{S}^{n+1} \subset \mathbb{R}^{n+2}$. Then M is mass-symmetric in $\mathbb{S}^{n+1}$.*

Proof. First, we use Theorem 2.7 to have (2.23), where both coefficients $|S|^2 + n$ and $|S|^2 - n\alpha^2$ are constant. Moreover, $|S|^2 - n\alpha^2 \neq 0$ because M is assumed to be of 2-type in $\mathbb{R}^{n+2}$ (notice that $|S|^2 = n\alpha^2$ implies M is a small hypersphere and so of 1-type in some hyperplane of $\mathbb{R}^{n+2}$ and then of 1-type in $\mathbb{R}^{n+2}$). Thus we have

$$0 = \int_M \Delta H dv = (|S|^2 + n) \int_M H dv + (|S|^2 - n\alpha^2) \int_M x dv,$$

and so

$$\int_M x dv = 0,$$

this means, the center of mass of M is nothing but the origin of $\mathbb{R}^{n+2}$. ∎

REMARK 2.10. We would like to point out that Theorem 2.7 and Corollaries 2.8 and 2.9 have been also obtained, simultaneously and independently, by Hasanis and Vlachos in [**32**], where they use a different method of proof.

2.2. Applications

A hypersurface M of $S^{n+1} \subset \mathbb{R}^{n+2}$ is called a Dupin hypersurface if the multiplicity of each principal curvature is constant on M and each principal curvature is constant along its associated principal directions. In [**6**] it is proved that compact embedded Dupin hypersurfaces are conformal images of isoparametric hypersurfaces when the number g of principal curvatures is $g \leq 2$, but this is not the case when $g \geq 3$. In [**46**], G. Thorbergsson proves that, in cohomology level, compact embedded Dupin hypersurfaces are isoparametric. That result leads to the Cecil-Ryan's conjecture [**7**]: *A compact embedded Dupin hypersurface is Lie equivalent to an isoparametric hypersurface.* That holds when $g \leq 3$, see [**6**] and [**37**]; otherwise, it can be found counterexamples to the conjecture in [**38**] and [**40**]. These facts suggest a close relation between compact embedded Dupin hypersurfaces and isoparametric ones.

It is a well-known fact that isoparametric hypersurfaces of $S^{n+1} \subset \mathbb{R}^{n+2}$ with $g \leq 2$ are spheres and Riemannian products of spheres. When $g = 3$, they were completely classified by E. Cartan [**5**]. They are all homogeneous spaces and the multiplicities of principal curvatures (m_1, m_2, m_3) and dimensions n are listed in the adjoint table:

M^n	(m_1, m_2, m_3)	n
$SO(3)/Z_2 + Z_2$	$(1,1,1)$	3
$SU(3)/T^2$	$(2,2,2)$	6
$SP(3)/SP(1)^3$	$(4,4,4)$	12
$F_4/Spin(8)$	$(8,8,8)$	24

Now, we are going to state and prove the following classification result.

THEOREM 2.11. *Let M be a Dupin hypersurface of S^{n+1} with at most three distinct principal curvatures which is not a small hypersphere of S^{n+1}. Then M is of 2-type if and only if one the following statements holds:*
1) M is an open piece of a Riemannian product $S^p \times S^{n-p}$.
2) M is an open piece of one of the hypersurfaces exhibited in the above table.

Proof. The sufficient condition follows easily from above results in this section. Now, let us suppose M is a 2-type hypersurface of S^{n+1}. Then from Theorems 2.7 and 2.6 we know that M has constant mean curvature and constant scalar curvature. Since M is a Dupin hypersurface it is not difficult to see that M is, in fact, an isoparametric hypersurface. Thus, we obtain the desired conclusion, because M cannot have only one principal curvature. ∎

As a consequence, we obtain the following.

COROLLARY 2.12. *Let M be a Dupin hypersurface of S^4 which is not a small hypersphere. Then M is of 2-type if and only if M is an open piece of one of the following hypersurfaces: $S^1 \times S^2$, $SO(3)/Z_2 + Z_2$.*

3. Hypersurfaces with a characteristic eigenvector field

In this section we will tackle the second extension of Takahashi's condition set in the Lorentz-Minkowski ambient. Before starting this task, it will be convenient recalling the pseudo-Riemannian version of Takahashi's theorem, which can be found in [**8**] and [**36**]. Let x be an isometric immersion of a submanifold M in a pseudo-Euclidean space $\mathbb{R}^{m}_{s}$. Then M satisfies equation (*) if and only if M is either minimal in $\mathbb{R}^{m}_{s}$, or minimal in a pseudo-hyperbolic space $H^{m-1}_{s-1}(r)$, or minimal in a pseudo-sphere $S^{m-1}_{s}(r)$. Furthermore, from here we obtain that minimal submanifolds of $\mathbb{R}^{m}_{s}$ are the only ones having harmonic coordinate functions and therefore there can be characterized by the equation $\Delta x = 0$.

As we pointed out in the Introduction, the condition (**) means that the coordinate functions of the mean curvature vector field H are eigenfunctions of the Laplacian associated to the same eigenvalue, thus that equation connects again with the spectral geometry of the submanifold.

We have already mentioned that information furnished by the equation (**) is different from that of (*). However, we wish to give an example in the Lorentzian ambient to ratify this fact.

Let $\mathbb{L}^{n+1}$ be the $(n+1)$-dimensional Lorentz-Minkowski space with the usual coordinates $(x_1, \ldots, x_{n+1})$ and the standard flat metric given by $ds^2 = -dx_1^2 + dx_2^2 + \cdots + dx_{n+1}^2$. Let us consider the differentiable function $f : \mathbb{L}^{n+1} \longrightarrow \mathbb{R}$ defined by

$$f(x_1, \ldots, x_{n+1}) = -x_1^2 + x_2^2 + \cdots + x_{k+1}^2 - r^2,$$

where $r > 0$ and $k \in \{1, 2, \ldots, n-1\}$. Then $M = f^{-1}(0)$ is a spacial hypersurface of $\mathbb{L}^{n+1}$, i.e., it is endowed with a Riemannian metric and furthermore isometric to the Riemannian product $H^k(r) \times \mathbb{R}^{n-k}$. Bearing in mind the relation between the second fundamental forms of $H^k(r) \times \mathbb{R}^{n-k}$ and $H^k(r)$, it is easy to see that both submanifolds satisfy equation (**), but we know that cylinders do not fall into Takahashi's family.

In analysing condition (**) we will study two cases separatedly, according to the curvature of the ambient space. To do that we would like to notice that no restriction on the causal character of the hypersurface is made.

3.1. Flat ambient space

Let M^n_s be a hypersurface, with index $s = 0, 1$, in $\mathbb{L}^{n+1}$ and let ∇f denote the gradient of a diferentiable function f. An easy computation allows us to get the following formula for the Laplacian of the mean curvature vector field H ([**8**],[**22**]):

$$\Delta H = 2S(\nabla\alpha) + \frac{n\varepsilon}{2}\nabla\alpha^2 + \{\Delta\alpha + \varepsilon\alpha\operatorname{tr}(S^2)\}N, \tag{3.1}$$

where S stands for the shape operator of M, α the mean curvature, N the unit normal vector field and $\varepsilon =< N, N >$.

Assuming that M satisfies the equation (**), we easily get from (3.1) the following equations:

$$S(\nabla\alpha) + \frac{n\varepsilon}{2}\alpha\nabla\alpha = 0, \tag{3.2}$$

$$\Delta\alpha + (\varepsilon \mathrm{tr}(S^2) - \lambda)\alpha = 0. \tag{3.3}$$

Now we wish to deduce some easy consequences from there. If α is a non-vanishing constant then (3.3) implies that $\mathrm{tr}(S^2)$ is constant and therefore M also has constant scalar curvature. On the other hand, if M has constant mean and scalar curvatures then equation (**) holds for the real constant $\lambda = \varepsilon\mathrm{tr}(S^2)$. Hence, the following problem arises in a natural way:

> **(P1)** *Are the non-vanishing constant mean curvature and constant scalar curvature hypersurfaces of the Lorentz-Minkowski space characterized by the equation $\Delta H = \lambda H$?*

Before beginning the study of this problem, we would like to remark that equation (3.2) can be obtained by supposing only that ΔH is a vector field normal to M. In this way, Garay and Romero, [**27**], have recently studied those hypersurfaces in $\mathbb{L}^{n+1}$ satisfying the condition $\Delta H = C$, where C is a constant vector in $\mathbb{L}^{n+1}$ which is normal to M at every point, and they show that C should vanish. Bearing in mind that minimal submanifolds are the only ones whose immersion is harmonic, i.e., $\Delta x = 0$, it seems natural to ask for the following geometric question:

> **(P2)** *Does the equation $\Delta H = 0$ characterize the vanishing mean curvature hypersurfaces of $\mathbb{L}^{n+1}$?*

Submanifolds satisfying the condition $\Delta H = 0$ are called *biharmonic*, because they satisfy $\Delta^2 x = 0$, and they have been handled, among others, in [**10**], [**17**], [**21**], in the Euclidean case, and in [**14**], in the pseudo-Euclidean case.

In dealing with problem (P1), we are going to find surfaces in $\mathbb{L}^3$ satisfying (**). At a first stage we know that minimal surfaces, hyperbolic $H^2(r)$ and de Sitter $S_1^2(r)$ planes are trivial solutions of (**), as well as the three cylinders appearing in this ambient space: $H^1(r) \times \mathbb{R}$, $\mathbb{L} \times S^1(r)$ and $S_1^1(r) \times \mathbb{R}$. Observe that all quoted examples, as spacial as Lorentzian ones, have diagonalizable shape operators. Therefore, there arises the following question: *Are there Lorentzian surfaces satisfying (**) and having non-diagonalizable shape operators?* To get an affirmative answer we present an example which was first given by Graves, [**28**].

Let $x(s)$ be a null curve in $\mathbb{L}^3$ with Cartan frame $\{A, B, C\}$, i.e., A, B and C are vector fields along $x(s)$ such that

$$\begin{aligned}
\dot{x}(s) &= A(s), \\
\dot{A}(s) &= k(s)C(s), \\
\dot{B}(s) &= w_0 C(s), \\
\dot{C}(s) &= w_0 A(s) + k(s)B(s),
\end{aligned}$$

where $k(s) \neq 0$ and w_0 is a nonzero constant. Then the map $\Phi : (s, u) \longrightarrow x(s) + uB(s)$ parametrizes a Lorentzian surface in $\mathbb{L}^3$, which Graves called a B-scroll. An easy computation shows that $\alpha = w_0$ and $\mathrm{tr}(S^2) = 2w_0^2$, and then from (3.1) we have $\Delta H = 2w_0^2 H$. Moreover, the shape operator S can be put, in the usual frame $\{\frac{\partial \Phi}{\partial s}, \frac{\partial \Phi}{\partial u}\}$, as

$$\begin{pmatrix} w_0 & 0 \\ k(s) & w_0 \end{pmatrix},$$

showing that the minimal polynomial $p_S(t)$ of S is given by $p_S(t) = (t - w_0)^2$ and S is not diagonalizable.

All above examples illustrating equation (**) satisfy a chiefly interesting geometric property: all of them have constant mean curvature. Then it is reasonable to ask for *surfaces in* $\mathbb{L}^3$ *satisfying (**) and having non-constant mean curvature.* However we are able to get a negative answer in [**24**]:

THEOREM 3.1. *Every surface* M_s^2 *in* $\mathbb{L}^3$ *satisfying the condition* $\Delta H = \lambda H$ *has constant mean curvature.*

As a consequence, from here and (3.3), it is easy to see that the only non-minimal surfaces in $\mathbb{L}^3$ satisfying (**) are the isoparametric ones, i.e., those whose shape operators have constant characteristic polynomial. Now bringing here the classification of such surfaces, given in [**35**] and [**39**], we get in [**24**] a complete answer to problem (P1):

THEOREM 3.2. *Let* M_s^2 *be a surface in* $\mathbb{L}^3$. *Then* $\Delta H = \lambda H$, *for a real constant* λ, *if and only if* M_s^2 *is either minimal or an open piece of one of the following surfaces:* $H^2(r)$, $S_1^2(r)$, $H^1(r) \times \mathbb{R}$, $\mathbb{L} \times S^1(r)$, $S_1^1(r) \times \mathbb{R}$ *and a* B*-scroll.*

Some consequences can be deduced from this theorem. On one hand, we find that minimal surfaces in $\mathbb{L}^3$ are characterized as the only ones having harmonic mean curvature vector field, solving problem (P2). On the other hand, paying attention on the causal character of the surface, we get that the only spacial surfaces satisfying (**) are either those having zero mean curvature (the so-called *maximal* ones) or open pieces of one of the following surfaces: a hyperbolic plane $H^2(r)$, a hyperbolic cylinder $H^1(r) \times \mathbb{R}$.

As for the n-dimensional case, problems (P1) and (P2) remain open and we have only found partial solutions. In **[22]** we deal with hypersurfaces in $\mathbb{L}^{n+1}$ satisfying (**) and such that the minimal polynomial of the shape operator is at most of degree two. *Under this additional hypothesis*, we are in a position to show the following.

PROPOSITION 3.3. *All hypersurfaces M_s^n in $\mathbb{L}^{n+1}$ satisfying $\Delta H = \lambda H$ have constant mean curvature.*

This result allows us to get an affirmative answer to (P1), under the above additional condition. Taking into account that theorem and the solution obtained for surfaces, we dare to state the following conjecture.

(C1) *The answer to problem (P1) is affirmative.*

Finally, we would like to remark that problem (P1) also involves the Euclidean case and thus we also guess that conjecture (C1) can be applied to Euclidean hypersurfaces.

3.2. Non-flat ambient space

Let $\bar{M}_\nu^{n+1}(c)$ denote the pseudo-Riemannian space form with index ν and constant sectional curvature $\mathrm{sgn}(c)/c^2$. Without loss of generality, we can assume $c = \pm 1$ and in what follows $\bar{M}_\nu^{n+1}$ will denote the pseudo-sphere S_ν^{n+1} or the pseudo-hyperbolic space H_ν^{n+1}, according to $c = 1$ or $c = -1$, respectively. Let x be an isometric immersion of a hypersurface M_s^n in $\bar{M}_\nu^{n+1}$ and let H denote the mean curvature vector field of M_s^n in the pseudo-Euclidean space $\mathbb{R}_t^{n+2}$ where $\bar{M}_\nu^{n+1}$ is lying. Then it is easy to show that H is given by

$$H = \alpha N - cx, \tag{3.4}$$

where N denotes a unit vector field normal to M_s^n in $\bar{M}_\nu^{n+1}$ and α the mean curvature of M_s^n in $\bar{M}_\nu^{n+1}$. An easy computation from (3.4) yields to the following nice formula for ΔH, **[8]**:

$$\begin{aligned}\Delta H &= 2S(\nabla\alpha) + n\varepsilon\alpha\nabla\alpha + \{\Delta\alpha + \varepsilon\alpha\mathrm{tr}(S^2) + nc\alpha\}N \\ &\quad -n(1+\varepsilon c\alpha^2)x,\end{aligned} \tag{3.5}$$

where S stands for the shape operator of the hypersurface.

If M_s^n is a hypersurface satisfying the condition (**), we can use equations (3.4) and (3.5) to obtain the following formulae:

$$\begin{aligned}&2S(\nabla\alpha) + n\varepsilon\alpha\nabla\alpha = 0,\\ &\Delta\alpha + (\varepsilon\mathrm{tr}(S^2) + nc - \lambda)\alpha = 0,\\ &n(1+\varepsilon c\alpha^2) - \lambda c = 0.\end{aligned}$$

Now we can combine those equations in order to get the following result, which can be considered as a first approximation to the solution of problem (P1) in the new ambient space.

THEOREM 3.4. *A hypersurface* M_s^n *of* $\bar{M}_\nu^{n+1}$ *satisfies the condition* $\Delta H = \lambda H$ *if and only if* M_s^n *is minimal in* $\bar{M}_\nu^{n+1}$ *or has nonzero constant mean curvature* α *and* $\mathrm{tr}(S^2) = (1/n)\mathrm{tr}(S)^2$.

As a first consequence we get the following. Let us suppose M_s^n has diagonalizable shape operator, for example provided M_s^n is a spacial hypersurface. Then the above result implies M_s^n is a totally umbilical hypersurface and we can use [**34**, Theorem 1.4] to find that M_s^n is an open piece either of $S_s^n(r)$, or $H_s^n(r)$, or $\mathbb{R}_s^n$ (in the last case, the immersion is constructed from a quadratic function). As a second consequence we deduce, taking into account the Gauss equation, that non-minimal hypersurfaces of $\bar{M}_\nu^{n+1}$ satisfying (**) are characterized by having constant mean α and scalar τ curvatures which satify the equation $\tau = n(n-1) < H, H >= n(n-1)(\varepsilon\alpha^2 + c)$.

With the aim of studying in depth the condition (**), we are going to deal with surfaces. Following [**15**], we construct B-scrolls over null curves to obtain some surfaces in $\bar{M}_1^3$ satisfying (**) and whose shape operators are non-diagonalizable, with minimal polynomials having only real roots. Nevertheless, it seems natural thinking of surfaces in $\bar{M}_1^3$ satisfying (**) and whose shape operator has a minimal polynomial with complex roots. However, that cannot happen because of the condition $\mathrm{tr}(S^2) = \frac{1}{2}\mathrm{tr}(S)^2$.

Let us suppose M_s^2 is a surface satisfying (**). Then M_s^2 has constant mean and scalar curvatures and thus M_s^2 is an isoparametric surface of $\bar{M}_1^3$. Now we may carry on an standard reasoning to obtain the following.

THEOREM 3.5. *Let* M_s^2 *be a non-minimal surface of* $\bar{M}_1^3$ *satisfying the condition* $\Delta H = \lambda H$. *Then* M_s^2 *is an open piece either of a totally umbilical surface or a* B*-scroll.*

This result leads to the characterization of biharmonic surfaces of $\bar{M}_1^3$.

COROLLARY 3.6. *A surface* M_s^2 *in* $\bar{M}_1^3$ *is biharmonic if and only if it is either a flat totally umbilical one or a flat* B*-scroll.*

In order to complete our study, we must consider hypersurfaces M_s^n in $\bar{M}_1^{n+1}$, where we guess more promising prospects than in the flat ambient space. We approach the problem by analising separatedly the different shape operators locally allowed for the hypersurface. In the diagonalizable case, the problem has been already solved. Let us suppose that the minimal polynomial of S is given, in an open set of M_s^n, by $p_S(t) = (t-\beta)^2(t-\mu_1)\cdots(t-\mu_k)$. Then the equatility $\mathrm{tr}(S^2) = \frac{1}{n}\mathrm{tr}(S)^2$ implies that

$\mu_1 = \cdots = \mu_k = \beta$ is constant and therefore M_s^n is an isoparametric hypersurface of $\bar{M}_1^{n+1}$ with $p_S(t) = (t-\beta)^2$. A standard reasoning on integral submanifolds leads us to an explicit description of the hypersurface M_s^n. The case $p_S(t) = (t-\beta)^3\,(t-\mu_1)\cdots(t-\mu_k)$ can be treated in a similar way. Finally, the situation when the minimal polynomial has complex roots, i.e., $p_S(t) = [(t-\beta)^2+\gamma^2](t-\mu_1)\cdots(t-\mu_k)$ with $\gamma \neq 0$, becomes more complicated, but at the present we think that cannot hold. In this way, we state the following conjecture.

> **(C2)** *There are no hypersurfaces in M_1^{n+1} satisfying (**) and whose shape operators have minimal polynomials with complex roots.*

To finish this section, we would like telling of that the results of this subsection, i.e., the non-flat ambient space case, are being purified in order to be published elsewhere.

4. Hypersurfaces satisfying the condition $\Delta x = Ax + B$

As we have pointed out in the Introduction, both conditions $\Delta x = Ax$, original from Garay, and (***), due to Dillen, Pas and Verstraelen, were only established for submanifolds and, particularly, hypersurfaces in the Euclidean space and, in this context, they have been recently studied by some authors, **[13]**, **[16]**, **[26]**, **[25]**, **[30]**, **[31]**. However, the pseudo-Riemannian case presents an own behaviour, mainly because the shape operator need not be diagonalizable, which plays a chief role in the Riemannian case.

In this section, we will study those pseudo-Riemannian hypersurfaces in pseudo-Riemannian space forms which are characterized by the matricial condition (***) in the Laplacian of the isometric immersion. A first step in this way was given by the authors in [**3**], where surfaces in the 3-dimensional Lorentz-Minkowski space satisfying the equation (***) were classified. The interesting changes found here with regard to the Euclidean case leaded us to consider that condition not only for hypersurfaces in a pseudo-Euclidean space, but also for hypersurfaces in a pseudo-spherical or pseudo-hyperbolic space.

4.1. Flat ambient space

Let $\mathbb{R}_\nu^{n+1}$ be the $(n+1)$-dimensional pseudo-Euclidean space of index ν with metric tensor, in the usual coordinates $(x_1, \ldots, x_{n+1})$, given by

$$ds^2 = -\sum_{i=1}^{\nu} dx_i \otimes dx_i + \sum_{j=\nu+1}^{n+1} dx_j \otimes dx_j.$$

Let M_s^n be a pseudo-Riemannian hypersurface in $\mathbb{R}_\nu^{n+1}$ with index $s = \nu - 1, \nu$ and let us write by H and N the mean curvature and the unit normal vector fields of M_s^n

in $\mathbb{R}^{n+1}_\nu$, respectively, so that $H = \alpha N$, being α the mean curvature in the direction of N.

Let $x : M^n_s \longrightarrow \mathbb{R}^{n+1}_\nu$ be an isometric immersion satisfying (***), where A is now an endomorphism of $\mathbb{R}^{n+1}_\nu$ and B is a constant vector. From here, the formula for ΔH given in Section 3 and the well known formula $\Delta x = -nH$, it is not difficult to see that

$$AX = n\alpha SX - nX(\alpha)N, \tag{4.1}$$

for any vector field X tangent to M^n_s, and

$$AH = 2S(\nabla\alpha) + n\varepsilon\alpha\nabla\alpha + \{\Delta\alpha + \varepsilon\alpha\mathrm{tr}(S^2)\}N. \tag{4.2}$$

If we suppose now M^n_s has non-zero constant mean curvature, then from (4.1) and (4.2) we have

$$AX = n\alpha SX, \tag{4.3}$$
$$AN = \varepsilon\mathrm{tr}(S^2)N, \tag{4.4}$$

from which we deduce $\mathrm{tr}(S^2)$ is also a constant and, taking covariant differentiation in (4.4), we find that the shape operator satisfies the polynomial equation

$$S(S - \lambda I) = 0,$$

where λ is the non-vanishing real constant given by $\lambda = \dfrac{\varepsilon\mathrm{tr}(S^2)}{n\alpha}$. That equation means that M^n_s is an isoparametric hypersurface in $\mathbb{R}^{n+1}_\nu$ with diagonalizable shape operator and having as principal curvatures zero, with multiplicity at most $n-1$, and $\lambda \neq 0$, with multiplicity at least one. Therefore, if M^n_s is totally umbilical in $\mathbb{R}^{n+1}_\nu$ then M^n_s is an open piece of a pseudo-sphere $S^n_\nu(r)$ or a pseudo-hyperbolic space $H^n_{\nu-1}(r)$. Otherwise, by using similar arguments as those in [**44**, Theorem 2.5] and [**33**, Lemma 2], M^n_s is an open piece of one of the pseudo-Riemannian products $\mathbb{R}^k_u \times S^{n-k}_{\nu-u}(r)$ and $\mathbb{R}^k_u \times H^{n-k}_{\nu-u-1}(r)$. We will refer to these four classes as the standard examples in $\mathbb{R}^{n+1}_\nu$.

On the other hand, besides the trivial case of minimal hypersurfaces, it is not difficult to show that the standard examples also satisfy equation (***).

Then it seems reasonable to state the following question.

> **(P3)** *Does the equation $\Delta x = Ax+B$ characterize to the family of minimal hypersurfaces and standard examples in $\mathbb{R}^{n+1}_\nu$?*

We have just obtained an affirmative answer to this question when the mean curvature is constant. The following result completely solves that problem, [**2**].

THEOREM 4.1. *The hypersurfaces in $\mathbb{R}^{n+1}_\nu$ satisfying the condition $\Delta x = Ax + B$ have constant mean curvature.*

Our result generalize those given, when the ambient space is $\mathbb{R}^{n+1}$, in [**16**], [**13**] and [**31**]. On the other hand, for hypersurfaces in the Lorentz-Minkowski space we have the following proposition.

PROPOSITION 4.2. *Let $x : M^n_s \longrightarrow \mathbb{L}^{n+1}$ be an isometric immersion. Then $\Delta x = Ax + B$ if and only if M^n_s is either minimal or an open piece of one of the following hypersurfaces: $S^n_1(r)$, $H^n(r)$, $S^k_1(r) \times \mathbb{R}^{n-k}$, $\mathbb{L}^k \times S^{n-k}(r)$, $H^k(r) \times \mathbb{R}^{n-k}$.*

REMARK 4.3. We wish to pointed out that both equations $\Delta H = \lambda H$ and $\Delta x = Ax + B$ characterize the same family of surfaces in the Euclidean case, but they make notably differences in Lorentzian ambient. In fact, we have seen that a B-scroll, which has constant mean curvature but non-diagonalizable shape operator, satisfies the former but not the latter.

4.2. Non-flat ambient space

Through this section, we will keep the notation fixed in Section 3.2. A hypersurface M^n_s in $\bar{M}^{n+1}_\nu$ can also be viewed as a codimension two submanifold in the corresponding pseudo-Euclidean space $\mathbb{R}^{n+2}_t$, where $\bar{M}^{n+1}_\nu$ is canonically immersed, and therefore we can ask ourselves for those hypersurfaces in $\bar{M}^{n+1}_\nu$ whose isometric immersion $x : M^n_s \longrightarrow \bar{M}^{n+1}_\nu \subset \mathbb{R}^{n+2}_t$ satisfies the condition (***), being A and endomorphism of $\mathbb{R}^{n+2}_t$ and B a constant vector in $\mathbb{R}^{n+2}_t$.

In order to guide our study, we are going to give some examples. A first trivial one is provided by minimal hypersurfaces in $\bar{M}^{n+1}_\nu$. Consider now a totally umbilical hypersurface M^n_s in $\bar{M}^{n+1}_\nu$. Bearing in mind the classification given in [**34**], we know that, according to the causal character of H, M^n_s is an open piece of a pseudo-Riemannian space form. It is not difficult to see that both $S^n_s(r)$ and $H^n_s(r)$ satisfy the asked condition, so that the most interesting situacion comes up in the flat case, where H is a null vector. Here the isometric immersion $x : \mathbb{R}^n_s \longrightarrow \bar{M}^{n+1}_\nu \subset \mathbb{R}^{n+2}_{s+1}$ is given by $x = f - x_0$, being x_0 a fixed vector in $\mathbb{R}^{n+2}_{s+1}$ and $f : \mathbb{R}^n_s \longrightarrow \mathbb{R}^{n+2}_{s+1}$ the function defined by $f(u_1, \ldots, u_n) = (q(u), u_1, \ldots, u_n, q(u))$, where $q(u) = a < u, u > + < b, u > +c$, $a \neq 0$. Now we have $\Delta x = (-2na, 0, \ldots, 0, -2na)$, showing that this hypersurface also satisfies equation (***) with $A = 0$ and $B = (-2na, 0, \ldots, 0, -2na)$. Finally, a straightforward computation shows that those hypersurfaces in $\bar{M}^{n+1}_\nu$ built up as the following pseudo-Riemannian products $S^k_u(r_1) \times S^{n-k}_{s-u}(r_2)$, $H^k_u(r_1) \times H^{n-k}_{s-u}(r_2)$ and $S^k_u(r_1) \times H^{n-k}_{s-u}(r_2)$ also satisfy that condition (see [**1**]). These hypersurfaces will be called the standard pseudo-Riemannian products in $\bar{M}^{n+1}_\nu$.

Now, let $x : M_s^n \longrightarrow \bar{M}_\nu^{n+1} \subset \mathbb{R}_t^{n+2}$ be an isometric immersion satisfying the condition (***). Then the well known formula $\Delta x = -nH$ and the expresions for H and ΔH obtained in Section 3.2 lead to

$$\begin{aligned} AX &= n\alpha SX + ncX - nX(\alpha)N, \\ \alpha AN &= 2S(\nabla\alpha) + n\varepsilon\alpha\nabla\alpha + \{\Delta\alpha + \varepsilon\alpha \operatorname{tr}(S^2)\}N - nc\varepsilon\alpha^2 x - cB, \\ Ax &= -n\alpha N + ncx - B, \end{aligned} \tag{4.5}$$

for any tangent vector field X.

Then we have $< AX, x >= 0$ and taking covariant derivative we deduce that

$$< A\sigma(X,Y), x >= - < AX, Y >,$$

that along with (4.5) leads to

$$< SX - \varepsilon\alpha X, Y >< B, x >= 0. \tag{4.6}$$

In a first aproximation to our problem, let us suppose that the mean curvature of M_s^n in $\bar{M}_\nu^{n+1}$ is a non-zero constant. If M_s^n is not totally umbilical (4.6) implies $< B, x >= 0$ and, reasoning as in [**1**, Lemma 3.1], $B = 0$. Now, (4.5) can be rewritten as

$$\begin{aligned} AX &= n\alpha SX + ncX, \\ AN &= \varepsilon \operatorname{tr}(S^2)N - nc\varepsilon\alpha x, \\ Ax &= -n\alpha N + ncx. \end{aligned} \tag{4.7}$$

From these equations we deduce M_s^n is an isoparametric hypersurface in $\bar{M}_\nu^{n+1}$ whose shape operator S satisfies the polynomial equation

$$S^2 + \frac{nc - \varepsilon \operatorname{tr}(S^2)}{n\alpha} S - c\varepsilon I = 0. \tag{4.8}$$

The last equation plays a key role in the following reasoning. If S is diagonalizable and M_s^n is not totally umbilical in $\bar{M}_\nu^{n+1}$ from (4.8) we get M_s^n is isoparametric with two principal curvatures and, by using similar arguments as in [**44**, Theorem 2.5] and [**33**, Lemma 2], it is an open piece of one of the standard pseudo-Riemannian products. In particular, we have got a first solution to the problem when the ambient space is either spherical or hyperbolic, that is, when $\nu = 0$. Otherwise, M_s^n could be endowed with an indefinite metric and then the shape operator needs not be diagonalizable.

Now, we will consider the simplest situation where one can find a non-diagonalizable shape operator, that is, a Lorentzian surface M_1^2 in $\bar{M}_1^3$, satisfying condition (***) and having non-zero constant mean curvature. To do that, we know M_1^2 is an isoparametric surface in $\bar{M}_1^3$ and the characteristic polynomial of S is given by

(4.8), being $n = 2$ and $\varepsilon = 1$. From here we find M_1^2 is a flat surface in $\bar{M}_1^3 \subset \mathbb{R}_t^4$ with non diagonalizable shape operator and parallel second fundamental form in $\mathbb{R}_t^4$. Therefore, by using [**34**, Theorem 1.15 and Theorem 1.17] we deduce such a situation only appears when $M_1^3 = H_1^3$ and, in that case, M_1^2 is an open piece either of a complex circle, [**34**, Example 1.12], or of the surface exhibited in [**34**, Example 1.13]. Now it is not difficult to see that both surfaces satisfy the asked condition.

In conclusion, we have found a first significant difference in studying the condition (***) in non-flat pseudo-Riemannian space forms with respect to the similar one in the flat case. In fact, we have seen in Section 4.1 that hypersurfaces in $\mathbb{R}_\nu^{n+1}$ satisfying that condition must have diagonalizable shape operators. However we have just obtained some examples of surfaces in H_1^3 satisfying (***) and having non-diagonalizable shape operators.

In order to generalize these examples to any $\bar{M}_\nu^{n+1}$ we profit by Hahn's ideas, [**29**]. Let L be a self-adjoint endomorphism of $\mathbb{R}_t^{n+2}$, consider the quadratic function $f : \bar{M}_\nu^{n+1} \longrightarrow \mathbb{R}$ defined by $f(x) = < Lx, x >$ and assume that the minimal polynomial of L is given by $p_L(t) = t^2 + at + b$, $a, b \in \mathbb{R}$. Then the level set $M = f^{-1}(r)$, where r is a real constant such that $p_L(cr) \neq 0$, is an isoparametric hypersurface in $\bar{M}_\nu^{n+1}$. A straightforward computation shows that the mean curvature vector field of M_s^n in $\bar{M}_\nu^{n+1}$ is given by

$$H' = \frac{a + \mathrm{tr}(L) - cnr}{cnp_L(cr)}(Lx - crx),$$

from which we deduce that $\Delta x = Ax$, A being the following matrix

$$A = \frac{cnr - a - \mathrm{tr}(L)}{cp_L(cr)}L + \frac{r\mathrm{tr}(L) + (n+1)ar + cnb}{p_L(cr)}I_{n+2}.$$

We will refer this example as a *quadratic hypersurface.* It is worth noticing that in the above family all possibilities for the shape operator can appear, depending on the sign of $a^2 - 4b$.

At this point, it seems reasonable to ask for non-constant mean curvature hypersurfaces satisfying (***). In this case, $\mathcal{U} = \{p \in M_s^n : \nabla\alpha^2(p) \neq 0\}$ is a non-empty open set and the equation (4.6) leads to $< B, x >= 0$ on $\mathcal{U}$. Taking covariant derivative here we deduce B should vanish. From equation (4.5) we have $< AX, Y >=< X, AY >$ which yields

$$< A\sigma(X, Z), Y > - < A\sigma(Y, Z), X >=$$
$$< \sigma(X, Z), AY > - < \sigma(Y, Z), AX > .$$

Finally, from (4.5) we obtain

$$TX(\alpha)SY = TY(\alpha)SX, \tag{4.9}$$

where T denotes the self-adjoint operator defined by $TX = n\alpha X + \varepsilon SX$. This equation is the key to show the following result, [**1**].

THEOREM 4.4. *Let $x : M_s^n \longrightarrow \bar{M}_\nu^{n+1}$ be an isometric immersion such that $\Delta x = Ax + B$. Then M_s^n has constant mean curvature.*

Proof. (Outline) We consider two cases.

(A) $T(\nabla\alpha) \neq 0$ on $\mathcal{U}$. Then the shape operator S has rank one on $\mathcal{U}$ and thus we can choose a local orthonormal frame $\{E_1, \ldots, E_n\}$ such that $SE_1 = n\varepsilon\alpha E_1$, $SE_i = 0$, $i = 2, \ldots, n$. Working on the characteristic polynomials of both A and S it can be deduced that α is a root of a polynomial with constant coefficients and therefore it is locally constant on $\mathcal{U}$, which is a contradiction.

(B) There exists a point $p \in \mathcal{U}$ such that $T(\nabla\alpha)(p) = 0$. Then from (4.5) we have A is a self-adjoint endomorphism of $\mathbb{R}_t^{n+2}$ and $T(\nabla\alpha) = 0$ on $\mathcal{U}$. Moreover $\nabla\alpha$ is a non-null vector which allows us to take a local orthonormal frame $\{E_1, \ldots, E_n\}$ with E_1 parallel to $\nabla\alpha$. Working again on the characteristic polynomials we obtain the same contradiction as in case (A). ∎

Now we are ready to state the main result of this subsection.

THEOREM 4.5. *Let $x : M_s^n \longrightarrow \bar{M}_\nu^{n+1}$ be an isometric immersion. Then $\Delta x = Ax + B$ if and only if one of the following statements holds:*

a) M_s^n is a minimal hypersurface,

b) M_s^n is a totally umbilical hypersurface,

c) M_s^n is one of the standard pseudo-Riemannian products.

d) M_s^n is a quadratic hypersurface with non-diagonalizable shape operator.

Proof. We know α is constant. If $\alpha = 0$ there is nothing to prove, so we can assume $\alpha \neq 0$. When S is diagonalizable, we have seen that either (b) or (c) holds. Otherwise, from (4.6) and (4.7) we get

$$A^2 - (\varepsilon\mathrm{tr}(S^2) + cn)A + n\varepsilon c(\mathrm{tr}(S^2) - n\alpha^2)I_{n+2} = 0,$$

and therefore the minimal polynomial $p_A(t)$ of A is given by $p_A(t) = t^2 + at + b$, where $a = -(\varepsilon\mathrm{tr}(S^2) + cn)$ and $b = n\varepsilon c(\mathrm{tr}(S^2) - n\alpha^2)$. From (4.5) we know A is a self-adjoint endomorphism of $\mathbb{R}_t^{n+2}$ and $< Ax, x >$ is constant on M_s^n. Hence, it is an open piece of a quadratic hypersurface with non-diagonalizable S. ∎

References

[1] L. J. Alías, A. Ferrández and P. Lucas. *On a certain class of hypersurfaces in the pseudo-Riemannian space forms*, 1991. Preprint.

[2] L. J. Alías, A. Ferrández and P. Lucas. *Submanifolds in pseudo-Euclidean spaces satisfying the condition $\Delta x = Ax + B$*, 1991. To appear in Geometriae Dedicata.

[3] L. J. Alías, A. Ferrández and P. Lucas. *Surfaces in the 3-dimensional Lorentz-Minkowski space satisfying* $\Delta x = Ax + B$, 1991. To appear in Pacific J. Math.

[4] M. Barros, A. Ferrández and P. Lucas. *Hipersuperficies esféricas de tipo 2.* Trabajo de Investigación 1, Departamento de Matemáticas, Universidad de Murcia, 1991. *In English.*

[5] E. Cartan. *Sur des familles remarquables d'hypersurfaces isoparamétriques dans les espaces sphériques.* Math. Z. **45** (1939), 335–367.

[6] T. Cecil and P. Ryan. *Focal sets, taut embeddings and the cyclides of Dupin.* Math. Ann. **236** (1978), 177–190.

[7] T. Cecil and P. Ryan. *Tight and taut immersions of manifolds.* Volume 107 of *Research Notes in Math.*, Pitman, London, 1985.

[8] B. Y. Chen. *Finite-type pseudo-Riemannian submanifolds.* Tamkang J. of Math. **17** (1986), 137–151.

[9] B. Y. Chen. *Some open problems and conjectures on finite type submanifolds,* 1989. (Revised version).

[10] B. Y. Chen. *Some open problems and conjectures on submanifolds of finite type,* 1991. University of Michigan.

[11] B. Y. Chen. *Total mean curvature and submanifolds of finite type.* Volume 1 of *Series in Pure Math.*, World Scientific, Singapur, 1984.

[12] B. Y. Chen, M. Barros and O. J. Garay. *Spherical finite type hypersurfaces.* Algebras, Groups and Geometries **4** (1987), 58–72.

[13] B. Y. Chen, F. Dillen, L. Verstraelen and L. Vrancken. *Submanifold of restricted type,* 1990. Preprint.

[14] B. Y. Chen and S. Ishikawa. *Biharmonic surfaces in pseudo-Euclidean spaces.* Special issue dedicated to Prof. T. Otsuki on the occasion of his 75th birthday.

[15] M. Dajczer and K. Nomizu. *On flat surfaces in* S_1^3 *and* H_1^3. In *Manifolds and Lie Groups*, pages 71–108, Univ. Notre Dame, Indiana, Birkhäuser, 1981.

[16] F. Dillen, J. Pas and L. Verstraelen. *On surfaces of finite type in Euclidean 3-space.* Kodai Math. J. **13** (1990), 10–21.

[17] I. Dimitric. *Quadric representation and submanifolds of finite type.* PhD thesis, Michigan State University, 1989.

[18] J. Douglas. *A general formulation of the problem of Plateau.* Bull. Amer. Math. Soc. **36** (1930), p. 50.

[**19**] J. Douglas. *The problem of Plateau.* Bull. Amer. Math. Soc. **39** (1933), 227–251.

[**20**] J. Douglas. *Solution of the problem of Plateau.* Trans. Amer. Math. Soc. **33** (1931), 263–321.

[**21**] A. Ferrández, O. J. Garay and P. Lucas. *On a certain class of conformally flat Euclidean hypersurfaces.* In Ferus, Pinkall, Simon and Wegner, editors, *Global Differential Geometry and Global Analysis, Berlin 1990*, pages 48–54, 1991. Lecture Notes in Mathematics, 1481.

[**22**] A. Ferrández and P. Lucas. *Classifying hypersurfaces in the Lorentz-Minkowski space with a characteristic eigenvector*, 1991. Preprint.

[**23**] A. Ferrández and P. Lucas. *Null finite type hypersurfaces in space forms.* Kodai Math. J. **14** (1991), 406–419.

[**24**] A. Ferrández and P. Lucas. *On surfaces in the 3-dimensional Lorentz-Minkowski space*, 1991. To appear in Pacific J. Math.

[**25**] O. J. Garay. *An extension of Takahashi's theorem.* Geometriae Dedicata **34** (1990), 105–112.

[**26**] O. J. Garay. *On a certain class of finite type surfaces of revolution.* Kodai Math. J. **11** (1988), 25–31.

[**27**] O. J. Garay and A. Romero. *An isometric embedding of the complex hyperbolic space in a pseudo- Euclidean space and its applications to the study of real hypersurfaces.* Tsukuba J. Math. **14** (1990), 293–313.

[**28**] L. Graves. *Codimension one isometric immersions between Lorentz spaces.* Trans. A.M.S. **252** (1979), 367–392.

[**29**] J. Hahn. *Isoparametric hypersurfaces in the pseudo-Riemannian space forms.* Math. Z. **187** (1984), 195–208.

[**30**] T. Hasanis and T. Vlachos. *Coordinate finite-type submanifolds.* Geometriae Dedicata **37** (1991), 155–165.

[**31**] T. Hasanis and T. Vlachos. *Hypersurfaces of E^{n+1} satisfying $\Delta x = Ax + B$*, 1990. Preprint.

[**32**] T. Hasanis and T. Vlachos. *Spherical 2-type hypersurfaces.* J. of Geom. (1991), 82–94.

[**33**] H. Lawson. *Local rigidity theorems for minimal hypersurfaces.* Ann. of Math. **89** (1969), 187–197.

[34] M. A. Magid. *Isometric immersions of Lorentz space with parallel second fundamental forms.* Tsukuba J. Math. **8** (1984), 31–54.

[35] M. A. Magid. *Lorentzian isoparametric hypersurfaces.* Pacific J. Math. **118** (1985), 165–197.

[36] S. Markvorsen. *A characteristic eigenfunction for minimal hypersurfaces in space forms.* Math. Z. **202** (1989), 375–382.

[37] R. Miyaoka. *Compact Dupin hypersurfaces with three principal curvatures.* Math. Z. **187** (1984), 433–452.

[38] R. Miyaoka and T. Ozawa. *Construction of taut embeddings and Cecil-Ryan conjecture.* In *Proc. of Symp. in Diff. Geom.*, 1988.

[39] K. Nomizu. *On isoparametric hypersurfaces in the Lorentzian space forms.* Japan J. Math. **7** (1981), 217–226.

[40] U. Pinkall and G. Thorbergsson. *Deformations of Dupin hypersurfaces.* Proc. A.M.S. **107** (1989), 1037–1043.

[41] T. Radó. *On Plateau's problem.* Ann. of Math. **31** (1930), 457–469.

[42] T. Radó. *On the functional of Mr. Douglas.* Ann. of Math. **32** (1931), 785–803.

[43] T. Radó. *The problem of the least area and the problem of Plateau.* Math. Z. **32** (1930), 763–796.

[44] P. J. Ryan. *Homogeneity and some curvature conditions for hypersurfaces.* Tôhoku Math. J. **21** (1969), 363–388.

[45] T. Takahashi. *Minimal immersions of Riemannian manifolds.* J. Math. Soc. Japan **18** (1966), 380–385.

[46] G. Thorbergsson. *Dupin hypersurfaces.* Bull. London Math. Soc. **15** (1983), 493–498.

The Problem of Plateau (pp. 76– 86)
ed. Th. M. Rassias

AFFINE MINIMAL HIGHER ORDER PARALLEL AFFINE SURFACES

FRANKI DILLEN[1]
Katholieke Universiteit Leuven, Department Wiskunde
Celestijnenlaan 200 B, B-3001 Leuven, Belgium

and

LUC VRANCKEN[2]
Katholieke Universiteit Leuven, Department Wiskunde
Celestijnenlaan 200 B, B-3001 Leuven, Belgium

ABSTRACT

We study affine minimal surfaces in the 3-dimensional affine space. We completely classify the affine minimal surfaces which have higher order parallel cubic form. We show that they are affine flat and either parallel (in which case we obtain the paraboloids), or affine equivalent to $z = xy + P(y)$, where P is an arbitrary polynomial in y.

1. Introduction

The study of affine differential geometry started with the work of W. Blaschke and his coworkers around 1920. The last twenty years there has been a resurgence of interest in affine differential geometry through the work of amongst others Calabi, Chern, Nomizu, Pinkall, Simon, Yau,...

Starting from the standard connection D on $\mathbb{R}^{n+1}$ and the parallel volume form ω on $\mathbb{R}^{n+1}$, given by the determinant, one can, on a nondegenerate hypersurface M, induce in a canonical way an affine connection ∇ and an affine volume form θ. This is done by determining a unique transversal vector field ξ, called the affine normal. More details about this construction are given in Section 2. Then, using the affine normal, just as in the Euclidean case, we can define a $(1,1)$-tensor field S. We call S the affine shape operator.

If trace$(S) = 0$, then M is called affine minimal. It is proved in [2] (for surfaces), in [3] (for convex hypersurfaces) and in [9] (for general nondegenerate hypersurfaces)

[1] Supported by a post-doctoral fellowship of the Research Council of the Katholieke Universiteit Leuven

[2] Research Assistant of the National Fund for Scientific Research (Belgium)

that affine minimal hypersurfaces are extremal hypersurfaces for the area functional. In that sense they are the affine analog of Euclidean minimal submanifolds. But there are also some remarkable differences between the affine case and the Euclidean case. For example, where in the Euclidean case minimal surfaces are locally area minimizing (cfr. Plateau's problem), more or less the opposite is true in the affine case. Indeed, E. Calabi showed that for an affine minimal convex surface in $\mathbb{R}^3$, the second variation of the area functional is negative definite. Hence affine minimal convex surfaces are locally volume maximizing. Therefore, as E. Calabi suggested in [3], a more appropriate name for these surfaces would be affine maximal surfaces. In the nonconvex case however, as is shown in [9], the situation is much more complicated. For example, for the helicoid, which is affine minimal, there exist deformations such that the second variation is positive as well as deformations such that the second variation is negative. On the other hand, see [6], the affine minimal surface with parameterization $x(u,v) = (u\cosh(v), u\sinh(v), v)$ is not convex but every deformation has negative second variation.

Let h be the second fundamental form associated with the immersion. We call ∇h the cubic form. We say that M has higher order parallel cubic form if $\nabla^n h = 0$, for some natural number n. In this paper we study affine minimal surfaces in $\mathbb{R}^3$ which have higher order parallel cubic form. It is proved by L. Berwald [1], that the vanishing of the cubic form characterizes the nondegenerate quadrics. Surfaces with $\nabla^n h = 0$ have been classified in [8] for $n = 2$ and in [5] for small values of n, i.e. for $n = 2, 3, 4, 5$. Here, we shall give a complete classification of affine minimal surfaces with higher order parallel cubic form. In particular, the whole of Section 3 will be dedicated to the proof of the following theorem.

Theorem. *Let M be an affine minimal surface in $\mathbb{R}^3$. Assume also that M has higher order parallel cubic form, i. e. $\nabla^n h = 0$, for some natural number n. Then M is affine equivalent to one of the following affine surfaces:*

(1) *the elliptic paraboloid $z = x^2 + y^2$,*

(2) $z = xy + P(y)$,

where P is a polynomial of degree m, with $m \leq n + 1$.

2. Preliminaries

Let M^n be a differentiable manifold and let $f : M \to \mathbb{R}^{n+1}$ be an immersion. On $\mathbb{R}^{n+1}$, we consider the standard connection D and a parallel volume form ω. Then, starting from an arbitrary transversal (i.e. nowhere tangent) vector field ξ one can introduce a torsion-free affine connection ∇ on M by

$$D_X Y = \nabla_X Y + h(X,Y)\xi,$$

where h is a symmetric $(0,2)$-tensor field on M. We call h the second fundamental form. We say that M is nondegenerate if h is nondegenerate. It is easy to show that the fact that M is nondegenerate is independent of the choice of transversal vector field ξ. From now on, we shall always assume that M is nondegenerate. One can also induce a volume form θ on M by

$$\theta(X_1, \ldots, X_n) = \omega(X_1, \ldots, X_n, \xi).$$

Since f is nondegenerate, we know (see [7]) that there is a unique choice of ξ such that the corresponding induced connection ∇, the nondegenerate second fundamental form h and the induced volume form θ satisfy the following conditions:

(1) $\nabla\theta = 0$

(2) $\theta = \nu_h$ (ν_h is the metric volume form of h)

Such an immersion is called a Blaschke immersion. We call h the affine metric. Condition (1) and (2) imply that $\nabla\nu_h = 0$. This condition is called the apolarity condition. Condition (1) implies that $D_X\xi$ is tangent to $f(M^n)$ for any tangent vector X to M^n. Therefore, we can define a $(1,1)$-tensor field S on M^n, called the shape operator, by

$$D_X\xi = -f_*(SX).$$

If S is a multiple of the identity, then we call M an affine sphere. An affine sphere with $S \neq 0$ is called a proper affine sphere. If $S = 0$, then M is called flat or an improper affine sphere. The mean curvature H is defined by

$$H = \text{trace}(S).$$

M is called affine minimal if H vanishes identically. We define ∇h by

$$(\nabla h)(X,Y,Z) = Xh(Y,Z) - h(\nabla_X Y, Z) - h(Y, \nabla_X Z).$$

Then ∇h is called the cubic form. The equations of Gauss, Codazzi (for h and for S) and Ricci are then respectively given by

$$\begin{aligned} R(X,Y)Z &= h(Y,Z)SX - h(X,Z)SY, \\ (\nabla h)(X,Y,Z) &= (\nabla h)(Y,X,Z), \\ (\nabla_X S)Y &= (\nabla_Y S)X, \\ h(X,SY) &= h(SX,Y). \end{aligned}$$

Notice that the induced connection ∇ is flat ($R = 0$) if and only if $S = 0$. Using the cubic form one can write down the apolarity condition in the following form:

$$\text{trace}_h\{(X,Y) \mapsto (\nabla h)(Z,X,Y)\} = 0 \tag{3}$$

for all vectors Z, where trace_h denotes the trace with respect to the affine metric h.

From now on, we shall assume that the dimension of M is 2. We say that M has higher order parallel cubic form, or shorter : is higher order parallel, if $\nabla^n h = 0$, for some natural number n. We shall need the following results from [5].

Lemma 2.1. *Let M be a Blaschke surface. If M has higher order parallel cubic form, then the shape operator S at any point p has one of the following forms :*

(a) $S = \lambda I$, $\lambda \in \mathbb{R}$,

(b) *There exists a basis $\{e_1, e_2\}$, with $h(e_i, e_j) = \epsilon_j \delta_{ij}$, where $\epsilon_j = \pm 1$ and $i, j \in \{1, 2\}$, such that the shape operator is given by*

$$Se_1 = \lambda e_1$$
$$Se_2 = 0.$$

(c) *There exists a basis $\{e_1, e_2\}$, with $h(e_1, e_1) = h(e_2, e_2) = 0$ and $h(e_1, e_2) = 1$ such that the shape operator is given by*

$$Se_1 = \lambda e_2$$
$$Se_2 = 0.$$

So if we also assume that M is affine minimal, we see that case (a) and case (b) can only occur if $\lambda = 0$, i.e. if $S = 0$. If M is flat, then we have the following result.

Theorem 2.1. *Let M be a flat Blaschke surface. Assume also that M has higher order parallel cubic form. Then, M is affine equivalent to one of the following surfaces*

(1) *the elliptic paraboloid $z = x^2 + y^2$*

(2) $z = xy + P(y)$,

where P is a polynomial in one variable.

So in order to prove our theorem, we only have to show that case (c) can not occur when $\lambda \neq 0$. We can remark that there do exist Blaschke surfaces in $\mathbb{R}^3$ such that the affine shape operator takes the form (c), but of course they are not higher order parallel. Examples are the so-called warped helicoids, introduced in [4], where they are introduced as complex surfaces in $\mathbb{C}^3$, but it is easy to make the translation into the real case, see also [6].

3. Proof of the theorem.

Let M be a higher order parallel minimal Blaschke surface in $\mathbb{R}^3$. As mentioned above, we only have to show that case (c) of Lemma 2.1 can not occur. If $p \in M$, then from Lemma 2.1 we know that there are only two possibilities, namely $S_p = 0$ or S takes the form (c) at p. We now suppose that M is not an improper affine sphere and we shall try to derive a contradiction. Since M is not an improper affine sphere, there exists a point p of M such that $S_p \neq 0$. Hence, we can take a neighbourhood around p on which S does not vanish. So, we can choose a basis $\{E_1, E_2\}$ on a neighbourhood U of p such that for each point $q \in U$, $\{E_1(q), E_2(q)\}$ is a basis as in (c). The basis can be chosen differentiably, cfr. the proof of Lemma 2.1 in [5]. Then we put

$$\nabla_{E_1} E_1 = a_1 E_1 + a_2 E_2, \qquad \nabla_{E_2} E_2 = a_3 E_1 + a_4 E_2,$$
$$\nabla_{E_1} E_2 = a_5 E_1 + a_6 E_2, \qquad \nabla_{E_2} E_1 = a_7 E_1 + a_8 E_2,$$

where a_1 up to a_8 are differentiable functions on U.

Lemma 3.1. *We have the following equations*

$$a_5 = a_3 = a_8 = 0,$$
$$a_7 + a_4 = 0,$$
$$a_6 + a_1 = 0.$$

Proof. The apolarity condition in this situation states that

$$0 = (\nabla h)(E_2, E_1, E_1) = -2h(\nabla_{E_2} E_1, E_1) = -2a_8.$$

By applying the Codazzi equation, we also find that

$$0 = (\nabla h)(E_1, E_2, E_1) = -h(\nabla_{E_1} E_2, E_1) - h(E_2, \nabla_{E_1} E_1) = -a_6 - a_1.$$

Similarly, starting from $(\nabla h)(E_1, E_2, E_2) = (\nabla h)(E_2, E_1, E_2) = 0$, we obtain that

$$a_7 + a_4 = 0 = a_5.$$

Finally, we obtain from the Codazzi equation for S that

$$0 = (\nabla_{E_1} S)E_2 - (\nabla_{E_2} S)E_1 = -E_2(\lambda)E_2 - \lambda a_3 E_1 - 2\lambda a_4 E_2.$$

Hence $a_3 = 0$ and $E_2(\lambda) = -2\lambda a_4$. □

Lemma 3.2. $E_1(a_4) + E_2(a_1) = -2a_1 a_4$.

Proof. By the Gauss equation, we know that $R(E_1, E_2)E_2 = 0$. On the other hand,

$$R(E_1, E_2)E_2 = \nabla_{E_1} \nabla_{E_2} E_2 - \nabla_{E_2} \nabla_{E_1} E_2 - \nabla_{-a_1 E_2 + a_4 E_1} E_2,$$
$$= E_1(a_4) + E_2(a_1) + 2a_1 a_4.$$

□

Lemma 3.3. *There exist coordinates $\{u, v\}$ on a neighbourhood of the point p such that*

$$h(x_u, x_u) = h(x_v, x_v) = 0, \text{ and } h(x_u, x_v) = 1,$$
$$Sx_u = f(u)x_v, \text{ and } Sx_v = 0,$$
$$\nabla_{x_v} x_v = \nabla_{x_u} x_v = \nabla_{x_v} x_u = 0, \text{ and } \nabla_{x_u} x_u = (g(u) - vf(u))x_v,$$

where f and g are differentiable functions, depending only on u, on a neighbourhood U of the point p. Furthermore, f is nonzero on U.

Proof. First, we consider the following system of differential equations on M :

$$E_1(\mu) = -a_1,$$
$$E_2(\mu) = a_4.$$

The integrability condition of this system is given by

$$E_1(a_4) + E_2(a_1) = -2a_1 a_4.$$

Hence, because of Lemma 3.2, we know that the system of differential equations has a solution. We put $\omega = \exp \mu$ and define,

$$F_1 = \omega E_1,$$
$$F_2 = \frac{1}{\omega} E_2.$$

Then $h(F_1, F_1) = h(F_2, F_2) = 0$ and $h(F_1, F_2) = 1$. Furthermore,

$$\nabla_{F_2} F_1 = -\frac{1}{\omega} E_2(\omega) E_1 - a_4 E_1 = 0.$$

Similarly, we also obtain that $\nabla_{F_1} F_2 = 0$. Therefore, by applying Lemma 3.1 for the frame field $\{F_1, F_2\}$ we see that also $\nabla_{F_2} F_2 = 0$ and that $\nabla_{F_1} F_1$ is a multiple of F_2. Hence, we can write

$$\nabla_{F_1} F_1 = \alpha F_2.$$

Also, since $\nabla_{F_1} F_2 - \nabla_{F_2} F_1 = 0$, there exist coordinates $\{u, v\}$ around p such that

$$F_1 = x_u \quad \text{and} \quad F_2 = x_v.$$

Furthermore we have by the definition of F_1 and F_2 that

$$Sx_u = \omega^2 \lambda x_v,$$
$$Sx_v = 0.$$

Hence, by applying the Codazzi equation on S for x_u and x_v, we obtain that

$$(\omega^2\lambda)_v = 0.$$

Hence $\omega^2\lambda$ depends only on u. Finally it follows from the Gauss equation that

$$(\omega^2\lambda)x_v = R(x_u, x_v)x_u = -\nabla_{x_v}(\alpha x_v) = -\alpha_v x_v.$$

Hence $\alpha_{vv} = 0$. So we can write $\alpha(u,v) = g(u) - v(f(u))$. But then, we also find that

$$(\omega^2\lambda) = -\alpha_v = f.$$

Since ω and λ are not zero on U, this implies that f is also not zero. This completes the proof of the lemma. □

For the remainder of this section, we shall always work with the coordinates given in Lemma 3.3. Before we can formulate the next lemma, we need two more definitions. The vector fields x_u and x_v are called special vector fields. Let $(x_1, \ldots, x_k)$ be a sequence of special vector fields. Then, we define the index $\mathrm{Ind}(x_1, \ldots, x_k)$ of $(x_1, \ldots, x_k)$ as the number of $i \in \{1, \ldots, k\}$ such that $x_i = x_v$. Then we have the following lemma.

Lemma 3.4. *Let m be a natural number and let $x_1, \ldots, x_{m+2}$ be special vector fields on a neighbourhood of p. If $\mathrm{Ind}(x_1, \ldots, x_{m+2}) \geq 2$, then*

$$(\nabla^m h)(x_1, \ldots, x_{m+2}) = 0,$$

and in the other cases we have

$$\begin{aligned}
&(\nabla^m h)(x_v, x_u, \ldots, x_u) = P_1^m(u),\\
&(\nabla^m h)(x_u, x_v, x_u, \ldots, x_u) = P_2^m(u),\\
&\ldots\\
&(\nabla^m h)(x_u, x_u, \ldots, x_v, x_u, x_u, x_u) = P_{m-1}^m(u),\\
&(\nabla^m h)(x_u, x_u, \ldots, x_u, x_v, x_u, x_u) = 0,\\
&(\nabla^m h)(x_u, x_u, \ldots, x_u, x_v, x_u) = 0,\\
&(\nabla^m h)(x_u, x_u, \ldots, x_u, x_u, x_v) = 0,\\
&(\nabla^m h)(x_u, \ldots, x_u) = P^m(u) + vQ^m(u),
\end{aligned}$$

where $P_1^m, \ldots, P_{m-1}^m, P^m, Q^m$ are functions depending only on u and are defined *inductively in the following way:*

$$\begin{aligned}
P^1(u) &= -2g(u),\\
Q^1(u) &= 2f(u),\\
P_1^{m+1}(u) &= Q^m(u),\\
P_i^{m+1}(u) &= (P_{i-1}^m(u))_u \qquad i \geq 2,\\
P^{m+1}(u) &= (P^m(u))_u - g(u)\sum_{i=1}^{m-1} P_i^m(u),\\
Q^{m+1}(u) &= (Q^m(u))_u + f(u)\sum_{i=1}^{m-1} P_i^m(u).
\end{aligned}$$

Proof. We shall prove this lemma by induction. If $m = 1$, we recall that the apolarity condition states that

$$(\nabla h)(x_u, x_v, x_v) = (\nabla h)(x_v, x_u, x_u) = 0.$$

Further, using the expression for ∇ given in Lemma 3.3, a straightforward computation shows that

$$(\nabla h)(x_v, x_v, x_v) = 0$$

and

$$(\nabla h)(x_u, x_u, x_u) = -2(g(u) - vf(u)).$$

Hence the lemma is true for $m = 1$. Next, we assume that the lemma holds for some natural number m and we show that it also holds for $m + 1$.

First, we take special vector fields x_1 up to x_{m+3} such that the index of $(x_1, \dots, {}_km + 3)$ is greater than or equal to 2. If $x_1 = x_v$, then, since $\nabla_{x_v} x_i = 0$ for all i, we have that

$$(\nabla^{m+1} h)(x_v, x_2, \dots, x_{m+3}) = ((\nabla^m h)(x_2, \dots, x_{m+3}))_v = 0,$$

using $\mathrm{Ind}(x_2, \dots, x_{m+3}) \geq 1$ and the induction hypothesis. If $x_1 = x_u$, then we have that $\mathrm{Ind}(x_2, \dots, x_{m+3}) \geq 2$. Then, using also the fact that $\nabla_{x_u} x_v = 0$ and that $\nabla_{x_u} x_u$ only has a component in the x_v direction, we obtain that

$$\begin{aligned}
(\nabla^{m+1} h)(x_u, x_2, \dots, x_{m+3}) = -(\nabla^m h)(\nabla_{x_u} x_2, \dots, x_{m+3}) \dots\\
- (\nabla^m h)(x_2, \dots, \nabla_{x_u} x_{m+3}) = 0.
\end{aligned}$$

This settles the matter if the index is greater than or equal to two.

Secondly, if the index is one, we immediately see that

$$(\nabla^{m+1} h)(x_v, x_u, \dots, x_u) = ((\nabla^m h)(x_u, x_u, \dots, x_u))_v = Q^m(u),$$

so that

$$P_1^{m+1}(u) = Q^m(u).$$

Using the induction hypothesis and the expression for ∇ given in Lemma 3.3, we obtain in a similar way that

$$(\nabla^{m+1}h)(x_u, x_v, \dots, x_u) = ((\nabla^m h)(x_v, x_u, \dots, x_u))_u = (P_1^m(u))_u,$$

so that $P_2^{m+1}(u) = (P_1^m(u))_u$. Similarly, we obtain that $P_i^{m+1}(u) = (P_{i-1}^m(u))_u$. This completes the proof if $\mathrm{Ind}(x_1, \dots, x_{m+3}) = 1$. Finally, we may assume that the index is zero. Then, we obtain that

$$\begin{aligned}(\nabla^{m+1}h)(x_u, x_u, \dots, x_u) &= ((\nabla^m h)(x_u, \dots, x_u))_u \\ &\quad - (\nabla^m h)(\nabla_{x_u} x_u, x_u, \dots, x_u) - \dots - (\nabla^m h)(x_u, \dots, x_u, \nabla_{x_u} x_u) \\ &= (P^m)_u + v(Q^m)_u - (g(u) - vf(u))(\sum_{i=1}^{m-1} P_i^m(u)).\end{aligned}$$

This completes the proof of the lemma. □

Lemma 3.5. *The function f is a nonzero polynomial in u of degree l, where $0 \le l \le n-3$.*

Proof. Because $\nabla^n h = 0$, we know that $(\nabla^n h)(x_u, x_u, \dots, x_u, x_v, x_u, x_u) = 0$. Therefore, if we use Lemma 3.4, we see that

$$0 = P_{n-1}^n(u) = (P_{n-2}^{n-1}(u))_u = (P_{n-3}^{n-2}(u))_{uu} = \dots = (P_1^2)_{uu\dots u}.$$

Hence the $(n-2)$-th derivative of $P_1^2 = 2f$ is zero. Thus f is a polynomial in u of degree less than or equal to $n-3$. □

By combining Lemma 3.4 and Lemma 3.5, we immediately see that for all natural numbers i and j such that $i > j$, P_j^i and Q_j are polynomials in u. Then, in the following lemma, we give some relations between the degrees of these polynomials. Here, we define the degree of the 0-polynomial to be $-\infty$. We denote by l the degree of f.

Lemma 3.6. *For every natural number m we have*

$$\deg(P_1^{2m}) = ml,$$

$$\deg(P_i^{2m}) < ml, \qquad i \ge 2$$

$$\deg(Q^{2m}) < ml,$$

for every natural number m.

Proof. We shall prove this lemma by induction. If $m = 1$, we have

$$P_1^2(u) = 2f(u),$$

$$Q^2(u) = 2(f(u))_u.$$

Now, we may assume that the lemma holds for a natural number m. Then, we have

$$\begin{aligned}
&\deg(P_1^{2m+1}(u)) = \deg(Q^{2m}(u)) < ml,\\
&\deg(P_i^{2m+1}(u)) = \deg((P_{i-1}^{2m}(u))_u) < ml-1, \qquad i \geq 2,\\
&\deg(Q^{2m+1}(u)) = \deg((Q^{2m}(u))_u + f(u)(\sum_{i=1}^{n} P_i^{2m}(u))) = (m+1)l.
\end{aligned}$$

Therefore, we also obtain that

$$\begin{aligned}
&\deg(P_1^{2m+2}(u)) = \deg(Q^{2m+1}(u)) < (m+1)l,\\
&\deg(P_i^{2m+2}(u)) = \deg((P_{i-1}^{2m+1}(u))_u) < ml-1 < (m+1)l, \qquad i \geq 2,\\
&\deg(Q^{2m+2}(u)) = \deg((Q^{2m+1}(u))_u + f(u)(\sum_{i=1}^{n} P_i^{2m+1}(u))) = (m+1)l.
\end{aligned}$$

This completes the proof of the lemma. □

Since M has higher order parallel second fundamental form, we know that $\nabla^n h = 0$ and hence $\nabla^{2n} h = 0$. So we also know that $P^{2n} = 0$. Hence $\deg(P_1^{2n}) = -\infty$. But on the other hand, by Lemma 3.6 we know that P_1^{2n} is a polynomial in u of degree nl where n is a natural number and $l \neq -\infty$. This is a contradiction, so that the theorem is proved.

4. References

1. L. Berwald, *Die Grundgleichungen der Hyperflächen im Euklidischen R_{n+1} gegenüber inhaltstreuen affinitäten*, Monatshefte für Math und Phys. **32** (1922), 89–106.
2. W. Blaschke, *Vorlesungen über Differentialgeometrie, II, Affine Differentialgeometrie*, Springer Verlag, Berlin, 1923.
3. E. Calabi, *Hypersurfaces with maximal affinely invariant area*, Amer. J. Math. **104** (1982), 91–126.
4. F. Dillen, *Locally symmetric hypersurfaces of the complex affine space*, J. Geom. **33** (1988), 27–38.
5. F. Dillen, L. Vrancken, *Generalized Cayley surfaces*, Proceedings of the Conference on Global Analysis and Global Differential Geometry, Berlin 1990, Lecture Notes in Mathematics, Springer Verlag, Berlin (to appear).
6. F. Dillen, L. Vrancken, *Affine differential geometry of hypersurfaces*, Geometry and Topology of Submanifolds II, World Scientific, Singapore, 1990, pp. 144–164.
7. K. Nomizu, *Introduction to affine differential geometry, part I*, MPI/88–37, Bonn.

8. K. Nomizu and U. Pinkall, *Cayley surfaces in affine differential geometry*, Tôhoku Math. J. **41** (1989), 589–596.
9. L. Verstraelen and L. Vrancken, *Affine Variation Formulas and Affine Minimal Surfaces*, Michigan Math. J. **36** (1989), 77–93.

The Problem of Plateau (pp. 87–110)
ed. Th. M. Rassias

CARTAN'S METHOD AND PLATEAU'S PROBLEM

JERRY DONATO

Institute for Mathematics and its Applications
University of Minnesota
206 Church Street S.E.
Minneapolis, Minnesota 55455-0436
U.S.A.

ABSTRACT

Cartan's method of moving frames is used to describe minimal surfaces where curvature, torsion and group properties are considered.

1. Introduction

Sections 2 and 3 review the following fundamental concepts of arithmetic and algebra

- (i) The relation between the coefficients of a polynomial and the corresponding roots are noted, and their relationship to the mean curvature and Gaussian curvature in classical differential geometry is mentioned.
- (ii) The coefficients and roots of an equation may vary and a transformation procedure may be used.
- (iv) the concepts of group and field and the solution of polynomial equations are mentioned.
- (v) The characteristic value problem is outlined and its relationship to mean and Gaussian curvature is noted.
- (vi) A review of linear algebra is made and the notion of the trace of a matrix is developed and its use in describing a minimal surface is noted.
- (vii) The following well-known theorem is stated: if A is skew-symmetric matrix, then $\exp A$ is orthogonal and its relationship to Cartan's method and the problem of Plateau is mentioned.

Section 4 outlines the fundamentals of curves in space using the Serret- Frenet equations of a curve where the tangent, normal and binormal vectors and the curvature and torsion parameters (along with a natural parameter) are presented. A skew-symmetric matrix appears from these equations. The properties of some curves are noted.

Section 5 reviews and comments on the following fundamental notions of classical surface theory:

(i) The mean curvature is the sum of the roots and the Gaussian curvature is the product of the roots of a polynomial equation; this result suggests that a geometric viewpoint is reflected in an algebraic viewpoint, that is, algebra gives information about geometry and geometry gives information about algebra— two viewpoints meet;

(ii) The previously mentioned theorem: if A is a skew-symmetric matrix, then $\exp A$ is orthogonal, sets the meeting of two different views of the matrix A on the stage of group theory; the exponential function suggests Lie group theory.

(iii) From an algebraic viewpoint, the theory of polynomial equations provides a procedure that may reduce the $(n-1)$st degree term to zero through a transformation process; from a geometric viewpoint this indicates a procedure to reduce mean curvature to zero, that is, describe a minimal surface.

(iv) A rotation of coordinates procedure results in a diagonalized second fundamental form.

(v) The Theoremea Egreguim of Gauss is stated and the intrinsic concept of a surface is noted.

(vi) The integrability conditions of surface theory are noted.

Section 6 presents Poincare's Lemma and it's Converse and notes that these two lemmas establish the integrability conditions of differential equations.

Section 7 presents the Frobenius integrability conditions for the Pfaffian equations. Section 8 presents the influence of the exponential form and related group property on the well-known first order ordinary differential equation.

Section 9 presents an overview of Plaţeau's problem and mentions some of the major investigators of this problem. Section 10 briefly reviews the following selected notions of the theory of Lie groups:

(i) presenting the mapping conditions of a Lie Group;

(ii) establishing that $G\ell(n, \mathbf{R})$ is a Lie group and so is $O(n)$ and $S\ell(n, \mathbf{R})$;

(iii) describing the action of a group on a set;

(iv) establishing that $G\ell(n, \mathbf{R})$ is a natural action on $\mathbf{R}^n$;

(v) establishing that every one-parameter subgroup of $G\ell(n, \mathbf{R})$ has an exponential form that is an orthogonal matrix which is related to a skew-symmetric matrix;

(vi) presenting a Lie Group adjoint representation;

(vii) presenting the Maurer-Cartan forms and their relationship to the First Fundamental Theory of Lie;

(viii) establishing the existence of a Lie Group having constants of structure provided that linearly independent one-forms can be found that satisfy a specified equation.

Section 11 outlines Cartan's method of moving frames with a given Riemannian manifold and the principal fiber bundle with the structure group $0(n)$, the bundle of orthonormal frames. The structure equations and the compatability (integrability) equations without curvature and with curvature are presented. Given arbitrary one-forms (called a basis or co-frame) and one-forms called connections certain integrability conditions must hold in order to establish the existence of a local subspace and a unique family of frames. The connection one-forms are skew-symmetric and there is no torsion term.

The equations without curvature and in matrix form are as follows:

$$\begin{aligned} dx &= \sigma e \\ de &= \Omega e \\ \Omega &= -\Omega^t \text{ where } t \text{ indicates a transpose} \\ d\sigma &= \Omega\sigma \\ d\Omega &= \Omega^2 \,. \end{aligned}$$

The above set of equations would describe a minimal surface provided the mean curvature is zero. With curvature, the last equation would be written as

$$d\Omega = \Omega^2 + \Theta$$

where Θ is the curvature term. Again a minimal surface would be described provided the mean curvature is zero. The general integrability conditions, the Bianchi identities and the torsion equations are presented in matrix form.

Section 12 briefly outlines Cartan's method of equivalence as given by Robert B. Gardner's 1989 monograph *The Method of Equivalence and Its Applications.* The equivalence problem of Élie Cartan is stated. The method considers a prescribed linear group in $G\ell(n, \mathbf{R})$, hence, the group is larger than the structure group $0(n)$ considered in section 11. The torsion concept is also incorporated into the method. There are also many other notions in the method; some of these concepts are mentioned in the description presented. A set of equations is given in index form where, in addition to curvature, there is a lifting of the linear group to G spaces and a mapping of intrinsic torsion. If torsion is not constant, and the group reduction and normalization proceses are not constant, then many other possibilities and problems appear. Hence, specifying a minimal surface–a surface with mean curvature zero–requires using broader postulates on which the subject of mathematics rests.

2. Arithmetic

Theorem 2.1. The Fundamental Theorem of Arithmetic states: *Every positive integer $a > 1$ can be expressed as a product of positive primes in one and only one way except for the order of the factors.*

Remark 2.2. The operations of multiplication with its corresponding commutativity or non-commutativity properties and related uniqueness are important.

3. Algebra

Theorem 3.1. The Fundamental Theorem of algebra states: *Every polynomial of degree n has exactly n roots, some or all of which may be complex.*

Example 3.2. Let $\lambda_1, \lambda_2, \lambda_3, \ldots, \lambda_n$ denote the n roots of the integral rational equation

$$a_0 x^n + a_1 x^{n-1} + a_2 x^{n-2} + \cdots + a_n = 0$$

or the equivalent equation

$$x^n + \frac{a_1}{a_0} x^{n-1} + \frac{a_2}{a_0} x^{n-2} + \cdots + \frac{a_n}{a_0} = 0 \, .$$

Then the following identity appears

$$\begin{aligned} x^n &+ \frac{a_1}{a_0} x^{n-1} + \frac{a_2}{a_0} x^{n-2} + \cdots + \frac{a_n}{a_0} = 0 \\ &= (x - \lambda_1)(x - \lambda_2)(x - \lambda_3) \cdots (x - \lambda_n) \\ &= x^n - (\lambda_1 + \lambda_2 + \lambda_3 + \cdots + \lambda_n) x^{n-1} \\ &+ (\lambda_1\lambda_2 + \lambda_1\lambda_3 + \cdots + \lambda_1\lambda_n + \cdots + \lambda_{n-1}\lambda_n) x^{n-2} \\ &- (\lambda_1\lambda_2\lambda_3 + \cdots + \lambda_1\lambda_2\lambda_3 + \cdots + \lambda_{n-2}\lambda_{n-2}\lambda_n) x^{n-3} \\ &\vdots \\ &+ (-1)^n (\lambda_1\lambda_2\lambda_3 \cdots \lambda_n) \, . \end{aligned}$$

Equate the coefficients of like powers of x in the two members of the above identity to obtain the following Vieta's relations between the coefficients

$a_0, a_1, a_2, \ldots, a_n$ and the roots $\lambda_1, \lambda_2, \lambda_3, \ldots, \lambda_n$

$$\frac{a_1}{a_0} = -(\lambda_1 + \lambda_2 + \lambda_3 + \cdots + \lambda_n),$$

$$\frac{a_2}{a_0} = +(\lambda_1\lambda_2 + \lambda_1\lambda_3 + \cdots + \lambda_1\lambda_n + \cdots + \lambda_{n-1}\lambda_n)$$

$$\frac{a_3}{a_0} = -(\lambda_1\lambda_2\lambda_3 + \lambda_1\lambda_2\lambda_4 + \cdots + \lambda_1\lambda_2\lambda_n + \cdots \lambda_{n-2}\lambda_{n-1}\lambda_n)$$

$$\cdots\cdots\cdots\cdots\cdots\cdots\cdots\cdots\cdots\cdots$$

$$\frac{a_n}{a_0} = (-1)^n(\lambda_1\lambda_2\lambda_3\cdots\lambda_n)\,.$$

Remark 3.3. The above theory of equations is used in describing the mean curvature and the Gaussian curvature of surfaces in classical differential geometry.

Remark 3.4. The coefficients $a_0, a_1, \ldots, a_n$ may be regarded as variable quantities. These variables may be independent or they may be in turn functions of other variables. As the coefficients vary, the roots of the equation will likewise vary and conversely.

Remark 3.5. It is frequently desirable to transform the following equation

$$a_0x^n + a_1x^{n-1} + a_2x^{n-2} + \cdots + a_n = 0$$

into another equation in which the term of the $(n-1)$st degree is missing.

Remark 3.6. The above algebraic procedure suggests that a mean curvature of zero value can be attained through a transformation process and a minimal surface can be described.

Remark 3.7. ([8]) A basic problem of algebra is the solution of polynomial equations. The broad question of which equations are solvable by algebraic operations were comprehensively answered by Niels Abel and Evariste Galois. The concepts of a group and a field emerged from their work. The group of an equation is a key to its solvability because the group expresses the degree of indistinguishability of the roots. Hence in a sense, a group measures the ignorance of the roots.

Example 3.8. Consider determining the solutions, if any, of the vector equation

$$Ax = \lambda x$$

where λ is a scalar parameter for which there exists vectors $x \neq 0$ and where A is an nth-order matrix. Such a problem is called a characteristic value or eigenvalue problem.

The above equation corresponds to the system of linear equations for the components

$$\sum_{j=1}^{n} a_{ij}x^j = \lambda x^i\,, \qquad i = 1, 2, \ldots, n\,.$$

This is a homogeneous system of n equations in the n unknowns $x^1, x^2, \ldots, x^n$ and maybe written as

$$\sum_{j=1}^{n}(a_{ij} - \lambda\delta_{ij})x^j = 0\,, \quad i = 1, 2, \ldots, n\,.$$

If $x \neq 0$ satisfies the above equation for a given λ, then A operating on x gives a vector which is a scalar multiple of x. Obviously $x = 0$ is one solution for any λ but this is a trivial solution. The more interesting cases occur when $x \neq 0$.

The above equation can also be written in the form

$$Ax = \lambda I x$$

or in matrix form as

$$(A - \lambda I)x = 0\,.$$

The above matrix has a determinant form as follows:

$$\begin{vmatrix} a_{11} - \lambda & a_{12} & \ldots & a_{1n} \\ a_{21} & a_{22} - \lambda & \ldots & a_{2n} \\ \vdots & & & \\ a_{n1} & a_{n2} & \ldots & a_{nn} \end{vmatrix} = (-1)^n \lambda^n + \cdots$$

which is a polynomial in λ of degree n. As the Fundamental Theorem of Algebra indicates, there are n values of λ which make the above determinant vanish. If λ is a real root of the polynomial equation

$$|A - \lambda I| = 0$$

then there is at least one solution $(x^1, x^2, \ldots, x^n)$ of the system which is not identically zero. Hence, for this same value of λ, there is a nonzero solution of the vector equation. It can also be proved that similar matrices have the same eigenvalues. If a matrix A is given and another matrix B can be found which has the same eigenvalues as A and which is diagonal, then the eigenvalues of A can be read off from those of B, provided the eigenvalues of A are all linearly independent.

Remark 3.9. The above formulation of the characteristic value problem is also applied in describing the mean curvature and the Gaussian curvature of surfaces in classical differential geometry.

Example 3.10. ([1]) The determinant of a linear operator $A : \mathbf{R}^n \to \mathbf{R}^n$, denoted by $|A|$, is defined to be the determinant of the matrix representation of A in any basis $e_1, e_2, \ldots, e_n$ of $\mathbf{R}^n$. The determinant of the matrix of the operator does not depend on the basis. If A is the matrix of the operator A in the basis $e_1, e_2, \ldots, e_n$, then the matrix of A with respect to another basis is of the form BAB^{-1}. But $|BAB^{-1}| = |A|$.

The trace of a matrix A denoted by trA is defined to be the sum of its diagonal elements, that is,

$$trA = \sum_{i=1}^{n} a_{ii} .$$

The trace of the matrix of an operator $A : \mathbf{R}^n \to \mathbf{R}^n$ does not depend on the basis of $\mathbf{R}^n$, but only on the operator itself. The trace of a matrix equals the sum of all n of its eigenvalues while the determinant equals the product of the eigenvalues. The trace of an operator A is the trace of its matrix in any (and hence in every) basis. For any linear operator $A : \mathbf{R}^n \to \mathbf{R}^n$, $|e^A| = e^{trA}$. The operator e^A is nonsingular. The operator e^A preserves the orientation of $\mathbf{R}^n$ (i.e. $|e^A| > 0$). If the trace of A equals zero, then the phase flow of

$$\dot{x} = Ax, x \in \mathbf{R}^n$$

preserves volume. Note that $e^0 = I$.

When considering the equation

$$\ddot{x} = -x + k\dot{x},\ x \in \mathbf{R}^n$$

of a pendulum with coefficient of friction $-k$, the behavior of area under transformations of the phase flow $\{g^t\}$ of the pendulum are as follows: if $k = 0$, area is preserved (g^t is a rotation through the angle t) if $k < 0$, the area is smaller and if $k > 0$, the area is larger.

Remark 3.11. The quadratic equation used in the theory of surfaces in classical differential geometry gives two roots (the characteristic values). The mean curvature of the surface is described as one-half of the sum of the roots. The trace is the sum of the roots. When the trace is zero, the mean curvature is zero and the surfaces are minimal. Hence the above algebraic operations are of great value in describing minimal surfaces.

Theorem 3.12. ([5]) *If A is a skew-symmetric matrix, then e^A is orthogonal where*

$$e^A = I + \sum_{n=1}^{\infty} \frac{A^n}{n!}$$

for the real matrix A.

Remark 3.13. A skew-symmetric matrix is a square matrix whose diagonal elements are zero, that is, the trace is zero and whose number of arbitrary elements in an nth-order skew-symmetric matrix is $n(n-1)/2$. This suggests that a skew-symmetric matrix can describe a minimal surface. The above Theorem 3.12 establishes that e^A is orthogonal, hence, a minimal surface can be viewed as e^A. There are now two different equivalent approaches to describing minimal surfaces: A as a skew-symmetric matrix and e^A as being orthogonal. Further investigations of this relationship suggests an application of Élie Cartan's method of equivalence to the problem of Plateau.

4. Curves

Theorem 4.1. The Fundamental existence and uniqueness theorem for space curves states: *Let $k(s)$ and $\tau(s)$ be arbitrary continuous functions on $a \leq s \leq b$, then there exists, except for position in space, one and only one space curve C, for which $k(s)$ is the curvature, $\tau(s)$ is the torsion and s is a natural parameter along C.*

Theorem 4.2. *The Serret-Frenet formulas of a curve are as follows:*

$$\begin{aligned} \dot{T} &= \quad\quad\quad +kN \\ \dot{N} &= -kN \quad\quad\quad -\tau B \\ \dot{B} &= \quad\quad\quad +\tau B \end{aligned}$$

where T, N and B represent the tangent, normal and binormal vectors respectively, k and τ represent the curvature and torsion parameters respectively and $\dot{T}$, $\dot{N}$ and $\dot{B}$ represent the derivatives with respect to the natural parameters.

Remark 4.3. Observe that the matrix is skew-symmetric. The three vectors T, N and B form a right-handed orthonormal triplet.

Theorem 4.4. (cf.[9]) *A regular curve of class ≥ 2 is a straight line if and only if its curvature is identically zero. A curve of class ≥ 3 along which N is of class C^1 is a plane curve if and only if its torsion vanishes identically. A curve is a general helix if and only if τ/k is constant where $k \neq 0$ and $\tau = 0$ whenever $k = 0$.*

5. Surfaces

Theorem 5.1. (cf.[9]) *The curvature of a normal section of a patch at a point P is equal to the normal curvature of the section at P.*

The normal curvature to the curve C at P depends only upon P and the direction of the tangent line to C at P. Hence the normal curvature, denoted as

k_n, at P in the direction $dv : du$, $du^2 + dv^2 \neq 0$ can be written as

$$k_n = \frac{L\,du^2 + 2M du\,dv + N\,dv^2}{E\,du^2 + 2\,F du\,dv + G\,dv^2} = \frac{II}{I}$$

where E, F and G are the coefficients of the first fundamental form, I, and where L, M and N are the coefficients of the second fundamental form, II.

Definition 5.2. (cf.[18]) *The direction numbers are denoted by $dv : du$. These direction numbers determine the same line if and only if they are proportional, that is,*

$$\frac{dv}{du} = \lambda$$

where there exists a $\lambda \neq 0$.

Definition 5.3. *The two perpendicular directions from which the values of k_n take on a maximum and minimum are called principal directions and the corresponding normal curvatures, k_1 and k_2, are called principal curvatures.*

Theorem 5.4. (cf.[9]) *A real number k is a principal curvature at P in the direction $dv : du$ if and only if k, du and dv satisfy*

$$(L - kE)du + (M - kF)dv = 0$$
$$(M - kF)du + (N - kG)dv = 0$$

where $du^2 + dv^2 \neq 0$.

The above is a homogeneous system of equations and will have a nontrivial solution dv, du if and only if

$$(EG - F^2)k^2 - (EN + GL - 2FM)k + (LN - M^2) = 0\,.$$

Theorem 5.5. (cf.[9]) *A number k is a principal curvature if and only if k is a solution of the equation*

$$(EG - F^2)k^2 - (EN + GL - 2FM)k + (LN - M^2) = 0\,.$$

Dividing by $EG - F^2$ it gives

$$k^2 - 2Hk + K = 0$$

where H and K are defined below.

Definition 5.6. *The mean curvature, H, is*

$$H = \frac{1}{2}(k_1 + k_2) = \frac{EN + GL - 2FM}{2(EG - F^2)}$$

which is the average of the roots k_1 and k_2.

Definition 5.7. *The Gaussian curvature K, is*

$$K = k_1 k_2 = \frac{LN - M^2}{EG - F^2}$$

which is the product of the roots.

Remark 5.8. Consider the following polynomial equation presented in Example 3.2:

$$a_0 x^n + a_1 x^{n-1} + a_2 x^{n-2} + \cdots + a_n = 0 \, .$$

Next consider the following equation presented in Theorem 5.5:

$$(EG - F^2)k^2 - (EN + GL - 2FM)k + (LN - M^2) = 0 \, .$$

This suggests the following analogies term by term:

$$\begin{aligned} a_0 &\sim (EG - F^2) \\ x^n &\sim k^2 \\ a_1 &\sim -(EN + GL - 2FM) \\ x^{n-1} &\sim k \\ a_n &\sim (LN - M^2) \end{aligned}$$

where $\sim$ indicates an analogy. The example in 3.2 also indicates that the above equation can be written in the following equivalent form:

$$x^n + \frac{a_1}{a_0} x^{n-1} + \frac{a_2}{a_0} x^{n-2} + \cdots + \frac{a_n}{a_0} = 0 \, .$$

A similar procedure is followed in Theorem 5.5, that is,

$$k^2 - \frac{(EN + GL - 2FM)}{(EG - F^2)} k + \frac{(LN - M^2)}{(EG - F^2)} = 0 \, .$$

The information in Example 3.2 also establishes the relationship between the coefficients and the roots from an algebraic viewpoint. The definitions in 5.6 and 5.7 establish that the relationship between the coefficients and the roots define mean curvature and Gaussian curvature, that is, a geometric viewpoint is reflected in an algebraic viewpoint. Algebra gives information about geometry and geometry gives information about algebra. A meeting of two viewpoints also has appeared in Theorem 3.12: if A is a skew-symmetric real matrix, then e^A is orthogonal. The appearance of the exponential function further expands the above relationships into group theory–a Lie group theory. Remark 3.7 states that Abel and Galois brought in the concepts of groups and fields in the solution of polynomial equations.

Remark 5.9. As Remarks 3.5 and 3.6 note, an integral rational equation may be transformed into another equation in which the term of the $(n-1)$st degree is missing from an algebraic viewpoint. From a geometric viewpoint, this indicates that the mean curvature is zero, that is, a minimal surface can be described. Observe that a transformation procedure is used.

Theorem 5.10. (cf.[9]) *The solutions to* $L\,du^2 + 2M\,du\,dv + N\,dv^2 = 0$ *form orthogonal families on a patch if and only if* $GL + EN - 2FM = 0$.

Remark 5.11. These are the conditions for zero mean curvature, that is, a minimal surface.

Theorem 5.12. *It is possible to rotate the coordinates in such a way that the cross terms,* xy, *in the original general equation of the second degree*

$$Ax^2 + Bxy + Cy^2 + Dx + Ey + F = 0$$

can be eliminated. This can be done by selecting the angle θ *so that*

$$\cot 2\theta = \frac{A - C}{B}$$

where the analogy with the equation in Theorem 5.10 would be ([7])

$$\tan 2\theta = \frac{2M}{L - N}\,.$$

Remark 5.13. Applying the above algebraic procedure to the following equation, the second fundamental form, in Theorem 5.10

$$Ldu^2 + 2Mdu\,dv + N\,dv^2 = 0$$

would result in a diagonalized second fundamental form ([7]).

Theorem 5.14. The Theorema Egreguim of Gauss states: *The Gaussian curvature on a surface of class* $\geqq 3$ *is a function only of the coefficients of the first fundamental form and their derivatives.*

Remark 5.15. ([18], [8]) Gaspard Monge saw in a surface primarily the boundary of a solid body and hence stressed the properties of a surface in relation to the surrounding space. Monge viewed problems involving curves led to ordinary differential equations and problems involving surfaces led to partial differential equations. Carl Gauss viewed the surface as a space itself, hence, intrinsic. The corresponding extrinsic view can be described as follows:

"Houston, Apollo 11 ... I've got the world in my window." Mike Collins, 28 hours and 7 minutes ground elapsed time quoted from *Carrying the Fire* by Michael Collins.

Theorem 5.16. (cf.[9]) The Fundamental Theorem of Surfaces states: *Let E, F and G be functions of u and v of class C^2 and let L, M and N be functions of u and v of class C^1 all defined on an open set containing (u_0, v_0) such that for all (u, v),*

(i) $EG - F^2 > 0, G > 0, \vec{G} > 0$

(ii) *E, F, G, L, M, N satisfy certain integrability conditions.*

Then there exists a patch $x = x(u, v)$ of class C^3 defined in the neighborhood of (u_0, v_0) for which E, F, G, L, M, N are the first and second fundamental coefficients. The surface represented by $x = x(u, v)$ is unique except for position in space.

Remark 5.17. The integrability (compatibility) conditions arise from the fact that if $x(u, v)$ is a function of class C^3 then the third order mixed partial derivatives of x are independent of the order of differentiation.

6. Poincaré's Lemma

Lemma 6.1. (cf.[9]) Poincaré's Lemma states: *If ω is a p-form on M (manifold) for which there exists a $(p-1)$-form α such that $d\alpha = \omega$, then $d\omega = 0$.*

Converse of Poincaré's Lemma 6.2. (cf.[9]) The converse of Poincaré's Lemma states: *If ω is a p-form on an open set $U \subset M$ (which is contractible to a point) such that $d\omega = 0$, then there exists a $(p-1)$-form α such that $\omega = d\alpha$ (Exceptions: if $p = 0$, then $\omega = f$ and the vanishing of df means f is constant).*

Remark 6.3. The above two lemmas establish the integrability conditions of differential equations; that is, the order of taking partial derivatives commutes.

7. Frobenius Conditions and Pfaffian Equations

Theorem 7.1. (cf.[19]) *The Frobenius condition $d\omega \wedge \omega = 0$ is called integrability condition for the Pfaffian equation $\omega = 0$.*

Example 7.2. Consider the following functions $f, g : U \to \mathbf{R}$ where U is a neighborhood of $\mathbf{R}^n$ satisfying

$$\omega = g df \, .$$

Use an inverse operation to rewrite the above as

$$df = g^{-1} \omega \, .$$

Apply the exterior algebraic operations to the first equation to get

$$d\omega = dg \wedge df \, .$$

Then substitute to get

$$d\omega = dg \wedge g^{-1}\omega = \frac{dg}{g} \wedge \omega \, .$$

Thus

$$d\omega = \theta \wedge \omega$$

where $\theta = g^{-1}dg = d\,\ell n|g|$. Hence

$$\omega \wedge d\omega = \omega \wedge \theta \wedge \omega = 0 \, .$$

Remark 7.3. Observe the appearance of the exponential form $-d\,\ell n|g|$.

8. Ordinary Differential Equations

Example 8.1. ([4,3]) Consider the following well-known ordinary differential equation (ODE)

$$\frac{dx}{dt} = Ax$$

and its corresponding solution

$$\phi(t) = e^{At}x(t)$$

where the initial conditions are specified as

$$\phi(0) = x(0) \, .$$

The usual procedure to solve the ODE requires the exponential form of the constant of integration be multiplied by the exponential form of the parameter, that is,

$$x = e^{At} + e^{B} = e^{At}e^{B}$$

where e^B represents the initial conditions. In other words the exponential product equals the sum of the exponentials, that is, multiplication commutes and addition is associative.

When $g^t : \mathbf{R}^n \to \mathbf{R}^n$ is a one-parameter group of linear transformation, then there exists a linear operator such that

$$g^t = e^{At} \, .$$

The group structure acts on the initial conditions and e^{At}.

When the linear operators $A, B : \mathbf{R}^n \to \mathbf{R}^n$ commute $AB = BA$, then

$$e^{A+B} = e^A e^B \, .$$

However, in general

$$e^{A+B} \neq e^A e^B \, .$$

Hence, addition is non-associative and multiplication is non-commutative, that is, the exponential operators do not commute. An obstruction has appeared.

Remark 8.2. The matrix methods for first order systems with constant coefficients use the above exponential form in constructing the solutions.

9. Overview of Plateau's Problem

I quote from Th. M. Rassias [14]: One of the most difficult problems in global variational analysis is Plateau's problem. Plateau dipped a wire in the form of a closed space curve into a soap solution and realized minimal surfaces as soap films. This is a problem of determining the surfaces of minimum area spanned in a given curve or subject to other boundary conditions.

Minimal surfaces belong to one of the best-studied surfaces in differential geometry. Their theory was initiated by Lagrange and followed by many mathematicians.

Jesse Douglas and Tibor Radó gave the very important existence proofs to the problem of Plateau.

M. Morse and C. Tompkins, M. Shiffman independently studied the existence of an unstable extremal in the theory of Plateau's problem. R. Palais and S. Smale found a condition called Condition C by them which allowed them to apply the Morse theory (critical point theory) to a wide variety of variational problems and deduce theorems related to the existence of solutions of problems in non-linear elliptic partial differential equations.

Th. M. Rassias ([14]) has shown his investigations of Morse theory on Hilbert manifolds to Plateau's problem. He has also presented a paper ([15]) on the Morse-Smale Index theorem for a global analysis of variational problems in several variables with new applications to the computation of the Morse-Smale index of the catenoid and Enneper's minimal surface; the Bessel equation is used. He has also presented ([16]) a research survey on uniqueness and classification theorems for minimal surfaces.

The Bessel functions and the squares of Bessel functions were used by J. D. Watson [20] in establishing a double helix structure for deoxyribose nuclei acid (DNA).

10. Lie Groups

The following is an overview of Lie Groups ([19]): Lie groups can be investigated locally and the local structures give rise to the concept of Lie algebra, the set of left invariant vector fields on a given Lie group G. The local properties of a Lie group can be read off from its Lie algebra $\mathfrak{G}$. A structure of Lie algebra on $\mathfrak{G}$ is given by a law of composition $[X_i, X_j] = \sum_k c_{ij}^k X_k$, $X_i, X_j, X_k \in \mathfrak{G}$ defined by the structure constants. The structure of the corresponding Lie group is determined by these structure constants as well. Cartan's criterion states that every closed subgroup of a Lie group is a Lie subgroup. Lie group theory may be viewed as the study of a given Lie group by means of its one-parameter subgroups in which case the corresponding lie algebra should be defined in terms of a set of one- parameter subgroups of G. This may be done via the exponential map $X \to \exp X$ of $\mathfrak{G}$ into G i.e. by means of the 1-parameter group $t \to \exp tX$. A Lie group admits a representation by linear transformations referred to as adjoint representations operating on homogeneous spaces of Lie Groups.

Definition 10.2. Following [2]. *G is a Lie group provided the mapping of $G \times G \to G$ defined by $(x, y) \to xy$ and the mapping $G \to G$ defined by $x \to x^{-1}$ are both C^∞ mappings.*

Example 10.3. $G\ell(n, \mathbf{R})$, the set of nonsingular matrices, is an open submanifold of $M_n(\mathbf{R})$, the set of $n \times n$ real matrices identified by $\mathbf{R}^{n^2}$. Also, $G\ell(n, \mathbf{R})$ is a group with respect to matrix multiplication. An $n \times n$ matrix A is nonsingular if and only if $\det A \neq 0$, but $\det(AB) = (\det A)(\det B)$ so if A and B are nonsingular, AB is also. An $n \times n$ matrix A is nonsingular, that is, $\det A \neq 0$ if and only if A has a multiplication inverse. Hence $G\ell(n, \mathbf{R})$ is a group. Both the maps $(A, B) \to AB$ and $A \to A^{-1}$ are C^ω. The product has entries which are polynomials in the entries of A and B and these entries are exactly the expression in local coordinates of the product map, which is C^ω (ω means real analytic), hence C^∞. The inverse of $A = (a_{ij})$ may be written as $A^{-1} = \left(\frac{1}{\det A}\right) \tilde{a}_{ij}$ where $\tilde{a}_{ij}$ are the cofactors of A (thus polynomials in the entries of A) and where $\det A$ is a polynomial in these entries which does not vanish on $G\ell(n, \mathbf{R})$. Hence the entries of A^{-1} are rational functions on $G\ell(n, \mathbf{R})$ with nonvanishing denominators, hence C^ω (and C^∞). Hence $G\ell(n, \mathbf{R})$ is a Lie group.

Example 10.4. $0(n) = \{X \in G\ell(n, \mathbf{R}), {}^tXX = I\}$, where the t indicates a transpose, the subgroup of orthogonal $n \times n$ matrices, is a regular submanifold and hence a Lie group. Also $S\ell(n, \mathbf{R}) = \{X \in G\ell(n, \mathbf{R}) | \det X = +1\}$ is a subgroup and regular submanifold of $G\ell(n, \mathbf{R})$ hence a Lie group.

Definition 10.5. *Let G be a group and X a set. Then G is said to act on X (on the left; a right action can also be defined) if there is a mapping $\theta : G \times X \to X$ satisfying the following two conditions:*

(1) *If e is the identity element of G, then $\theta(e, x) = x$ for all $x \in X$ and*
(2) *If $g_1, g_2 \in G$, then $\theta(g_1, \theta(g_2, x)) = \theta(g, g_2, x)$ for all $x \in X$. When G is a topological group and X is a topological space and θ is continuous, then the action is called continuous.*

Example 10.6. The natural action of $G\ell(n, \mathbf{R})$ on $\mathbf{R}^n$ is as follows: Let $G = G\ell(n, \mathbf{R})$ and $X = \mathbf{R}^n$ and define $\theta : G \times \mathbf{R}^n \to \mathbf{R}^n$ by $\theta(A, x) = Ax$, this being multiplication of $n \times n$ matrix A by the $n \times 1$ column vector obtained by writing $x \in \mathbf{R}^n$ vertically.

Remark 10.7. The one-parameter subgroups of a Lie group G are in one-to-one correspondence with the elements $T_e(G)$, the tangent space to G at the identity element. This information will assist to determine all one-parameter subgroups of various matrix groups.

Example 10.8. Consider $G = G\ell(n, \mathbf{R})$. The matrix entries $x_{ij}, 1 \le i, j \le n$ for any $X = (x_{ij}) \in G\ell(n, \mathbf{R})$ are coordinates on a single neighborhood covering the group which is an open subset of $M_n(\mathbf{R})$, the $n \times n$ matrices over $\mathbf{R}$. Hence $\frac{\partial}{\partial x_{ij}}$, $1 \le i, j \le n$, is a field of frames on G and relative to these frames as a basis at e, there is an isomorphism of $M_n(\mathbf{R})$ as a vector space onto $T_e(G)$ given by $A = (a_{ij}) \to \sum_{i,j} a_{ij} \left(\frac{\partial}{\partial x_{ij}}\right)_e$. Note that when $G = G\ell(n, \mathbf{R})$, e is the $n \times n$ identity matrix I.

Definition 10.9. *The exponential e^X of a matrix $X \in M_n(\mathbf{R})$ is defined to be the matrix given by*

$$e^X = I + X + \frac{1}{2!}X^2 + \frac{1}{3!}X^3 + \cdots \qquad (*)$$

if the series converges.

Theorem 10.10. *The series (*) converges absolutely for all $X \in M_n(\mathbf{R})$ and uniformly on compact subsets. The mapping $M_n(\mathbf{R}) \to M_n(\mathbf{R})$ defined by $X \in e^X$ is C^∞ and has nonsingular Jacobian at $X = 0$. Its image lies in $G\ell(n, \mathbf{R})$. If $A, B \in M_n(\mathbf{R})$ such that $AB = BA$, then $e^{A+B} = e^A e^B$.*

Remark 10.11. The above Theorem implies $e^A e^{-A} = e^0 = I$, hence e^A is nonsingular, that is, $e^A \in GL(n, \mathbf{R})$ for any $a \in M_n(\mathbf{R})$.

Theorem 10.12. *$t \to e^{tA}$ is the one-parameter subgroup of $G\ell(n, \mathbf{R})$ whose corresponding left-invariant vector field has the value $\sum_{ij} a_{ij} \left(\frac{\partial}{\partial x_{ij}}\right)_e$. All one-parameter subgroups of $GL(n, \mathbf{R})$ are of this form.*

Theorem 10.13. *If H is a Lie subgroup of G, then the one-parameter subgroups of H are exactly those one-parameter subgroups $t \to F(t)$ of G such that $\dot{F}(0) \in T_e(H)$ considered as a subspace of $T_e(G)$.*

Remark 10.14. ([2]) Use the mapping $X \to e^X$ to determine the one-parameter subgroups of other matrix groups, for example, $0(n)$, $S\ell(n, \mathbf{R})$. Consider $G\ell(n, \mathbf{R})$ as described above, then the following application can be presented.

Theorem 10.15. *The one-parameter subgroups of a subgroup $H \supseteq G\ell(n, \mathbf{R})$ are all of the form $t \to e^{tA}$ where $A = (a_{ij})$ are the components of a vector $\sum_{i,j} a_{ij} \left(\frac{\partial}{\partial x_{ij}}\right)_e \in T_e(G)$ which is tangent to H at e, that is, is $T_e(H) \in T_e(G)$.*

Remark 10.16. This is a consequence of the theorem and the fact that every one-parameter subgroup of $G = G\ell(n, \mathbf{R})$ is of the form $F(t) = e^{tA}$.

Theorem 10.17. *The homomorphism $t \to e^{tA}$ is a one-parameter subgroup of $0(n)$ if and only if $A^t = -A$ which is the necessary and sufficient condition on $A = (a_{ij})$ in order that the tangent vector $\sum_{i,j} a_{ij} \left(\frac{\partial}{\partial x_{ij}}\right)_e$ to $G\ell(n, \mathbf{R})$ at the identity be tangent to the subgroup $0(n)$.*

Example 10.18. Let $H = 0(n)$, $G = G\ell(n, \mathbf{R})$ and determine the one-parameter subgroups of H. If $e^{tA} \in H$ for all t, then $(e^{tA})(e^{tA})' = I$ where the prime indicates the transpose. Hence $(e^{tA})' = e^{tA'}$ and $(e^{tA})^{-1} = e^{-tA}$. Hence $e^{tA} \in H$ implies $e^{tA} = e^{-tA}$. Furthermore, $X \to e^X$ maps the linear submanifold of $M_n(\mathbf{R})$ of skew-symmetric matrices to the submanifold $0(n)$ of G; both manifolds have the same dimension and the Jacobian of the mapping is non-singular at $X = 0$. Hence, some neighborhood of the 0 matrix, $X \to e^X$ is a diffeomorphism. Hence, there is a δ such that if $|t| < \delta$, then $tA' = -tA$. Hence A is skew-symmetric. Conversely, if $A = -A$, then $e^{tA}(e^{tA})' = e^{tA}e^{tA'} = e^{tA}e^{-tA} = I$ which means that e^{tA} is an orthogonal matrix.

Remark 10.19. The comments made in Remark 3.13 should be noted when Example 10.18 is being considered. Also, note Remark 4.3 and Remark 5.8 and Example 8.1.

Example 10.20. Let $G = S0(n)$ be the group of $n \times n$ orthogonal matrices of determinant $+1$. The tangent space $T_e(G)$, $e = I$, the $n \times n$ identity matrix, may be identified with the skew-symmetric matrices $A = (a_{ij}) = -A'$ in the sense that $X_e = \sum_{i,j} a_{ij} \left(\frac{\partial}{\partial x_{ij}}\right)$ is tangent at I to $S0(n)$ considered as a submanifold of $G\ell(n, \mathbf{R}) \in \mathbf{R}^{n^2}$. The one-parameter subgroups are of the form $Z(t) = e^{tA}$. Hence compute $Ad\ B = T_e(G) \to T_e(G)$, where Ad denotes adjoint, as follows: verify from the definition of e^{tA} that $Be^{tA}B^{-1} = e^{tBAB^{-1}}$. Since $Ad\ B$ acting on $T_e(G)$ is just the linear map of the tangent space induced by the mapping $Z \to BZB^{-1}$ on $S0(n)$ this indicates that $Ad\ B$ takes the component matrix $A = (a_{ij})$ of X_e to BAB^{-1}. Next define on $T_e(G)$ an inner product (X_e, Y_e) for $X_e = \sum a_{ij} \left(\frac{\partial}{\partial x_{ij}}\right)$ and $Y_e = \sum c_{ij} \left(\frac{\partial}{\partial x_{ij}}\right)$ by $(X_e, Y_e) = tr\ A'C = \sum_{i,j=1}^{n} a_{ij}c_{ij}$. This is bilinear and symmetric since $(X_e, X_e) = tr\ A'A = \sum_{ij} a_{ij}a_{ij} = a_{ij}^2$, it is positive definite. Finally for $B \in S0(n)$, $(Ad(B)X_e, Ad(B)Y_e) = tr((BAB^{-1})'BCB^{-1}) = tr(BACB^{-1}) = tr\ AC = (X_e, Y_e)$. Hence, the inner product determines a bi-invariant Riemannian metric on G and G is a symmetric space with this Riemannian metric.

Theorem 10.21. ([6]) *Let G be a Lie group and choose a basis $\{\omega^i\big|_e\}$ of T_e^*, the cotangent space at the identity e and let the basis be for the right invariant* *one-forms of the group*

$$d\omega^i = \frac{1}{2}\sum c^i_{jk}\omega^j \wedge \omega^k$$

where $c^i_{jk} = -c^i_{kj}$ *are real constants.*

Remark 10.22. When the above functions are in fact constant, then this is a description of the First Fundamental Theorem of Lie. These constants are called the structure constants of G relative to the choice of Mauer-Cartan forms.

Theorem 10.23. ([19]) *Given* n^3 *constants,* c^k_{ij} *subject to the following constraints:*

$$c^k_{ij} + c^k_{ji} = 0 \text{ and}$$
$$\sum (c^h_{ij}c^\ell_{hk} + c^h_{jk}c^\ell_{hj} + c^h_{ki}c^\ell_{kj}) = 0$$

there does exist a Lie group having c^k_{ij} *as constants of structure provided linearly independent one-forms can be found such that*

$$d\omega^k = -\frac{1}{2}c^k_{ij}\omega^j \wedge \omega^i\,.$$

Remark 10.24. ([19]) The following relationship appears using the properties of exterior derivatives:

$$d^2\omega = 0 = \frac{1}{2}\sum c^i_{jk}(d\omega^j \wedge \omega^k - \omega^j \wedge d\omega^k)$$
$$= \sum_{j<k} c^i_{jk}\left(\sum_{\ell<m} c^j_{\ell m}\omega^\ell \wedge \omega^m \wedge \omega^k - \sum_{\ell<m} c^k_{\ell m}\omega^j \wedge \omega^\ell \wedge \omega^m\right)$$

which, in turn, implies the following Jacobi identity to hold

$$\sum_h (c^h_{jk}c^s_{h\ell} + c^h_{k\ell}c^s_{hj} + c^h_{\ell j}c^s_{hk}) = 0\,.$$

Theorem 10.25. ([5]) *If* A *is an orthogonal matrix whose elements are functions of any number of variables, then* $(dA)A^{-1}$ *is a skew-symmetric matrix of one-forms.*
Let ${}^tAA = I$ *where* t *is a transpose then*

$${}^tdAA + {}^tAdA = 0$$
$${}^tA^{-1}dA + dAA^{-1} = 0$$
$${}^t(dAA^{-1}) + dAA^{-1} = 0\,.$$

The converse can also be established.

Remark 10.26. Theorem 3.12 established that if A is skew-symmetric, then e^A is orthogonal. Theorem 10.25 established a similar relationship but without using the exponential directly. See also Remark 3.13. See Theorem 10.17 and Example 10.18.

11. Cartan's Method–Orthonormal Frames

Cartan's method of moving frames has the following features ([19], see also [5]): Given a Riemannian manifold M and the principal fiber bundle $P(M, 0(n))$ over M with structural group $0(n)$– the bundle of orthonormal frames, use the method of moving frames to look at the geometries of submanifolds of $\mathbf{R}^n$, that is, choose moving frames tangent to the submanifold and study the properties which can be obtained from the structure equations of these frames.

Example 11.2. Consider the Euclidean case $M = \mathbf{R}^n$ and regard the vector fields as $\mathbf{R}^n$-valued functions. The basis σ^i is dual to e_i hence write

$$dx = \sum_{i=1}^{n} \sigma^i e_i$$

or in matrix notation

$$dx = \sigma e\ .$$

Because $x : \mathbf{R}^n \to \mathbf{R}^n$ (an identity map) is a $\mathbf{R}^n$ valued 0-form; the above equation is a $\mathbf{R}^n$-valued one-form.

Example 11.3. Consider one-forms ω^i_j called connection forms described as

$$de_i = \sum_{j=1}^{n} \omega^j_i e_j$$

where

$$\omega^j_i = \sum \{jk, i\} \sigma^k$$

or in matrix form

$$de = \Omega e\ .$$

Example 11.4. The matrix of one-forms is skew-symmetric, that is, whenever $\{e_1, \ldots, e_n\}$ is a moving frame in $\mathbf{R}^n$, the following relationship holds

$$\omega^i_j = -\omega^j_i$$

or in matrix form

$$\Omega = -\Omega^t$$

where t indicates a transpose.

Remark 11.5. Given arbitrary one forms σ and Ω, there is, in general, no local surface or submanifold of $\mathbf{R}^n$. For existence of a local subspace and a unique family of frames, integrability conditions must hold.

Example 11.6. Without curvature, the one forms σ and Ω satisfy the following Cartan equations:

$$d\sigma^i = \sum_j \sigma^j \wedge \omega_j^i \qquad 1 \leq i, j \leq n$$

or in matrix form

$$d\sigma = \Omega\sigma$$

and

$$d\omega_j^i = -\sum \omega_k^i \wedge \omega_j^k$$

or in matrix form

$$d\Omega = \Omega^2 \, .$$

Example 11.7. With curvature, the above equations are

$$d\omega_j^i = -\sum \omega_k^i \wedge \omega_j^k + \frac{1}{2} R_{jk\ell}^i \sigma^k \wedge \sigma^\ell$$

or in matrix form

$$d\Omega = \Omega^2 + \Theta$$

where Θ denotes the curvature two form.

Remark 11.8. Further basic exterior algebraic operations give the following properties:

Example 11.9. The integrability conditions are

$$\Theta\sigma = 0 \, .$$

Example 11.10. The Bianchi identities are

$$d\Theta = \Theta\Omega - \Theta\Omega \, .$$

Example 11.11. Torsion is described as

$$\tau = d\sigma - \Omega\sigma \, .$$

Remark 11.12. The structure equations and integrability equations without curvature would describe a minimal surface provided the mean curvature is zero. With curvature, a minimal surface would also be described, provided the mean curvature is zero.

12. Cartan's Method of Equivalence

12.1. The presentation in this section is based on Robert B. Gardner's 1989 monograph *The Method of Equivalence and Its Applications* [6].

12.2. The purpose of the method of equivalence is to find the necessary and sufficient conditions so that geometric objects be equivalent, that is, the geometric objects should be mapped onto each other by a class of diffeomorphisms characterized as the set of solutions of a system of differential equations. The necessary and sufficient conditions are found in the form of differential invariants of the geometric object under the class of diffeomorphisms. The following presentation will be restricted to those classes of diffeomorphisms which can be described as solutions of a first-order system of differential equations or equivalently by conditions on their Jacobians.

Problem 12.3. *The equivalence problem of Élie Cartan is as follows: Let* $\Omega_V = {}^t(\Omega_V^1, \ldots, \Omega_V^n)$ *be a coframe on an open set* $V \subset \mathbf{R}^n$, *and let* $\omega_U = {}^t(\omega_U^1, \ldots, \omega_U^n)$ *be a coframe on* $U \subset \mathbf{R}^n$, *and let* G *be a prescribed linear group in* $G\ell(n, \mathbf{R})$, *then find the necessary and sufficient conditions that there exists a diffeomorphism* $\Phi = U \to V$ *such that for each* $u \in V$

$$\Phi^* \Omega_V \big|_{\Phi(u)} = \gamma_{VU}(u) \omega_U \big|_u ,$$

where $\gamma_{VU}(u) \in G$.

Remark 12.4. The ten lectures in Gardner's presentation [6] was summarized as a flow chart. The following description is based on that flow chart:

Begin with a group, coframes and open set, compute the Maurer-Cartan forms and defining relations, perform a principal component decomposition and a Lie algebra compatible absorption, compute infinitesimal action on structure tensor. Then determine if there is a trivial action.

If there is a trivial action, then determine if there is an identity structure; if there is an identity structure, then the problem is solved.

If there is no trivial action, then perform a normalization and group reduction procedure. Then determine if the procedure is a constant type.

If the procedure is a constant type, then change the group and coframe and begin the process again.

If the procedure is not a constant type, then other possibilities have to be considered. If there is no identity structure, then determine if the system is in involution.

If the system is not in involution, perform a prolongation procedure and change the group, coframe and open set and begin the process again. If the system is in involution, determine if the torsion is constant. If the torsion is constant, then the problem is solved.

If the torsion is not constant, then "wild things" will appear.

Proposition 12.5. *Cartan's method of equivalence can be summarized in the* *following set of equations using Gardner's symbols:*

$$d\omega^i = \sum \phi^i_k \wedge \omega^k + \pi \wedge \omega^i$$
$$d\phi^i_j = \sum \phi^i_k \wedge \phi^k_j - \frac{1}{2} w^i_{jk\ell}\omega^\ell \wedge \omega^k + (\tau_i\delta_{jk} - \tau_j\delta_{ik}) \wedge \omega^k$$
$$d\pi = \sum \tau_i \wedge \omega^i$$
$$d\tau_k = \tau_k \wedge \pi - \sum \tau_j \wedge \phi^j_k - \sum \lambda_{k\ell} \wedge \omega^\ell$$

with

$$\sum w^i_{ji\ell} = 0 \text{ and } \lambda_{k\ell} = \lambda_{\ell k} \; ,$$

Remark 12.6. A comparison of the symbols used in Section 11 with those presented in Problem 12.5 above reveal the following new symbols

$$\pi, \; \tau, \; \lambda, \; \omega$$

which reflects a more general approach to the method of equivalence than outlined in Section 11; recall that Definition 12.3 notes that the equivalence problem of E. Cartan is prescribing a linear group in $G\ell(n, \mathbf{R})$.

Remark 12.7. The following very brief comments are made on the above symbols and their corresponding concept and/or operations.

(1) The symbol π suggests that the equivalence problem is being lifted to the G-spaces where the connected linear group $G \subset G\ell(n, \mathbf{R})$ is being considered.
(2) The symbol τ suggests that the intrinsic torsion is being considered in the analysis.
(3) The symbol $\omega^i_{jk\ell}$ is a type of geometric curvature in the analysis and
(4) The symbol λ suggests the influence of the ubiquitous Cartan Lemma.

When considering Cartan's method of equivalence, the notions of the lifting of the linear group to G spaces and the related mapping of intrinsic torsion have to be considered in addition to curvature. If torsion is not constant and the group reduction and normalization processes are not constant, then many other possibilities and problem appear. Consequently, specifying a minimal surface–a surface with mean curvature zero–requires using broader postulates on which the subject of mathematic rests.

REFERENCES

1. V.I. Arnold, *Ordinary Differential Equations*, (MIT Press, Cambridge, MA, 1973), 93–118.
2. W.M. Boothby, *An Introduction to Differentiable Manifolds and Riemannian Geometry*, (Academic Press, New York, 1975), 81–95, 145–149, 349–350.
3. J. Donato, *Donato's Research on Economics, Social Choice, Statistics and Control Theory Through 1986*, (MC Printing Company, Syracuse, New York, April, 1987), copyright 1987, Jerry Donato.
4. J. Donato, "Path Dependent Analysis", in: *Topics in Mathematical Analysis, A Volume Dedicated to the Memory of A.L. Cauchy*, Ed. Th. M. Rassias, World Scientific Publishing Company, Singapore, 1989), 210–230.
5. H. Flanders, *Differential Forms with Applications to the Physical Sciences*, (Academic Press, Inc., New York, 1963), 32–48, 82–162.
6. R.B. Gardner, *The Method of Equivalence and Its Applications*, (Society for Industrial and Applied Mathematics, Philadelphia, Pennsylvania, 1989).
7. H.W. Guggenheimer, *Differential Geometry*, (Dover Publications, Inc., New York, 1977), 206–260.
8. M. Kline, *Mathematical Though from Ancient to Modern Times*, (Oxford University Press, New York, 1972), 536, 753.
9. M.M. Lipschutz, *Differential Geometry*, (McGraw-Hill Book Company, New York, 1969), 43–226.
10. N.H. McCoy, *Introduction to Modern Algebra*, (Allyn and Bacon, Inc., Boston, 1960), 48–111.
11. F.H. Miller, *College Algebra and Trigonometry*, (John Wiley and Sons, Inc., New York, 1945), 254–282.
12. M.H. Protter and C.B. Morrey, Jr., *Modern Mathematical Analysis*, (Addison-Wesley Publishing Company, Inc., Reading, Massachusetts, 1964), 359–395.
13. M.H. Protter and C.B. Morrey, Jr., *College Calculus with Analytic Geometry*, (Addison-Wesley Publishing Company, Inc., Reading, Massachusetts, 1964), 286–330.
14. Th.M. Rassias, "Morse Theory and Plateau's Problem", in: *Selected Studies: Physics-Astro-physics, Mathematics, History of Science, A Volume Dedicated to the Memory of Albert Einstein*, eds: Th. M. Rassias and G. M. Rassias, (North Holland Publishing Company, Amsterdam, 1982), 261–292.
15. Th.M. Rassias, "On the Morse-Smale Index Theorem for Minimal Surfaces", in: *Differential Geometry, Calculus of Variations and Their Applications*, eds: G.M. Rassias and Th. M. Rassias, (Marcel Dekker, Inc., 1985, Vol. 100), 429–453.
16. Th.M. Rassias, "Survey on Uniqueness and Classification Theorems for Minimal Surfaces", in: *Nonlinear Analysis*, ed. Th. M. Rassias, (World Scientific Publishing Co., Singapore, 1987), 493–511.
17. J.B. Rosenbach and E.A. Whitman, *College Algebra*, third edition, (Ginn and Company, New York, 1949), 278–332.

18. D.J. Struik, *Lectures on Classical Differential Geometry*, second edition, (Dover Publications, Inc., New York, 1961), **80**, 105.
19. C. Von Westenholz, *Differential Forms in Mathematical Physics*, revised edition, (North-Holland Publishing Company, 1981), 84–256.
20. J.D. Watson, *The Double Helix* in a Norton Critical Edition, ed. G.S. Stent, (W. W. Norton and Company, New York, 1980), 237–292.

The Problem of Plateau (pp. 111–128)
ed. Th. M. Rassias

CRITICAL POINT THEORY AND MULTIPLE PERIODIC SOLUTIONS OF CONSERVATIVE SYSTEMS WITH PERIODIC NONLINEARITY

ALESSANDRO FONDA
Université de Louvain, Institut Mathématique, Chemin du cyclotron, 2,
B-1348 Louvain-la-Neuve, Belgium

and

JEAN MAWHIN
Université de Louvain, Institut Mathématique, Chemin du cyclotron, 2,
B-1348 Louvain-la-Neuve, Belgium

ABSTRACT

We use a cohomological minimax theorem of Chang to prove multiplicity results for the periodic solutions of some Hamiltonian systems of the second and or the first order with nonlinearity verifying some periodicity condition. Applications are given to systems of coupled pendulums, discretization of Josephson equations and extensions of the Conley-Zehnder results about Arnold's conjecture.

1. Introduction

The first aim of this paper is to extend the results obtained by Mawhin [19] concerning the multiplicity of periodic solutions for a second order system of the form

$$(M(t)u')' + Au + D_uF(t,u) = h(t), \tag{1.1}$$

where $M(t)$ and A are symmetric matrices, $M(t)$ being positive definite, F and D_uF are bounded and satisfy a periodicity condition along the directions of the null-space of A, and h belongs to a suitable subspace of L^1.

The special case $M(t) \equiv Id$ and $A = 0$ has been studied in [21]. The existence of two distinct solutions was proved, generalizing a result of [20] for the pendulum equation (see also [17], [22]).

In [19], Mawhin proved,whenever A is semi-negative definite, the existence of (m + 1) distinct periodic solutions of (1), where m is the dimension of the null-space of A. His result unifies and completes previous existence theorems for the satellite equation (cf. [18]), the linearly coupled pendulum (cf. [15]) and the Josephson multipoint system (cf. [14]). See also [12].

Here we will prove the existence of at least (m + 1) distinct periodic solutions of (1), without requiring A to be semi-negative definite. As already shown in [19], if all the solutions are non-degenerate, then there are at least 2^m of them. The proof consists in applying an abstract minimax theorem of Chang [6], which is based on algebraic topology methods and Morse theory. Minimax techniques of this type have been applied by Morse-Tompkins and Shiffman, to the problem of the multiplicity of solutions of the Plateau problem , a few years after the fundamental contributions of Jesse Douglas and Tibor Rado to the existence question for this problem. Nowadays, the Plateau problem continues to be a source of inspiration for the development of modern tools in critical point theory (see e.g. [29]).

Our second objective is to give a multiplicity result for the periodic solutions of a Hamiltonian system of the form

$$J\dot{u} + Au + D_uH(t,u) = h(t), \tag{1.2}$$

where J is the standard symplectic matrix, A is a symmetric matrix, H, D_uH and $D_{uu}H$ are bounded and satisfy a periodicity condition along the directions of the null-space of A, and h belongs to a suitable subspace of L^2.

The situation differs from the above one for the fact that the functional associated to (1.2) is strongly indefinite. A finite dimensional reduction will be used in order to overpass this difficulty.

The special case A = 0 has been studied by Conley and Zehnder [9, 10] and Chang [6]. Here we extend their results to the case $A \neq 0$, and prove a theorem which is the perfect analogous to the one we have for system (1.1). When the paper (which is summarized in [28]) was written, we have received a preprint of Chang [27] which contains very close results.

The paper is organized as follows. In section 2 we recall some concepts of algebraic topology, two deformation lemmas and some results of Morse Theory. In section 3 we present two abstract existence theorems by Chang [6, 7]. The proofs are also carried out for the reader's convenience. In sections 4 and 5 we apply the abstract results of section 3 to prove multiplicity results for periodic solutions of (1.1) and (1.2), respectively.

2. Some preliminaries

Given a topological space X and a subspace $A \subset X$, we denote by $H_n(X,A)$ $[H^n(X,A)]$ the nth singular homology [cohomology] vector space of X relative to A, with respect to a given field (e.g. $\mathbb{R}$).

We recall that the elements of $H_n(X,A)$ are equivalence classes of singular chains having zero boundary. These elements are invariant under every continuous deformation $\tau : X \to X$ such that $\tau|_A = id_A$, in the sense that τ induces an isomorphism τ_* of $H_n(X,A)$ into itself, and we identify each element of $H_n(X,A)$ with its image under τ_*. The analogous is true for cohomologies, as well.

We state the following two properties of homologies, which hold for cohomologies as well.

(a) If X' is a strong deformation retract of X and $A' \subset X'$ is a strong deformation retract of A, then $H_n(X,A) \approx H_n(X',A')$.

(b) (Künneth formula) For any topological space Y,

$$H_n(X \times Y, A \times Y) \approx \oplus_{p+q=n} [H_p(X,A) \otimes H_q(Y)].$$

Let us now briefly recall the concepts of cup and cap products. Suppose A and B are subspaces of X such that, for example, one of the following three situations is true :

$$A = \emptyset, \quad A = B, \quad B = \emptyset.$$

Then there exist an operation, called the "cup product"

$$\begin{aligned} H^n(X,B) \times H^m(X,A) &\to H^{n+m}(X, A \cup B) \\ (\omega,\rho) &\mapsto \omega \cup \rho \end{aligned} \tag{2.1}$$

and an operation, called the "cap product"

$$\begin{aligned} H_{n+m}(X, A \cup B) \times H^n(X,B) &\to H_m(X,A) \\ (z,\omega) &\mapsto z \cap \omega, \end{aligned} \tag{2.2}$$

which are bilinear and invariant under continuous deformations leaving $A \cup B$ fixed (in the sense we saw above).

The cup and cap products are naturally induced by two operators, defined on the corresponding spaces of singular chains and cochains, which are denoted by the same symbols $\cup$ and $\cap$. If $z \in H_{n+m}(X, A \cup B)$, $\omega \in H^n(X,B)$ and $\rho \in H^m(X,A)$, then for all $\eta \in z$, $c \in \omega$ and $d \in \rho$ one has

$$\langle \eta, c \cup d \rangle = \langle \eta \cap c, d \rangle, \tag{2.3}$$

where $\langle .,. \rangle$ denotes the pairing between singular chains and cochains. Moreover, denoting by $|\eta|$ the support of the chain η, one has

$$|\eta \cap c| \subset |\eta|. \tag{2.4}$$

We now consider a Riemannian manifold $\mathcal{M}$ of class C^2 and a C^1 functional $f : \mathcal{M} \to \mathbb{R}$. We will use the following notations.

$$f^a = \{x \in \mathcal{M} : f(x) \le a\}$$

$$K_c = \{x \in \mathcal{M} : f(x) = c, df(x) = 0\}.$$

That is, K_c is the set of critical points at the level c.

The Palais-Smale condition, in short (PS), plays a fundamental role in the following deformation lemmas. Recall that (PS) holds iff every sequence (x_n) in $\mathcal{M}$ such that $f(x_n)$ is bounded and $df(x_n) \to 0$ possesses a convergent subsequence.

Assume that at every point x of the boundary of $\mathcal{M}$, df(x) points inwards in $\mathcal{M}$. Then the following lemmas hold.

First Deformation Lemma. *Assume* (PS) *holds for* f. *Fix* $c \in \mathbb{R}$ *and let* N *be a closed neighborhood of* K_c. *Then there is a continuous map* $\eta : [0,1] \times \mathcal{M} \to \mathcal{M}$, *as well as numbers* $\bar{\varepsilon} > \varepsilon > 0$ *such that*

(1) $\eta|_{\mathcal{M} \setminus f^{-1}([c-\bar{\varepsilon}, c+\bar{\varepsilon}])} = id,$

(2) $\eta(0,\cdot) = id,$

(3) $\eta(1, f^{c+\varepsilon} \setminus N) \subset f^{c-\varepsilon},$

(4) $\forall\, t \in [0,1]$, $\eta(t,\cdot)$ *is a homeomorphism.*

Second Deformation Lemma. *Assume* (PS) *holds for* f *and that* df *is locally Lipschitzian. If there are no critical values in the open interval* (a, b), *then* f^a *is a strong deformation retract of* $(f^b \setminus K_b)$.

For the proofs, cf. [22], [8], [24], [6]. Since $\mathcal{M}$ is a Riemannian manifold, we can consider the flow determined by the gradient of f. The deformations are then constructed along this flow.

Let $a < b$, and suppose that $x_1, \ldots, x_i$ are the only critical points of f in $f^{-1}([a,b])$. Let $C_n(f, x_i)$ denote the nth critical group of x_i, and suppose that all these critical groups (which are vector spaces) are finite dimensional, and that they are trivial for n sufficiently large. Then we can define the Morse polynomial

$$M(t,f^b,f^a) = \sum_{n=0}^{\infty} \left(\sum_{i=1}^{j} \dim C_n(f,x_i)\right) t^n.$$

Moreover, the Poincaré polynomial

$$P(t,f^b,f^a) = \sum_{n=0}^{\infty} \dim H_n(f^b,f^a)\, t^n$$

is also well defined and we have that

$$M(t,f^b,f^a) = P(t,f^b,f^a) + (1+t)\, Q(t), \tag{2.5}$$

where Q is a polynomial with nonnegative integer coefficients (cf. [22]).

As a consequence of (2.5), if for every $i \in \{1, \ldots, j\}$ we have that $\sum_{n=0}^{\infty} \dim C_n(f,x_i) \leq 1$, then

$$j \geq \sum_{n=0}^{\infty} \dim H_n(f^b,f^a). \tag{2.6}$$

This is the case if all the critical points $x_1, \ldots, x_j$ are nondegenerate. Indeed, in that case one has

$$C_n(f,x_i) = \delta_{n,m_i}\, \mathbb{R}$$

where m_i is the Morse index of the point x_i.

3. An abstract multiplicity result

In this section we expose some results of Chang [6].

Definition. *Let* X *be a topological space and* $A \subset X$. *Consider two non-zero singular homology classes*

$$z_1 \in H_m(X,A), \quad z_2 \in H_{n+m}(X,A).$$

We set $z_1 \prec z_2$ *whenever* $n > 0$ *and there exists* $\omega \in H^n(X)$ *such that*

$$z_1 = z_2 \cap \omega$$

(the cap product is as in (2.2), *with* $B = \emptyset$*).*

Theorem 1. *Assume* (PS) *holds for* f. *Let* a < b *be two real numbers such that* f *has only a finite number of isolated critical points in* $f^{-1}[a,b]$. *If there are k non-zero singular homology classes* $z_1 \in H_{n_1}(f^b, f^a), \ldots, z_k \in H_{n_k}(f^b, f^a)$ *with* $z_1 \prec z_2 \prec \ldots \prec z_k$, *then* f *has at least* k *distinct critical values.*

Proof. Define the following quantities :

$$c_i = \inf_{\eta \in z_i} \sup_{x \in |\eta|} f(x) \qquad (i = 1, 2, \ldots, k).$$

The Minimax Principle tells us that whenever c_i is finite and the family of sets z_i is invariant under homeomorphisms, c_i is a critical value of f. Since $z_i \in H_{n_i}(f^b, f^a)$, for all $\eta \in z_i$ we have that $|\eta| \subset f^b$. Hence $c_i \leq b$. Moreover, since z_i is supposed to be non-zero in $H_{n_i}(f^b, f^a)$, this means that for any $\eta \in z_i$ we have that $|\eta|$ has non empty intersection with $f^b \setminus f^a$. Hence $c_i \geq a$. So c_i is finite; z_i is invariant under homeomorphisms, as we said in section 2.

So the c_i's are critical values of f. We want to prove now that we have

$$c_1 < c_2 < \ldots < c_k.$$

Let us concentrate on c_1 and c_2, the reasoning being the same for the others. We know, since $z_1 \prec z_2$ that there exists $\omega \in H^{n_2-n_1}(f^b)$, $n_2 - n_1 > 0$ such that $z_1 = z_2 \cap \omega$. This means that for any $\eta_2 \in z_2$ and any $c \in \omega$ we can define $\eta_1 = \eta_2 \cap c \in z_1$, and by (2.4) we have that $|\eta_1| \subset |\eta_2|$. Hence

$$\forall \eta_2 \in z_2 \, \exists \, \eta_1 \in z_1 : \sup_{x \in |\eta_1|} f(x) \leq \sup_{x \in |\eta_2|} f(x).$$

This immediately implies $c_1 \leq c_2$.

Suppose by contradiction that $c_1 = c_2$, and denote by c this common value. Then for every $\varepsilon > 0$ there must exist a $\eta_2 \in z_2$ such that $|\eta_2| \subset f^{c+\varepsilon}$. Since K_c is a set of isolated critical points, we choose two contractible neighborhoods of K_c in $f^b \setminus f^a$: $N \subset N'$. We can then write $\eta_2 = \eta'_2 + \eta''_2$, where $|\eta'_2| \subset N'$ and $|\eta''_2| \subset f^{c+\varepsilon} \setminus N$.
We can consider η'_2 as a cycle of f^b relative to $f^b \setminus N$. Hence $[\eta'_2] \in H_{n_2}(f^b, f^b \setminus N)$ and $[\eta'_2] \cap \omega \in H_{n_1}(f^b, f^b \setminus N)$. The cap product does not change if we shrink N', and hence η'_2, to a point. Since $\omega \in H^{n_2-n_1}(f^b)$ and $n_2 - n_1 > 0$, there exists $c \in \omega$ which, applied to any chain having support in N', gives 0. In particular, by the definition of the cap product, $\eta'_2 \cap c = 0$. Set $\eta_1 = \eta_2 \cap c \in z_1$; then

$$\eta_1 = \eta_2 \cap c = \eta'_2 \cap c + \eta''_2 \cap c = \eta''_2 \cap c.$$

Hence $|\eta_1| \subset |\eta''_2 \cap c| \subset |\eta''_2| \subset f^{c+\varepsilon} \setminus N$.

Consider now the homeomorphism $\tau : f^{c+\varepsilon} \setminus N \to f^{c-\varepsilon}$ given by the First Deformation Lemma. We have that $\tau(|\eta_1|) \subset f^{c-\varepsilon}$. But $\tau \circ \eta_1 \in z_1$ because of the invariance under homeomorphisms, and this contradicts the definition of c_1 (= c).

Given a compact manifold $\mathcal{V}$, let us define cuplength ($\mathcal{V}$) as the greatest natural number ℓ such that

$$(\forall\, i \in \{1, \ldots, \ell\})\ (\exists\, k_i > 0)\ (\exists\, \omega_i \in H^{k_i}(\mathcal{V})) : \omega_1 \cup \ldots \cup \omega_\ell \neq 0.$$

It can be seen that for a compact manifold such a number indeed exists.

Theorem 2. *Let* L *be a bounded self-adjoint operator with a bounded inverse, defined on a Hilbert space* H. *Suppose that the negative space determined by* L *is finite dimensional. Let* $\mathcal{V}$ *be a* C^2*-compact manifold without boundary. Let* $g \in C^2(H \times \mathcal{V}, \mathbb{R})$ *be a function having bounded and compact differential* dg. *Then the function*

$$f(x, v) = \frac{1}{2}(Lx, x) + g(x, v)$$

has at least [cuplength ($\mathcal{V}$) + 1] *critical points.*

Moreover, if all the critical points of f *are non degenerate, there are at least* $\sum_{n=0}^{\infty} \dim H_n(\mathcal{V})$ *of them.*

<u>Proof.</u> 1) In order to apply the previous theorem, let us verify that (PS) holds for f on $H \times \mathcal{V}$. Take a sequence (x_n, v_n) in $H \times \mathcal{V}$ such that $(f(x_n, v_n))$ is bounded and

$$df(x_n,v_n) = Lx_n + dg(x_n,v_n) \to 0. \tag{3.1}$$

Since $(dg(x_n,v_n))$ is bounded by hypothesis, it follows that (Lx_n) is also bounded. We deduce that, L having a bounded inverse, (x_n) is bounded, and since $\mathcal{V}$ is compact, the sequence (x_n,v_n) is bounded. Since dg is supposed to be compact and $\mathcal{V}$ is compact, there exists a subsequence (x_{n_k},v_{n_k}) such that $(dg(x_{n_k},v_{n_k}))$ and (v_{n_k}) are convergent. From (3.1) and the boundedness of L^{-1}, it then follows that (x_{n_k},v_{n_k}) itself is convergent.

2) Let $H = H^+ \oplus H^-$, where H^+ and H^- are the invariant subspaces corresponding to the positive and negative spectrum of L, respectively. Accordingly, every $x \in H$ can be written as $x = x^+ + x^-$, with $x^+ \in H^+$ and $x^- \in H^-$. Let $\gamma = \dim H^-$. Set $\varepsilon_\pm = \inf\{\|Lx^\pm\| : \|x^\pm\| = 1\}$. These are positive numbers. If $\bar{m}$ is such that $\|dg(x,v)\| \le \bar{m}$ $\forall\, (x,v) \in H \times \mathcal{V}$, set $R_+ = (\bar{m} + 1) / \varepsilon_+$. From now on it will be convenient to work on the manifold $\mathcal{M} = (H^+ \cap B_{R_+}) \times H^- \times \mathcal{V}$.

In order to be sure that the Deformation Lemmas can be used, we have to check that $-df$ points inward to $\mathcal{M}$ on each point of the boundary $\partial\mathcal{M}$, i.e. for every (x, v) such that $\|x^+\| = R_+$. Indeed in such a case we have

$$\begin{aligned}(-df(x,v), x^+) &= -(Lx^+,x^+) - (dg(x,v), x^+)\\ &\le -\varepsilon_+ \|x^+\|^2 + \bar{m}\|x^+\|\\ &= -R_+(\varepsilon_+ R_+ - \bar{m}) = -R_+ < 0.\end{aligned}$$

Since (on $\mathcal{M}$)

$$f(x,v) \le \frac{1}{2}\|L\| R_+^2 - \frac{1}{2}\varepsilon_-\|x^-\|^2 + \bar{m}(R_+ + \|x^-\| + \|v\|)$$

and since $\mathcal{V}$ is compact, we have that

$$f(x, v) \to -\infty \text{ as } \|x^-\| \to \infty \text{ uniformly in } x^+ \text{ and } v. \tag{3.2}$$

It is not restrictive to suppose there exist only a finite number of critical points, which are isolated and contained in $f^b \setminus f^a$, for certain fixed $a < b$.

From (3.2) it follows that there exists an $R_1 > 0$ such that

$$(H^+ \cap B_{R_+}) \times (H^- \setminus B_{R_1}) \times \mathcal{V} \subset f^a.$$

Since

$$f(x,v) \ge \frac{1}{2}\varepsilon_+ R_+^2 - \frac{1}{2}\|L\|\,\|x^-\|^2 - \bar{m}(R_+ + \|x^-\| + \|v\|)$$

this implies f is bounded below on $(H^+ \cap B_{R_+}) \times (H^- \cap B_{R_1}) \times \mathcal{V}$, and hence there exists an $a' < a$ such that

$$f^{a'} \subset (H^+ \cap B_{R_+}) \times (H^- \setminus B_{R_1}) \times \mathcal{V}$$

and, again, there exists an $R_2 > R_1$ such that

$$(H^+ \cap B_{R_+}) \times (H^- \setminus B_{R_2}) \times \mathcal{V} \subset f^{a'}.$$

The Second Deformation Lemma gives us a strong deformation retraction $\tau_1 : f^a \to f^{a'}$. Moreover a strong deformation retraction

$$\tau_2 : (H^+ \cap B_{R_+}) \times (H^- \setminus B_{R_1}) \times \mathcal{V} \to (H^+ \cap B_{R_-}) \times (H^- \setminus B_{R_2}) \times \mathcal{V}$$

can be constructed by hand (see [6]). So $\tau = \tau_2 \circ \tau_1$ is a strong deformation retraction from f^a to $(H^+ \cap B_{R_+}) \times (H^- \setminus B_{R_2}) \times \mathcal{V}$, and we have

$$
\begin{aligned}
H_n(f^b, f^a) &\approx H_n(\mathcal{M}, f^a) && \text{(deformation)}\\
&\approx H_n(\mathcal{M}, (H^+ \cap B_{R_+}) \times (H^- \setminus B_{R_2}) \times \mathcal{V}) && \text{(deformation)}\\
&\approx H_n(H^- \times \mathcal{V}, (H^- \setminus B_{R_2}) \times \mathcal{V}) && \text{(Künneth)}\\
&\approx \oplus_{p+q=n} [H_p(H^-, H^- \setminus B_{R_2}) \otimes H_q(\mathcal{V})] && \text{(Künneth)}\\
&\approx \oplus_{p+q=n} [H_p(B_{R_2}, \partial B_{R_2}) \otimes H_q(\mathcal{V})] && \text{(deformation)}\\
&\approx H_{n-\gamma}(\mathcal{V}).
\end{aligned}
$$

Moreover,

$$
\begin{aligned}
H^n(f^b) &\approx H^n(\mathcal{M})\\
&\approx H^n((H^+ \cap B_{R_+}) \times H^- \times \mathcal{V})\\
&\approx \oplus_{p+q=n} [H^p((H^+ \cap B_{R_+}) \times H^-) \otimes H^q(\mathcal{V})]\\
&\approx H^n(\mathcal{V}).
\end{aligned}
$$

Let ℓ = cuplength ($\mathcal{V}$). Then, according to the above isomorphisms, for every $i \in \{1, \ldots, \ell\}$ there exists $\omega_i \in H^{k_i}(f^b)$, $k_i > 0$ such that $\omega_1 \cup \ldots \cup \omega_\ell \neq 0$. Hence there exists $z_1 \in H_{k_1+\ldots+k_\ell+\gamma}(f^b, f^a)$ such that

$$(z_1, \omega_1 \cup \ldots \cup \omega_\ell) \neq 0.$$

Then we can define recursively $z_{i+1} \in H_{k_{j+1}+\ldots+k_\ell+\gamma}(f^b, f^a)$ by

$$z_{j+1} = z_j \cap \omega_j,$$

$j = 1, \ldots, \ell$. We thus obtain (ℓ+1) non-zero homology classes such that $z_{\ell+1} \prec z_\ell \prec \ldots \prec z_1$. Theorem 1 then proves the first part of the theorem.

As for the second part, it is immediate from (2.6) and the above isomorphisms.

4. Periodic solutions of second order systems

In this section we will apply the abstract multiplicity results of section 3 to the following second order system.

$$
\begin{aligned}
&(M(t)u')' + Au + D_uF(t,u) = h(t)\\
&u(0) - u(T) = u'(0) - u'(T) = 0
\end{aligned}
\tag{4.1}
$$

We assume $T > 0$, $F = [0,T] \times \mathbb{R}^n \to \mathbb{R}$ is continuous, bounded and such that

(i) $F(t + T,u) = F(t,u)$ for all $(t,u) \in [0,T] \times \mathbb{R}^n$.

(ii) D_uF exists, is continuous and bounded, and $D_{uu}F$ exists and is continuous.

Moreover, $S(\mathbb{R}^n,\mathbb{R}^n)$ being the space of symmetric real (n x n)-matrices, we have $A \in S(\mathbb{R}^n,\mathbb{R}^n)$, $M : [0,T] \to S(\mathbb{R}^n,\mathbb{R}^n)$ is continuous and such that, for some $\mu > 0$ and all $(t,v) \in [0,T] \times \mathbb{R}^n$,

$$(M(t)v \mid v) \geq \mu\, |v|^2, \tag{4.2}$$

and $h \in L^1(0,T; \mathbb{R}^n)$.

Using Schauder's fixed point theorem, one can prove the existence of at least one solution to (4.1) whenever the "linearized" problem

$$\begin{gathered}(M(t)u')' + Au = 0 \\ u(0) - u(T) = u'(0) - u'(T) = 0\end{gathered} \tag{4.3}$$

has only the zero solution.

We will consider the situation described by the following assumptions, where N(A) denotes the null-space of A.

(A1) $N(A) = \text{span}\,\{a_1, \ldots, a_m\}$, $1 \leq m \leq n$, *and problem* (4.3) *has as solutions only the elements of* N(A).

(A2) *For every* $v \in N(A)$,

$$\int_0^T (h(t) \mid v)\, dt = 0.$$

(A3) *There are positive numbers* $T_1, \ldots, T_m$ *such that*

$$F(t, u + T_ia_i) = F(t,u)$$

for every $(t,u) \in [0,T] \times \mathbb{R}^n$ *and* $j \in \{1, \ldots, m\}$.

Theorem 3. *Under the above assumptions, problem* (4.1) *has at least* (m + 1) *geometrically distinct solutions. If moreover all the solutions of* (4.1) *are nondegenerate, then there are at least* 2^m *of them.*

Remark. Theorem 3 generalizes previous results by Mawhin [19] where A was supposed to be semi-negative definite.

Proof. Let us consider the Hilbert space

$$H^1_T = \{u \in H^1(0,T;\mathbb{R}^n) : u(0) = u(T)\}$$

equipped with the inner product

$$\langle u \mid v\rangle = \int_0^T [(M(t)u'(t) \mid v'(t)) + (u(t) \mid v(t))]\, dt.$$

The corresponding norm $\|u\| = \langle u \mid u\rangle^{1/2}$ is by (4.2) equivalent to the classical norm of $H^1(0,T;\mathbb{R}^n)$.

Let us define the operator $L : H^1_T \to H^1_T$ such that

$$\langle Lu \mid v\rangle = \int_0^T [(M(t)u'(t) \mid v'(t)) - (Au(t) \mid v(t))]\, dt.$$

It can easily be seen that L is a self-adjoint operator on H^1_T. Because of the compact imbedding of H^1_T into $C([0,T],\mathbb{R}^n)$, $(I - L)^{-1}$ is compact. This implies that, writing $H^1_T = H^- \oplus H^\circ \oplus H^+$, where H^-, H° and H^+ are the invariant subspaces corresponding to the negative, zero and positive spectrum of L, respectively, the space H^- is finite dimensional. Moreover, by (A1), $H^\circ = N(A)$.

Let us consider the space $H = H^- \oplus H^+$. Then L can be considered as a bounded self-adjoint operator on H with a bounded inverse.

Let $T^m = \mathbb{R}^m/\mathbb{Z}^m$ be the m-fold torus, and define on $H \times T^m$ the following functional

$$g(u,(v_1,\ldots,v_m)) = \int_0^T \Big(-F\big(t, u(t) + \sum_{i=1}^m v_i a_i\big) + (h(t) \mid u(t))\Big)\, dt.$$

By (ii), g is of class C^2. Because of (A2), (A3) and classical arguments, the critical points of the C^2-functional f defined by

$$f(u,v) = \frac{1}{2}(Lu,u) + g(u,v)$$

correspond to geometrically distinct solutions of (4.1). It is easy to see that dg is bounded and compact, because of the compact imbedding of H^1_T into the space of continuous functions. So all the assumptions of Theorem 2 are satisfied, and the result follows from the following well known facts :

$$\text{cuplength}\,(T^m) = m + 1 \tag{4.4}$$

$$\dim H_n(T^m) = \binom{m}{n}. \tag{4.5}$$

5. Periodic solutions of Hamiltonian systems

In this section we will give a multiplicity result for the following Hamiltonian system

$$\begin{aligned} J\dot{u} + Au + D_uH(t,u) &= h(t) \\ u(0) &= u(T) \end{aligned} \tag{5.1}$$

We assume $T > 0$, $H = [0,T] \times \mathbb{R}^{2n} \to \mathbb{R}$ is continuous, bounded and such that

(i) $H(t + T, u) = H(t,u)$ for all $(t,u) \in [0,T] \times \mathbb{R}^{2n}$;

(ii) D_uH and $D_{uu}H$ exist, they are continuous and bounded.

Moreover, A is a symmetric real (2n x 2n)-matrix with null-space N(A), $h \in L^2(0,T; \mathbb{R}^{2n})$ and $J = \begin{pmatrix} 0 & -I_n \\ I_n & 0 \end{pmatrix}$ is the standard symplectic matrix.

It is not difficult to see, by Schauder's fixed point theorem, that if the "linearized" problem

$$\begin{aligned} J\dot{u} + Au &= 0 \\ u(0) &= u(T) \end{aligned} \tag{5.2}$$

has only the zero solution, then (5.1) has at least one solution.

We will consider the situation described by the following assumptions.

(A1) $N(A) = \text{span}\,\{a_1, \ldots, a_m\}$, $1 \le m \le 2n$, *and problem* (5.2) *has as solutions only the elements of* N(A).

(A2) *For every* $v \in N(A)$, $\int_0^T (h(t) \mid v)\, dt = 0.$

(A3) *There are positive numbers* $T_1, \ldots, T_m$ *such that*

$$H(t,u + T_j a_j) = H(t,u)$$

for every $(t,u) \in [0,T] \times \mathbb{R}^{2n}$ *and* $j \in \{1, \ldots, m\}$.

Theorem 4. *Under the above assumptions, problem* (5.1) *has at least* (m + 1) *geometrically distinct solutions.*

If moreover all the solutions of (5.1) *are nondegenerate, then there are at least* 2^m *of them.*

Remark. Theorem 4 generalizes previous results obtained by Conley and Zehnder [9, 10] and Chang [6] . They all consider the case $A = 0$ and hence $N(A) = \mathbb{R}^{2n}$. In [6] and [9] a conjecture of Arnold was proved (cf. [3], [4]).

Proof. We define a self-adjoint operator L on the Hilbert space $X = L^2(0,T; \mathbb{R}^{2n})$:

$$D(L) = \{u \in H^1(0,T; \mathbb{R}^{2n}) : u(0) = u(T)\}, \qquad Lu = J\dot{u} + Au.$$

It is well known that L is self-adjoint, has closed range and a discrete spectrum $\sigma(L) = \{\ldots < \lambda_{-1} < \lambda_0 < \lambda_1 < \ldots\}$ unbounded from below and from above, made of eigenvalues of finite multiplicity which do not accumulate at any finite point.

We define the operator $N : X \to X$ by

$$(Nu)(t) = - D_uH(t,u(t)) + h(t).$$

By (ii), N is Lipschitz continuous and possesses a symmetric Gateaux derivative given by

$$[N'(u)\eta](t) = - D_{uu}H(t,u(t))\eta(t) \tag{5.3}$$

for every $\eta \in X$.

Let α be the Lipshitz constant of N. By the properties of $\sigma(L)$ we can choose $\alpha' \geq \alpha$ and $\varepsilon > 0$ such that $[-(\alpha' + \varepsilon), - \alpha'] \cap \sigma(L)$ and $[\alpha', \alpha' + \varepsilon] \cap \sigma(L)$ are both empty.

By the above definitions of the operators L and N, it is clear that problem (5.1) is equivalent to the equation

$$Lu = Nu. \tag{5.4}$$

In order to be able to apply the abstract results of section 3 we need a reduction to a finite dimensional equation. To this aim, let us consider $\{E_\lambda : \lambda \in \mathbb{R}\}$, the spectral resolution of L, and define the following orthogonal projector in X :

$$P = \int_{-(\alpha'+\varepsilon)}^{\alpha'+\varepsilon} d\,E_\lambda.$$

For any $u \in D(L)$, we will write $u = v + w$, where $v = Pu$ and $w = (I - P)u$. Equation (5.4) is then equivalent to the following system.

$$Lw = (I - P)\, N\, (v + w) \tag{5.5}$$
$$Lv = P\, N\, (v + w). \tag{5.6}$$

Remark that, if V denotes the range of P and W the range of (I - P), then V is finite dimensional, and for every $u \in D(L)$, Pu can be expressed as a finite sum of terms in the spaces $\ker(\lambda I - L) \subset D(L)$, with $\lambda \in [-(\alpha' + \varepsilon), \alpha' + \varepsilon]$. Hence, if we pose $E = C([0,T], \mathbb{R}^{2n})$, we have that $V \subset D(L) \hookrightarrow E$. From (5.3) it follows moreover that $N'|_E$ is continuous.

We will now prove the following

Claim. *For each fixed* $v \in V$ *there exists a unique* $w \in W$ *which solves* (5.5). *Further,* $w \in E$ *and the map* ξ *associating* w *to* v *is of class* C^1 *from* V *to* E. *Moreover, for every* $v \in V$ *and* $j \in \{1, \ldots, m\}$,

$$\xi(v + T_j a_j) = \xi(v). \tag{5.7}$$

First of all, notice that, taking $\tau \in\,]-(\alpha' + \varepsilon), \alpha' + \varepsilon[\, \setminus \sigma(L)$, (5.5) becomes equivalent to the following fixed point problem :

$$w = (L - \tau I)^{-1} (I - P) [N(v + w) - \tau w] := T_v(w).$$

We want to show that, if τ is appropriately chosen, the map T_v is a contraction, for all v. Since

$$(L - \tau I)^{-1}(I - P) \; = \; \int_{-\infty}^{-(\alpha'+\varepsilon)} (\lambda - \tau)^{-1}\, d E_\lambda \; + \; \int_{\alpha'+\varepsilon}^{+\infty} (\lambda - \tau)^{-1}\, d E_\lambda,$$

we have that

$$|(L - \tau I)^{-1}(I - P)| \leq (\alpha' + \varepsilon - |\tau|)^{-1}.$$

Hence, since N is Lipshitzian of constant α,

$$|T_v(w) - T_v(\tilde{w})| \leq (\alpha' + \varepsilon - |\tau|)^{-1}(\alpha + |\tau|)\, |w - \tilde{w}|.$$

This shows that T_v is a contraction whenever $|\tau| < \frac{1}{2}(\alpha' + \varepsilon - \alpha)$, and such a choice is always possible because of the structure of $\sigma(L)$. Then T_v has a unique fixed point w, and we set

$$\xi(v) = w.$$

Hence we have that, for all $v \in V$,

$$\xi(v) = (L - \tau I)^{-1}(I - P) [N(v + \xi(v)) - \tau\xi(v)] \tag{5.8}$$

It can be shown (see [16]) that ξ is Lipshitz continuous from V to W. Moreover, for every $v \in H$,

$$[(\tau - L)^{-1} v](t) = \exp [t\, J(A - \tau)] u_0 + \int_0^t \exp [(t - s)J(A - \tau)]\, Jv(s)\, ds$$

where

$$u_0 = [I_{2n} - \exp[T J(A - \tau)]]^{-1} \int_0^T \exp[(T - s)J(A - \tau)]\, Jv(s)\, ds.$$

This implies

$$(\tau I - L)^{-1} \in \mathcal{L}(X,E).$$

Hence from (5.8) we have that ξ is continuous from V to E. Now consider the function

$$\varphi : V \times E \to E$$
$$\varphi(v,w) = w - T_v(w).$$

Recalling the fact that $N'|_E$ is continuous, since T_v is also a contraction as a map from E to E, we have that the implicit function theorem can be applied to φ. As a consequence, we have that ξ is of class C^1 from V to E. Finally, (5.7) holds since, by (A3), equation (5.5) does not change if we substitute v with $v + T_i a_i$. The Claim is then proved.

By the Claim proved above, we have that equation (5.4) is reduced to equation (5.6), with $w = \xi(v)$, i.e. to

$$Lv = PN(v + \xi(v)) \tag{5.9}$$

where v varies in the finite dimensional space V. By the spectral decomposition of L we can write $V = V^- \oplus V^\circ \oplus V^+$, where V^-, V° and V^+ are the invariant subspaces of V corresponding to the negative, zero and positive spectrum of L. By (A1), $V^\circ = N(A)$.Let us consider the Hilbert space $H = V^- \oplus V^+$. Then L can be considered as a bounded self-adjoint operator on H with a bounded inverse.

Let $T^m = \mathbb{R}^m/\mathbb{Z}^m$ be the m-fold torus, and define on $H \times T^m$ the following map

$$g(v, (\tau_1,\dots,\tau_m)) = \int_0^T [-H(t,v(t) + \xi(v)(t) + \sum_{i=1}^m \tau_i a_i) + (h(t), v(t) + \xi(v)(t))]\, dt.$$

Since $\xi \in C^1(H,E)$, $H \hookrightarrow E$ and $N'|_E$ is continuous, g is of class C^2. Set

$$f(v,\tau) = \frac{1}{2}(Lv,v) + g(v,\tau).$$

By (A2), (A3) and classical arguments, the critical points of the functional f correspond to geometrically distinct solutions of (5.9).

It is easy to see that dg is bounded and continuous. Since g is defined on a finite dimensional space, this implies dg is also compact. So all the assumptions of Theorem 2 are satisfied, and the result follows from (4.4) and (4.5).

References

[1] H. Amann, *Saddle points and multiple solutions of differential equations*, Math. Z. **169** (1979), 127-166.

[2] H. Amann and E. Zehnder, *Nontrivial solutions for a class of non-resonance problems and applications to nonlinear differential equations*, Ann. Scuola Norm. Sup. Pisa Cl. Sci. (4) **7** (1980), 539-603.

[3] V.I. Arnold, *Problems in present day mathematics : fixed points of symplectic diffeomorphisms*, Proc. Symp. Pure Math. vol. **28**, Amer. Math. Soc., Providence 1976, p. 66.

[4] V.I. Arnold, *Mathematical Methods of Classical Mechanics*, Springer, Berlin-Heidelberg-New York, 1978.

[5] R. Bott, *Lectures on Morse theory, old and new*, Bull. Amer. Math. Soc. **7** (1982), 331-358.

[6] K.C. Chang, *Infinite Dimensional Morse Theory and its Applications*, Séminaire de Mathématiques Supérieures, Presses Univ. Montréal, 1985.

[7] K.C. Chang, *Applications of homology theory to some problems in differential equations*, in : Nonlinear Functional Analysis and its Applications (F.E. Browder ed.), Proc. Symp. Pure Math., Amer. Math. Soc., Providence, 1985.

[8] D.C. Clark, *A variant of Ljusternik-Schnirelmann theory*, Indiana Math. J. 22 (1972), 65-74.

[9] C.C. Conley and E. Zehnder, *The Birkhoff-Lewis fixed point theorem and a conjecture of V. Arnold*, Invent. Math. **73** (1983), 33-49.

[10] C.C. Conley and E. Zehnder, *Morse type index theory for flows and periodic solutions for Hamiltonian equations*, Comm. Pure Appl. Math. **37** (1984), 207-253.

[11] A. Dold, *Algebraic Topology*, Springer Verlag, Berlin-Heidelberg-New York, 1972.

[12] P. Drabek and S. Invernizzi, *Periodic solutions for systems of forced coupled pendulum-like equations*, Quaderni Mat. n° **127**, Univ. Trieste, 1987.

[13] M.J. Greenberg, *Lectures on Algebraic Topology*, Benjamin, New York, 1977.

[14] M. Levi, F.C. Hoffenstaedt and W.L. Miranker, *Dynamics of the Josephson junction*, Quaterly Appl. Math. 36 (1978), 167-188.

[15] J.A. Marlin, *Periodic motions of coupled simple pendulums with periodic disturbances*, Int. J. Nonlinear Mech. **3** (1968), 439-447.

[16] J. Mawhin, *Semilinear equations of gradient type in Hilbert spaces and applications to differential equations*, in Nonlinear Differential Equations, stability, invariance and bifurcation, Academic Press, New York 1981, 269-282.

[17] J. Mawhin, *Points fixes, points critiques et problèmes aux limites*, Séminaire de Mathématiques Supérieures, Presses Univ. Montréal, 1985.

[18] J. Mawhin, *On a differential equation for the periodic motions of a satellite around its center of mass*, in Asymptotic Methods of mathematical physics, Nauka Dumka, Kiev, 1988, 150-157.

[19] J. Mawhin, *Forced second order conservative systems with periodic nonlinearity*, Anal. non linéaire, Inst. H. Poincaré, special issue dedicated to J.J. Moreau, 1989.

[20] J. Mawhin and M. Willem, *Multiple solutions of the periodic boundary value problem for some forced pendulum-type equations* , J. Diff. Equations **52** (1984), 264-287.

[21] J. Mawhin and M. Willem, *Variational methods and boundary value problems for vector second order differential equations and applications to the pendulum equation*, in Nonlinear Analysis and Optimization, ed. C. Vinti, Lect. Notes in Math. n° 1107, 1984.

[22] J. Mawhin and M. Willem, *Critical Point Theory and Hamiltonian Systems*, Springer-Verlag,, New York-Berlin, 1989.

[23] R.S. Palais, *Morse theory on Hilbert manifolds*, Topology **2** (1963), 299-340.

[24] R.S. Palais, *Critical point theory and the minimax principle, Global Analysis*, Proc. Sym. Pure Math. **15** (ed. S.S. Chern), Amer. Math. Soc., Providence, 1970, 185-202.

[25] P.H. Rabinowitz, *Variational methods for nonlinear eigenvalue problems, Eigenvalues of Nonlinear Problems*, Edizioni Cremonese, Roma, 1974, 141-195.

[26] P.H. Rabinowitz, *On a class of functionals invariant under a* Z^n *action*, Trans. Amer. Math. Soc. **310** (1988) 303-311.

SUPPLEMENTARY REFERENCES

[27] K.C. Chang, *On the periodic nonlinearity and the multiplicity of solutions*, Nonlinear Analysis, TMA, **13** (1989) 527-537.

[28] A. Fonda and J. Mawhin, *Multiple periodic solutions of conservative systems with periodic nonlinearity*, in Proceed. Intern Confer. Theory Applic. Differential Equations, Columbus, Ohio, 1988, to appear.

[29] M. Struwe, *Plateau's problem and the Calculus of Variations*, Princeton Univ. Press, Princeton, 1988.

The Problem of Plateau (pp. 129–137)
ed. Th. M. Rassias

ON THE UNIQUENESS FOR HYPERSURFACES WITH CONSTANT MEAN CURVATURE IN $\mathbf{R}^{n+1}$ BOUNDED BY A ROUND $(n-1)$ -SPHERE

MIYUKI KOISO
Department of Mathematics
Osaka University
Toyonaka, Osaka 560
Japan

ABSTRACT

It is very likely that spherical caps are the only compact embedded hypersurfaces with non-zero constant mean curvature in $\mathbf{R}^{n+1}$ bounded by a round $(n-1)$-sphere. Under certain additional assumptions on the considered hypersurfaces, the above conjecture is proved by using the reflection methods of A. D. Alexandrov.

1. Introduction

Let Γ be an $(n-1)$-dimensional round sphere with radius 1 in $\mathbf{R}^{n+1}$. For convenience, we choose orthogonal coordinates $(x^1, \cdots, x^{n+1})$ of $\mathbf{R}^{n+1}$ so that $\Gamma = \{(x^1, \cdots, x^n, 0); (x^1)^2 + \cdots + (x^n)^2 = 1\}$. We consider the problem of finding all compact hypersurfaces with constant mean curvature H bounded by Γ. It is known that in the case that $H = 0$, the n-dimensional closed ball $\overline{D} = \{(x^1, \cdots, x^n, 0); (x^1) + \cdots + (x^n)^2 \leq 1\}$ is the only compact hypersurface of mean curvature zero with Γ as its boundary. Therefore, we may devote our attention to the case that $H \neq 0$. Well-known examples are n-dimensional spherical caps with radius $1/|H|$ $(0 < |H| \leq 1)$ bounded by Γ. It has long been conjectured that these examples are the only possible ones. But recently, for $n = 2$, Kapouleas[1] proved the existence of compact surfaces with constant mean curvature H $(0 < |H| < 1)$ bounded by Γ with genus ≥ 3. However, the following two conjectures are still likely to be valid.

CONJECTURE 1. *A compact embedded hypersurface with non-zero constant mean curvature bounded by Γ must be a spherical cap.*

CONJECTURE 2. *In the case that $n = 2$, a compact immersed surface with non-zero constant mean curvature of genus zero bounded by Γ must be a spherical cap.*

There are several partial results relating to the above conjectures.[2,3,4,5] In this paper we prove a new partial result concerning Conjecture 1.

Let M be an n-dimensional compact connected smooth manifold with smooth boundary ∂M.

DEFINITION 1. Let H be a real constant. An embedding $\mathcal{X} : M \longrightarrow \mathbf{R}^{n+1}$ is called an *embedded H-hypersurface bounded by* Γ if $\mathcal{X}$ satisfies the following conditions (i) and (ii).

(i) The mean curvature of $\mathcal{X}$ is constant H.

(ii) The restriction of $\mathcal{X}$ to ∂M is a diffeomorphism of ∂M onto Γ.

If $\mathcal{X} : M \longrightarrow \mathbf{R}^{n+1}$ is an embedded H-hypersurface bounded by Γ and if $H \neq 0$, then M must be orientable and the sign of H depends only on the orientation of M. Hence, we may assume that H is positive without loss of generality. Moreover, we can choose the globally defined unit normal vector field $\mathcal{N} : M \longrightarrow \mathbf{R}^{n+1}$ of $\mathcal{X}$. If we further assume that the image $\mathcal{X}(M)$ does not intersect the open n-ball $D = \{(x^1, \cdots, x^n, 0) \in \mathbf{R}^{n+1}; (x^1)^2 + \cdots + (x^n)^2 < 1\}$, then, by the Alexander duality theorem, $\mathcal{X}(M) \cup D$ bounds a bounded domain G of $\mathbf{R}^{n+1}$. Our main result is as follows.

THEOREM 1. *Let $\mathcal{X} : M \longrightarrow \mathbf{R}^{n+1}$ be an embedded H-hypersurface ($H > 0$) bounded by Γ, and let $\mathcal{N} : M \longrightarrow \mathbf{R}^{n+1}$ be the unit normal vector field of $\mathcal{X}$. Assume that $\mathcal{X}$ satisfies the following conditions* (i) *and* (ii).

(i) *$\mathcal{X}(M) \cap D = \emptyset$. The domain G bounded by $\mathcal{X}(M) \cup D$ lies in the upper half space $\mathbf{R}^{n+1}_+ = \{(x^1, \cdots, x^n, x^{n+1}) \in \mathbf{R}^{n+1}; x^{n+1} > 0\}$ near D.*

(ii) *For any $\zeta \in \partial M$, $\mathcal{X}(\zeta) + \mathcal{N}(\zeta)$ is contained in the half open solid cylinder $\tilde{C} = C^\circ \cup (\partial C \cap \mathbf{R}^{n+1}_+)$, where $C = \{(x^1, \cdots, x^n, x^{n+1}) \in \mathbf{R}^{n+1}; (x^1)^2 + \cdots + (x^n)^2 \leq 1\}$ and C° is the interior of C.*

Then the image $\mathcal{X}(M)$ is a spherical cap.

Remark that the second statement of condition (i) in Theorem 1 is not any restriction.

Theorem 1 is proved by using the reflection method of A. D. Alexandrov.[6]

We should remark that the similar arguments of the proof of Theorem 1 with some further discussions enables us to prove a more general result, that is the symmetry of embedded H-hypersurfaces with symmetric boundaries. We will discuss such a generalization in another paper in the near future.

2. Preliminaries

In this section we give some definitions and known results which will be used to prove Theorem 1. An n-dimensional connected C^r-submanifold ($r \geq 1$) with or without boundary of $\mathbf{R}^{n+1}$ is called a hypersurface in $\mathbf{R}^{n+1}$.

Definition 2. Let S be a hypersurface in $\mathbf{R}^{n+1}$ and l a straight line. Then, we say that *S is a rotational hypersurface with axis l* if, for each hyperplane π perpendicular to l, $S \cap \pi$ is invariant under the rotation group $SO(n)$ acting on π with origin $\pi \cap l$.

Theorem A (c.f. E. Artin[7]). *A hypersurface S in $\mathbf{R}^{n+1}$ is a rotational hypersurface with axis l if and only if S is symmetric with respect to every hyperplane containing l.*

W.-Y. Hsiang and W.-C. Yu[8] classifies rotational hypersurfaces of constant mean curvature in $\mathbf{R}^{n+1}$. As an immediate consequence of their results we obtain the following

Theorem B (c.f. W.-Y. Hsiang and W.-C. Yu[8]). *Suppose that S is a rotational C^2-hypersurface with a straight line l as its axis. If S is of non-zero constant mean curvature H and intersects with l, then S is the n-dimensional round sphere or a spherical cap with its center on l and $1/|H|$ as its radius.*

Using Theorem A and Theorem B, we can reduce Theorem 1 to the following

Proposition 1. *Let $\mathcal{X} : M \longrightarrow \mathbf{R}^{n+1}$ satisfy the assumption of Theorem 1. Then the image $\mathcal{X}(M)$ is symmetric with respect to every hyperplane which contains the x^{n+1}-axis.*

To prove the symmetry of hypersurfaces, the reflection method of A.D.Alexandrov [6] is very effective. Now we give the fundamental definition and the key theorem for this method.

Definition 3 (A.D.Alexandrov[6]). Let S_1, S_2 be two oriented C^r-hypersurfaces $(r \geq 1)$ in $\mathbf{R}^{n+1}$. We say that *they touch each other from one side at a point p*, if the following conditions (i) $\sim$ (iii) are fulfilled.

(i) Either p is a common interior point or a common boundary point of S_1 and S_2.

(ii) S_1 and S_2 have a common normal at p.

(iii) If we introduce the coordinates $y^1, \cdots, y^n, y^{n+1}$ in a neighbourhood U of p so that $(0, \cdots, 0, 1)$ is the common normal vector as in (ii) and represent S_1 and S_2 by the equations

$$y^{n+1} = f_1(y^1, \cdots, y^n), \qquad y^{n+1} = f_2(y^1, \cdots, y^n)$$

respectively in U, then either $f_1 \geq f_2$ in U or $f_1 \leq f_2$ in U.

The following theorem is an immediate consequence of the comparison principle for solutions of strictly elliptic partial differential equations.

THEOREM C (A.D.Alexandrov[6]). *Let S_1, S_2 be two oriented C^2-hypersurfaces of constant mean curvature H in $\mathbf{R}^{n+1}$. Suppose that they touch each other from one side at a point p. Then they coincide in a neighbourhood of p.*

3. Proof of main results

As we mentioned in §2, Theorem 1 is reduced to Proposition 1. Before proving Proposition 1, we prepare a lemma.

LEMMA 1. *Let $\mathcal{X}$, $\mathcal{N}$ satisfy the assumtions in Theorem 1. Then for any point $p = (p^1, \cdots p^n, 0) = \mathcal{X}(\zeta)$, $\zeta \in \partial M$,*

$$\mathcal{N}(\zeta) = (\xi p^1, \cdots, \xi p^n, \eta),$$

where

$$\xi \in [-1, 0]$$

and

$$\eta = \sqrt{1-\xi^2} \qquad \text{or} \qquad \eta = -\sqrt{1-\xi^2}.$$

Moreover, if $\xi = 0$, then $\eta = 1$.

PROOF: Set $\mathcal{N}(\zeta) = (\nu^1, \cdots, \nu^{n+1})$. Denote the tangent spaces of Γ, $\mathcal{X}(M)$ at p by $T_p(\Gamma)$, $T_p(\mathcal{X}(M))$, respectively. Because $T_p(\Gamma) \subset T_p(\mathcal{X}(M))$, $\mathcal{N}(\zeta)$ is orthogonal to $T_p(\Gamma) = \{(q^1, \cdots q^n, 0) \in \mathbf{R}^{n+1}; p^1q^1 + \cdots + p^nq^n = 0\}$. Therefore, $(\nu^1, \cdots, \nu^n) = \xi(p^1, \cdots, p^n)$ for some $\xi \in \mathbf{R}$. Since $|\mathcal{N}(\zeta)| = 1$, $\mathcal{N}(\zeta)$ is represented as follows.

$$\mathcal{N}(\zeta) = (\xi p^1, \cdots, \xi p^n, \eta),$$

where

$$|\xi| \le 1, \qquad |\eta| = \sqrt{1-\xi^2}.$$

Therefore,

$$\mathcal{X}(\zeta) + \mathcal{N}(\zeta) = ((1+\xi)p^1, \cdots, (1+\xi)p^n, \eta).$$

By the assumption (ii) in Theorem 1, regarding $p \in \Gamma$, we see

$$(1+\xi)^2 \le 1.$$

Hence

$$\xi \le 0.$$

If $\xi = 0$, then $\eta = \pm 1$ and $\mathcal{X}(\zeta) + \mathcal{N}(\zeta) = (p^1, \cdots, p^n, \eta) \in \partial C$. Since $\mathcal{X}(\zeta) + \mathcal{N}(\zeta) \in \partial C \cap \mathbf{R}^{n+1}_+$, η must be 1. ∎

PROOF OF PROPOSITION 1: Set $S = \mathcal{X}(M)$. For any real number a, denote the hyperplane $\{x^1 = a\}$ by P_a. Set $a_0 = \min\{x^1; (x^1, \cdots, x^{n+1}) \in S\}$. And define a subset H_a of S by

$$H_a = \{(x^1, \cdots, x^{n+1}) \in S; x^1 < a\}$$

for $a > a_0$. Then H_a is not empty. Denote by $\tilde{H}_a$ the reflection of H_a with respect to P_a. If $a - a_0$ is sufficiently small, $\tilde{H}_a$ is contained in $G \cup D$. Set

$$d = \sup\{b \in \mathbf{R}; \tilde{H}_a \subset G \cup D \quad \text{for all} \quad a \in (a_0, b)\}.$$

Then $d > a_0$.

At first we assume that $d < 0$ and derive a contradiction. Since $\tilde{H}_d \subset \overline{G}$, there are three possibilities as follows:

(I) $\tilde{H}_d \subset G \cup D$.
(II) $\tilde{H}_d \not\subset G \cup D$ and $\tilde{H}_d \subset \overline{G} - \Gamma$.
(III) $\tilde{H}_d \cap \Gamma \neq \emptyset$.

Now we assume case (I). Let p be a point in $\partial H_d \cap \partial \tilde{H}_d = \{(d, x^2, \cdots, x^{n+1}) \in S\}$. Suppose that the tangent space $T_p(S)$ of S at p is not perpendicular to P_d. Then, for any sufficiently small $\delta > 0$, $\tilde{H}_{d+\delta}$ is contained in $G \cup D$ in some neighbourhood of p. Moreover, if p is in $\partial H_d \cap \partial \tilde{H}_d \cap \Gamma$ and if $T_p(\Gamma)$ coincides with $\{x^{n+1} = 0\}$, $\tilde{H}_{d+\delta}$ is contained in $G \cup D$ in some neighbourhood of p for sufficiently small $\delta > 0$. Therefore, the compactness of $\partial H_d \cap \partial \tilde{H}_d$ and the smoothness of S imply that $\tilde{H}_{d+\delta}$ is globally contained in $G \cup D$ for small $\delta > 0$ as soon as the following conditions (i) and (ii) are fulfilled:

(i) For any point $p \in (\partial H_d \cap \partial \tilde{H}_d) - \Gamma$, $T_p(S)$ is not perpendicular to P_d.

(ii) For any point $p \in \partial H_d \cap \partial \tilde{H}_d \cap \Gamma$, $T_p(S)$ is not perpendicular to P_d or it coincides with $\{x^{n+1} = 0\}$.

Hence, from the definition of d, either we have

$(\tilde{i})$ There exists some $p \in (\partial H_d \cap \partial \tilde{H}_d) - \Gamma$ such that $T_p(S)$ is perpendicular to P_d ;

or

$(\tilde{ii})$ Thre exists some $p \in \partial H_d \cap \partial \tilde{H}_d \cap \Gamma$ such that $T_p(S)$ is perpendicular to P_d and that it does not coincide with $\{x^{n+1} = 0\}$.

However, we can prove that $(\tilde{ii})$ can not occur as follows. Assume that $T_p(S)$ is perpendicular to P_d at a point $p = \mathcal{X}(\zeta_1) = (d, p^2, \cdots, p^n, 0) \in \partial H_d \cap \partial \tilde{H}_d \cap \Gamma$, $\zeta_1 \in \partial M$. By Lemma 1, $\mathcal{N}(\zeta_1)$ is represented as $\mathcal{N}(\zeta_1) = (\xi d, \xi p^2, \cdots, \xi p^n, \eta)$. Since $\mathcal{N}(\zeta_1)$ is orthogonal to $(1, 0, \cdots, 0)$, $\xi d = 0$. Since $d < 0$, we get $\xi = 0$ and $\mathcal{N}(\zeta_1) = (0, \cdots, 0, 1)$, which implies that $T_p(S)$ coincides with $\{x^{n+1} = 0\}$.

Consequently, $(\widetilde{ii})$ cannot happen. Now assume $(\widetilde{i})$. Then $S - H_d$ and $\tilde{H}_d$ touch each other from one side at p. Therefore, by Theorem C, $S - H_d$ and $\tilde{H}_d$ coincide in a neighbouhood of p, which contradicts the assumption of case (I).

Next we assume case (II). Then $\tilde{H}_d \cap (S - \Gamma)$ is not empty. Take any point p of $\tilde{H}_d \cap (S - \Gamma)$. Since p is an interior point of both $\tilde{H}_d$ and $S - H_d$, $\tilde{H}_d$ and $S - H_d$ touch each other from one side at p. Therefore, again by Theorem C, $\tilde{H}_d$ and $S - H_d$ coincide in a neighbourhood of p. Now, denote by $\tilde{K}$ the connected component of $\tilde{H}_d$ which contains p. Then $\tilde{K}$ is the reflection of a certain component K of H_d with respect to P_d. If we note that $\tilde{H}_d \cap \Gamma$ is empty and use Theorem C repeatedly, then we see that $\tilde{K} \cap S$ is an open and closed subset of $\tilde{K}$. Since $\tilde{K}$ is connected and $\tilde{K} \cap S$ is not empty, $\tilde{K} \cap S$ is all of $\tilde{K}$. Hence $\tilde{K}$ is contained in S, from which, together with the fact that $\tilde{H}_d \cap \Gamma$ is empty, it follows that $\tilde{K}$ is included in $S - \Gamma$. Therefore, ∂K coincides with $\partial \tilde{K}$. Set $K^* = \overline{K \cup \tilde{K}}$. Then, from the above consideration, K^* is an n-dimensional connected topological manifold without boundary which is contained in S. Therefore, K^* is open and closed in S. Since S is connected and K^* is not empty, K^* coincides with S. However, S has the non-empty boundary $\partial S = \Gamma$,which is a contradiction. Consequently, case (II) is also impossible.

Assume case (III). Let $\mathcal{X}(\zeta_2) = p = (p^1, \cdots, p^n, 0) \in \tilde{H}_d \cap \Gamma$. Let $q = \mathcal{X}(w)$ be the reflection of p with respect to P_d. Then

$$q = (q^1, p^2, \cdots, p^n, 0),$$

where

$$q^1 = 2d - p^1.$$

By the definition of $\tilde{H}_d$,

$$q^1 < d < p^1. \tag{1}$$

Since $d < 0$,

$$q \notin \Gamma. \tag{2}$$

By the definition of d,

$$\{(x^1, p^2, \cdots, p^n, 0); q^1 < x^1 < p^1\} \subset G \cup D. \tag{3}$$

We will prove that $\mathcal{N}(w) = \mathcal{N}(\zeta_2) = (0, \cdots, 0, 1)$ and that S and $\tilde{H}_d \cap \overline{\text{Ext}C}$ touch each other from one side at p, where

$$\overline{\text{Ext}C} = \left\{ (x^1, \cdots, x^n, x^{n+1}) \in \mathbf{R}^{n+1}; (x^1)^2 + \cdots + (x^n)^2 \geq 1 \right\}.$$

First, assume that $p^1 > 0$. Then, by the definition of d, near q, S must be contained in $\overline{\mathbf{R}^{n+1}_+}$. Therefore, the unit normal $\mathcal{N}(w)$ of S at q is $(0,\cdots,0,1)$. From this fact, Lemma 1, and the fact that $\tilde{H}_d \subset \overline{G}$, we see that $\mathcal{N}(\zeta_2) = (0,\cdots,0,1)$ and that S and $\tilde{H}_d \cap \overline{\mathrm{Ext}C}$ touch each other from one side at p.

Next assume that $p^1 < 0$. By (3) and Lemma 1, we see that $\mathcal{N}(\zeta_2) = (0,\cdots,0,1)$. Since $\tilde{H}_d \subset \overline{G}$, we know that $\mathcal{N}(w) = (0,\cdots,0,1)$ and that S and $\tilde{H}_d \cap \overline{\mathrm{Ext}C}$ touch each other from one side at p.

At last, we assume that $p^1 = 0$. By (3) and the fact that $\tilde{H}_d \subset \overline{G}$, we see that $T_q(S)$ must contain $\{(x^1,0,\cdots,0) \in \mathbf{R}^{n+1}; x^1 \in \mathbf{R}\}$. Therefore, $\mathcal{N}(w)$ is orthogonal to $(1,0,\cdots,0)$. Since $\tilde{H}_d \subset \overline{G}$, by the assumption (i) of Theorem 1, $\mathcal{N}(w)$ must be represented as follows.

$$\mathcal{N}(w) = \left(0, \eta^2 p^2, \cdots, \eta^n p^n, \eta^{n+1}\right), \tag{4}$$

where

$$\eta^j \geq 0, \qquad j = 2,\cdots,n. \tag{5}$$

By (4) and the assumption (ii) of Theorem 1, regarding $\tilde{H}_d \subset \overline{G}$, we get

$$\left(p^1\right)^2 + \left(p^2 + \eta^2 p^2\right)^2 + \cdots + \left(p^n + \eta^n p^n\right)^2 \leq 1. \tag{6}$$

Since $\left(p^1\right)^2 + \cdots + \left(p^n\right)^2 = 1$ and $\left(\eta^2 p^2\right)^2 + \cdots + \left(\eta^n p^n\right)^2 + \left(\eta^{n+1}\right)^2 = 1$, by (6), we obtain

$$2\left\{\eta^2\left(p^2\right)^2 + \cdots + \eta^n\left(p^n\right)^2\right\} \leq -\left\{1 - \left(\eta^{n+1}\right)^2\right\} \leq 0. \tag{7}$$

(5) and (7) imply

$$\eta^2 = \cdots = \eta^n = 0.$$

Therefore,

$$\mathcal{N}(w) = (0,\cdots,0,\eta^{n+1}).$$

Again by the assumption (ii) of Theorem 1 and $\tilde{H}_d \subset \overline{G}$, we see

$$\mathcal{N}(w) = (0,\cdots,0,1).$$

Therefore, since $\tilde{H}_d \subset \overline{G}$, we know that $\mathcal{N}(\zeta_2) = (0,\cdots,0,1)$ and that S and $\tilde{H}_d \cap \overline{\mathrm{Ext}C}$ touch each other from one side at p.

Therefore, by Theorem C, S and $\tilde{H}_d \cap \overline{\mathrm{Ext}C}$ coincide in a neighbourhood of p irrespective of sign of p^1. Hence, $\tilde{H}_d$ includes an open subset Γ_1 of Γ which contains p. Next we show the existence of a point $p{*}{*} = (p{*}{*}^1,\cdots,p{*}{*}^n,0) \in \tilde{H}_d \cap \Gamma$ such that $p{*}{*}^1 > 0$. When $p^1 > 0$, we can take p as $p{*}{*}$. Assume that $p^1 \leq 0$. Then, we can find a boundary point $p{*} = (p{*}^1, \cdots, p{*}^n, 0)$ of Γ_1 so that

$$(8) \qquad p{*}^1 > p^1.$$

Sinse $d < 0$, $\partial \tilde{H}_d \subset P_d \cup D$. Moreover, recall that $p^1 > d$ (c.f. (1)). Then, iequality (8) implies that $p{*}$ is an interior point of $\tilde{H}_d$. Therefore $p{*} \in \tilde{H}_d \cap \Gamma$. The similar discussion on $p{*}$ to the above on p leads us to the conclusion that $\tilde{H}_d$ includes an open subset Γ_2 of Γ which contains $p{*}$. By using this argument repeatedly, we see that there exists some point $p{*}{*} = (p{*}{*}^1, \cdots, p{*}{*}^n, 0) \in \tilde{H}_d \cap \Gamma$ such that $p{*}{*}^1 > 0$.

Therefore, we may assume that $p^1 > 0$. We saw that $\tilde{H}_d$ includes an open subset Γ_1 of Γ which contains $p = (p^1, \cdots, p^n, 0)$. Let $\hat{p} = (\hat{p}^1, \cdots, \hat{p}^n, 0)$ be any point in Γ_1. Let $\hat{q}$ be the reflection of $\hat{p}$ with respect to P_d. Then $\hat{q} \in H_d$ and

$$\hat{q} = (\hat{q}^1, \hat{p}^2, \cdots, \hat{p}^n, 0),$$

where

$$\hat{q}^1 = 2d - \hat{p}^1.$$

By the definitions of $\tilde{H}_d$ and d, we get

$$\hat{q}^1 < d < \hat{p}^1,$$

$$\{(\hat{x}^1, \hat{p}^2, \cdots, \hat{p}^n, 0); \hat{q}^1 < \hat{x}^1 < \hat{p}^1\} \subset G \cup D.$$

Since $\hat{p}$ is an arbitrary point of Γ_1 and since Γ_1 is an non-empty open subset of Γ, by regarding (2), we see that there exists an non-empty open subset Ω of hyperplane $\{x^{n+1} = 0\}$ which satisfies the following properties (i) $\sim$ (iv).

(i) $\Omega \subset \{(\hat{x}^1, \hat{p}^2, \cdots, \hat{p}^n, 0); (\hat{p}^1, \cdots, \hat{p}^n, 0) \in \Gamma_1, \quad 2d - \hat{p}^1 < \hat{x}^1 < \hat{p}^1\}$.
(ii) The boundary of Ω contains q.
(iii) $\Omega \subset G$.
(iv) Ω is contained in the reflection of $D = \{(x^1, \cdots, x^n, 0) \in \mathbf{R}^{n+1}; (x^1)^2 + \cdots + (x^n)^2 < 1\}$ with respect to P_d.

Since $p^1 > 0$, by the definition of $\tilde{H}_d$ and the assumption (i) of Theorem 1, we see that

$$\Omega \subset S,$$

which contradicts the assumption that the mean curvature of S is not zero.

Consequently, $d \geq 0$. Therefore,

$$(9) \qquad \tilde{H}_0 \subset \overline{G}.$$

Next, set $a_1 = \max\{x^1; (x^1, \cdots, x^{n+1}) \in S\}$. For $a < a_1$, define a subset F_a of S by

$$F_a = \{(x^1, \cdots, x^{n+1}) \in S; x^1 > a\},$$

and denote by $\tilde{F}_a$ the reflection of F_a with respect to P_a. Set

$$e = \inf\{b \in \mathbf{R}; \tilde{F}_a \subset G \cup D \quad \text{for all} \quad a \in (b, a_1)\}.$$

By making the similar discussion on F_a and $\tilde{F}_a$ to the above discussion on H_a and $\tilde{H}_a$, we conclude that $e \leq 0$. Therefore,

$$\tilde{F}_0 \subset \overline{G}. \tag{10}$$

(9) and (10) leads us to the symmetry of S with respect to P_0.[9] By rotating $\mathbf{R}^{n+1}$ around the x^{n+1}-axis, we get the symmetry of S with respect to every hyperplane containing the x^{n+1}-axis. ∎

4. References

1. N. Kapouleas, Compact constant mean curvature surfaces in Euclidean three-space, *J. Diff. Geom.* **33** (1991), pp. 683-715.
2. M. Koiso, Symmetry of hypersurfaces of constant mean curvature with symmetric boundary, *Math. Zeit.* **191** (1986), pp. 567-574.
3. J. L. M. Barbosa, Hypersurfaces of constant mean curvature on $\mathbf{R}^{n+1}$ bounded by an Euclean sphere, *preprint*.
4. R. Earp, F. Brito, W. H. Meeks, III and H. Rosenberg, Structure theorems for constant mean curvature surfaces bounded by a planar curve, *Indiana Univ. Math. J.* **40** (1991), pp. 333-343.
5. F. Brito and R. S. Earp, Geometric configurations of constant mean curvature surfaces with planar boundary, *An. Acad. bras. Ci.* **63** (1991), pp. 5-19.
6. A. D. Alexandrov, A characteristic property of spheres, *Ann. Mat. Pura Appl.* **58** (1962), pp. 303-315.
7. E. Artin, *Geometric Algebra* (Interscience, New York, 1957), p.129.
8. W.-Y. Hsiang and W.-C. Yu, A generalization of a theorem of Delaunay, *J. Diff. Geom.* **16** (1981), pp. 161-177.
9. The same paper as in 2, pp. 573-574.

The Problem of Plateau (pp. 138–164)
ed. Th. M. Rassias

ON THE THEORY OF MINIMAL SURFACES

ERWIN KREYSZIG
Department of Mathematics & Statistics
Carleton University, Ottawa, Canada K1S 5B6

Abstract

The evolution of the theory of minimal surfaces had two culmination points, one in the publications of general representation formulas relating to complex analysis, by Enneper (1864), Weierstrass (1866), and Riemann (1868, posthumous), and the second in the solution of Plateau's problem for general Jordan curves, by Garnier (1928), Radó (1930), Douglas (1931), and McShane (1933). In our paper we consider basic ideas and aspects in the development of minimal surface theory, particularly emphasizing those that were essential to the solution by Radó. This is based on extensive discussions between Prof. Tibor Radó and the author from 1956 until 1962 at The Ohio State University.

1. Introduction

Minimal surfaces in E^3 constitute a large and interesting class of surfaces that has attracted attention over a long period of time and for various theoretical and practical reasons. This development culminated in 1930-31 with the general solution of Plateau's problem by Tibor Radó[36] and independently by Jesse Douglas[11], to both of whom this volume is dedicated.

For his work, Douglas became one of the first two recipients of the Fields Medal (at the Oslo International Congress of Mathematicians 1936, the other medalist being Lars Ahlfors). In his laudatio, Carathéodory[7] emphasized the novelty of Douglas's method "which uses only very few elements of the traditional theory and leads to consequences of unexpected range [von ungeahnter Tragweite]." Carathéodory went on by saying that Douglas related Plateau's problem to the Dirichlet integral, showed that it suffices to consider harmonic surfaces, and transformed the Dirichlet integral into a functional $A(g)$ which was central to his approach and had a "surprisingly simple geometric meaning."

Carathéodory also contrasted this approach with the solution by Radó, saying [p. 311]:

> "After a rather long development, in which many mathematicians participated, in 1930 Tibor Radó was able to solve the boundary value problem of minimal surfaces and Plateau's problem connected with it for all closed Jordan curves in space which bound a surface of finite area."

It is with some main ideas of that rich and rather heterogeneous long development in differential geometry, as just mentioned by Carathéodory, that this paper will be concerned. It will be shown how these ideas evolved from the very beginning and contributed to Radó's paper[36] of 1930, a landmark of the whole evolution.

For reasons of space we shall discuss neither Douglas's work nor later solutions of Plateau's problem, such as that by McShane[28] of 1933. We mention that the latter was based on initial ideas by Lebesgue in his famous thesis[24] and a later paper[25]. Of his solution, McShane claimed [p. 717] "that the distinctive feature of the present method is not its elegance (in which respect it is inferior to its predecessors) but the directness of the line of thought," and a few lines later he added "that whatever familiarity I may have with this branch of mathematics is due in large part to my conversations with Professor Radó."

2. Evolution of Differential-Geometric Surface Theory

In this paper, "differential geometry" will mean classical differential geometry in three-dimensional Euclidean space E^3, and "surface theory" the differential-geometric theory of surfaces in E^3.

From the standpoint of the evolution of Analysis, the two main subjects of differential geometry, namely, curve theory and surface theory, have played entirely different roles. Indeed, one observes that curves had a much more fundamental impact on that development than surfaces. A reason is that curves were instrumental to the development of the concept of function (see[21], Sec. 3). Another reason is that, whereas the (local) differential-geometric curve theory can be obtained more or less from the theory of the Frenet formulas, it is well known that surface theory is much more complicated and diverse (see, for instance [10]). This entailed a rather heterogeneous study of various special classes of surfaces, each one providing different aspects, specific problems of interest, and special methods of approach, minimal surfaces being a particularly attractive class, as can be seen from standard texts (such as the monographies[12],[22],[46]).

In view of the facts just stated, it is not surprising that so far the development of surface theory as a whole has never been presented in detail, but many ideas appeared just marginally in other contexts, aimed in different directions. For orientation we briefly mention some of them. An authoritative résumé on minimal surfaces until about 1875 was given by G. Darboux[10],*I*,319–331. The state of differential geometry around the turn of the century is reflected in articles in the Encyklopädie[14]. An outline leading to that state was sketched by Struik[47]. However, that article paid little attention to minimal surfaces. The period from the publication of Gauss's "Disquisitiones"[17] in 1828 to Riemann's "Hypothesen" in 1868 was discussed by Reich[39], who included a rich bibliography. A wealth of relevant information is contained in Nitsche's classic[34]. For developments on minimal surfaces after 1930 – not our present concern – we refer to Nitsche[33].

3. Terminology and Notations

We say "*surface*" to mean a whole surface as well as a portion of it. x, y, z denote Cartesian coordinates in space E^3. u, v denote Gaussian parameters ("coordinates on a surface") in parametric representations of a surface S, which we write (Ω denoting some domain in the uv-plane)

$$r = [x(u,v),\ y(u,v),\ z(u,v)] \qquad (u,v) \in \Omega. \tag{3.1}$$

[Conforming to original papers, we shall sometimes also use the Gauss notation p, q for u, v; this should cause no confusion with the familiar notation for first partial derivatives.] For *nonparametric representations* we write

$$z = z(x,y). \tag{3.2}$$

Existence of partial derivatives will be assumed as needed. Also the surface normal vector

$$n = r_u \times r_v$$

will be assumed to be not the zero vector, perhaps except at finitely many points. For the coefficients of the fundamental forms we use the standard notation

$$\begin{aligned} E &= r_u \cdot r_u, & F &= r_u \cdot r_v, & G &= r_v \cdot r_v \\ L &= r_{uu} \cdot n, & M &= r_{uv} \cdot n, & N &= r_{vv} \cdot n. \end{aligned} \tag{3.3}$$

The (classical) surface area is

$$A(S) = \iint_\Omega (EG - F^2)^{\frac{1}{2}}\, du\, dv. \tag{3.4}$$

The *Gaussian curvature* is

$$K = \kappa_1 \kappa_2 = \frac{LN - M^2}{EG - F^2}, \tag{3.5}$$

where κ_1 and κ_2 denote the principal curvatures. A *minimal surface* S is a surface with identically vanishing *mean curvature*

$$H = \frac{1}{2}(\kappa_1 + \kappa_2) = \frac{1}{2}\left(\frac{GL - 2FM + EN}{EG - F^2}\right). \tag{3.6}$$

If S is represented nonparametrically by (3.2), then $H = 0$ is equivalent to the *minimal surface equation*

$$\left(1 + z_y^2\right) z_{xx} - 2 z_x z_y z_{xy} + \left(1 + z_x^2\right) z_{yy} = 0, \tag{3.7}$$

which is nonlinear and elliptic (but not uniformly elliptic).

By *Plateau's problem* we mean the determination of a minimal surface S in E^3 homeomorphic to a disk and bounded by a given Jordan curve in space. We call the problem *nonparametric* if S is given nonparametrically, say, by (3.2).

This problem is named after the Belgian physicist, J. Plateau (1801-1883), who realized minimal surfaces experimentally by dipping wires (the boundary curves) into soapy solutions, minimum area corresponding to minimum energy. We cannot enter into this fascinating theory of surface tension and capillarity, the latter discovered by Leonardo da Vinci, but mention that many famous mathematicians have contributed to it, among them Monge (1787), Laplace (1806), Gauss (1830), Poisson (1831), Betti (1867), Maxwell (1875), Mathieu (1883), and Poincaré (1895).

By the *problem of least area* we mean the determination of a surface S : $r(u, v)$ that is bounded by a given Jordan curve C, is *continuous* [i.e., $r(u, v)$ is continuous for $u^2 + v^2 \leq 1$] and has area $A(S)$ in the sense of Lebesgue equal to the infimum of the areas of all continuous surfaces bounded by C, and $r(u, v)$ is assumed to map the unit circle in the uv-plane topologically onto C.

4. Euler's Discovery of the Catenoid

Almost any mathematical theory is preceded by the discovery of special examples, but in the case of minimal surface theory this fact is particularly significant, because the *catenoid*, the earliest nontrivial minimal surface discovered, by Euler (1707-1783) in 1744[15], remained the only known nontrivial minimal surface for over thirty years (until Meusnier found the right helicoid), and it took almost one hundred years until Scherk discovered further minimal surfaces.

The catenoid is obtained by rotating a catenary, say,

$$\rho(x) = A\cosh(kx - c), \quad (A \neq 0,\ k \neq 0) \tag{4.1}$$

about the x-axis. This curve describes the form of a chain or flexible cable under the attraction of gravity, as was shown in 1691 independently by Huygens, Leibniz, and Johann Bernoulli.

The title of [15] shows that Euler obtained the catenoid as an application of the developing *calculus of variations*, so it seems worthwhile to recall that at that time, one was concerned with the minimization (or maximization) of a functional (integral)

$$J[y] = \int_a^b F(x, y, y')\, dx, \quad y(a) = y_0, \quad y(b) = y_1,$$

in a given class of functions ("admissible functions") satisfying the given boundary conditions. The impetus to the development had come in 1696 from Johann Bernoulli's famous problem of the brachistochrone (curve of fastest descent). It was Euler who gave the first systematic treatment of the calculus of variations, in a paper of 1736 and again in his book of 1744; this included the Euler (or Euler-Lagrange) equation

$$F_y - \frac{d}{dx} F_{y'} = 0 \tag{4.2}$$

as a necessary condition for $y = y(x)$ to be a solution of the problem $J[y] =$ minimum.

It is also worthwhile noting that Euler's study of developability and isometry of surfaces led him to investigations of *partial differential equations*, then a rather novel area about which but little was known; this was of significance to the development of surface theory in general. For some details, see Cantor[6],IV,551.

5. Lagrange: Beginning of the Theory

It is interesting that it took less than twenty years that Euler's discovery was followed up by the beginning of a systematic theory of minimal surfaces, again in the context of the calculus of variations, in a famous Memoir[23] of 1760-61 by Lagrange (1736-1813). At the time of publication, Lagrange was only twenty-four years old, but even more remarkable is the fact that he was only nineteen when he communicated his method in 1755 to Euler, who applauded his results; Lagrange [Œuvres II, 37] wrote:

> "Cette méthode, qu'on peut très-bien appeler, d'après M. Euler, *méthode des variations*, avait déjà été communiquée dès 1755 à ce grand Géomètre, qui l'avait jugée digne de son attention et de son suffrage..."

This Memoir became of great significance to the evolution of the calculus of variations and its applications to mechanics, where various minimum principles, such as that of least action, were guidelines in the further development.

Although Lagrange also arrived at minimal surfaces by way of the calculus of variations, hẹ took a big step forward by deriving the minimal surface equation, thus breaking ground for a general theory. He obtained the equation by the now familiar argument of considering, along with a surface $S : z = z(x, y)$, written [p. 354]

$$dz = p\,dx + q\,dy \tag{5.1}$$

and assumed to minimize the area functional A, a neighboring surface

$$\widehat{S} : \hat{z}(x, y) = z(x, y) + \epsilon\eta(x, y)$$

with $|\epsilon|$ small and $\eta = 0$ on the boundary curve C and the corresponding functional

$$A(\epsilon) = \iint_{\Omega} (1 + \hat{z}_x^2 + \hat{z}_y^2)^{\frac{1}{2}}\,dx\,dy.$$

Setting $A'(0) = 0$, integrating by parts, using $\eta(\partial\Omega) = 0$ and the fact that η is arbitrary otherwise, he got as a necessary condition for A to be minimum the *minimal surface equation*, which he wrote [p. 356] in the form

$$\left(\frac{dP}{dx}\right) + \left(\frac{dQ}{dy}\right) = 0 \tag{5.2}$$

where

$$P = \frac{p}{\sqrt{1 + p^2 + q^2}}, \quad Q = \frac{q}{\sqrt{1 + p^2 + q^2}},$$

with $p = z_x$ and $q = z_y$ and the parentheses denoted partial differentiation, as was common at that time. Lagrange did not perform these differentiations, which would have given him (3.7), but regarded (5.2) as integrability conditions:

> "Le problème se réduit donc à chercher p et q par ces conditions que
>
> $$p\,dx + q\,dy \quad \text{et} \quad \frac{p\,dy - q\,dx}{\sqrt{1 + p^2 + q^2}}$$
>
> [misprint p for q in the numerator corrected] soient l'une et l'autre des différentielles exactes."

6. Meusnier: Relation to Mean Curvature

That a solution of the minimal surface equation must be a surface with identically vanishing mean curvature

$$H = \frac{1}{2}(\kappa_1 + \kappa_2) \tag{6.1}$$

[cf. (3.6)] was discovered by J. Meusnier (1754-1793), an officer in the French Army, who died of wounds received during the siege of Mainz by Prussia. Meusnier presented his results to the Paris Academy in 1776, when he was only twenty-two years old, and had it published in 1785[29]. This is the same paper which also contained the well-known Meusnier's theorem (see [46],II or[22]). Thus $H = 0$ is necessary for the area of a surface to be minimum, but not sufficient; nevertheless, it became common to call any surface with $H = 0$ a *minimal surface.*

The term "mean curvature" (coubure moyenne) was introduced in 1831 by S. Germain [J. reine angew. Math. **7**, 7], three years after the Gauss curvature K had been given its fundamental position, by Gauss in his "Disquisitiones"[17].

The paper under discussion is the only work by Meusnier devoted to differential geometry. In it, Meusnier rediscovered the catenoid (unaware of Euler's earlier result) and found as a new minimal surface the *right helicoid.* This was a substantial accomplishment because it should take almost seventy years before further special minimal surfaces were found (by Scherk). Moreover, Meusnier's *method* by which he discovered the right helicoid as a minimal surface later became of importance to the study of particular solutions of partial differential equations.

Meusnier was a student of G. Monge (1746-1818), when the latter was working on the curvature of surfaces, and this undoubtedly influenced Meusnier's direction of mathematical interests.

Quite generally, the second half of the eighteenth century was a period of much work and great progress in surface theory. Tinseau (tangent surfaces, 1774), Euler (geodesics, 1732; Euler's theorem, 1760; parametric representations of special surfaces, 1771; spherical mapping, 1782 – but shown in its basic importance only much later by Gauss, who established various concepts as *intrinsic* properties of surfaces–), Lagrange (Sec. 5), Lambert (cartography, 1772), and Monge must be mentioned here. Monge's "Feuilles d'Analyse" of 1792 contained as novelties families of surfaces and their envelopes, characteristics, and lines of curvature (being the highlight of that work), with an application to surfaces for which the principal curvatures are equal and of opposite sign, leading to the equation which the "citoyen Lagrange" had found for minimal surfaces. Somewhat strange is that Monge did not mention Meusnier – but those times around 1789 were full of unrest, and perhaps Monge had really not seen Meusnier's finished work.

7. Monge and Legendre: General Solution Formulas

Quite in the spirit of the times, after the study of those two nontrivial minimal surfaces, one attempted next to find a "formula for the general solution" of the minimal surface equation. This began in 1784, led to deeper and deeper geometrical insight and analytical rigor, and culminated eighty years later after the completion of classical complex analysis. The first such formulas were obtained by Monge[31], the founder of Descriptive Geometry, whom we have just mentioned, and it seems not impossible that the inspiration to this work came from his collaboration with Meusnier. His general solution formulas were of the form

$$\begin{aligned} x &= a + b \\ y &= \phi(a) + \psi(b) \\ z &= \int \sqrt{-1 - \phi'(a)^2}\, da + \int \sqrt{-1 - \psi'(b)^2}\, db. \end{aligned} \tag{7.1}$$

In these formulas, ϕ and ψ are arbitrary analytic functions of a and b, respectively.

Monge read his paper to the Paris Academy in 1784, and it appeared in print three years later. He based his work on his theory of *characteristics* (which are imaginary for an elliptic equation, such as the minimal surface equation).

Contemporaries, however, regarded Monge's idea of generating solutions of partial differential equations from envelopes as unsatisfactory, if not mysterious – also, what should *complex* expressions be good for a *real* geometric problem!

This work was followed up by A.M. Legendre (1752-1833), who criticized it (for some errors on integrability conditions that were easily corrected). Legendre read his results to the Paris Academy in 1787, and they appeared in print in 1789[26]. In his paper he used his Legendre transformation suggested by the geometric interpretation of a differential equation if the solution surface is represented by its tangent plane coordinates. Thus, Legendre replaced x, y, z by p, q, $v = px + qy - z$, v being regarded as a function of p and q, and obtained as the transform of (3.7) the *linear* equation

$$(1 + p^2)v_{pp} + 2pqv_{pq} + (1 + q^2)v_{qq} = 0$$

which, together with $x = v_p$, $y = v_q$, $z = pv_p + qv_q - v$ determined minimal surfaces. He could have solved his new equation by the method of characteristics, but he used an elegant shortcut, writing the resulting formulas in the form

$$\begin{aligned} x &= \phi'(a) + \psi'(b) \\ y &= \phi(a) - a\phi'(a) + \psi(b) - b\psi'(b) \\ z &= \int \left(\sqrt{-1 - a^2}\right) \phi''(a)\, da + \int \left(\sqrt{-1 - b^2}\right) \psi''(b)\, db \end{aligned} \tag{7.2}$$

which was closely related to (7.1). He claimed that in his much simpler derivation (compared to that of Monge) he had replaced "certain metaphysical principles by the ordinary way." Legendre also transformed his formulas into integral-free form, an idea which we shall meet again in Weierstrass's work.

Since at that time, complex analysis had not yet been developed sufficiently far, the *geometric* significance of such complex-valued expressions was not obvious and became the subject of attempts to shed light onto the geometric aspects of the formalism only at a much later time, in 1832 and 1835 by Scherk, to whose work we turn next, and still later, in 1844 by E. Björling [Archiv Math. Phys. 4, 290-315], in 1853 and 1855 by O. Bonnet [Comptes Rendus Paris **37**, 529-532, and **40**, 1107-1110] and others.

One can cast Monge's representation into a locally equivalent complex form

$$\begin{aligned} x &= \operatorname{Re} w \\ y &= \operatorname{Re} f(w) \\ z &= \operatorname{Re} i \int_{w_0}^{w} \sqrt{1+f'^2(\tilde{w})}\, d\tilde{w}, \end{aligned} \tag{7.3}$$

but the systematic utilization of complex analysis for minimal surfaces had to wait for over seventy years, until Enneper, Riemann and, above all, Weierstrass introduced their famous representation formulas.

8. Scherk's Progress

The situation of minimal surface theory during the early nineteenth century did not look too promising, due to the facts that complex analysis was still in its infancy, rigor was lacking, for instance, in defining the notion of a surface and related concepts, and no further minimal surfaces had been found after Euler's and Meusnier's discoveries. Nevertheless, the period of attempts on *useful* general representations just discussed caused progress in terms of the discovery of special minimal surfaces, resulting in strong encouragement of further work in the area.

It was H. Scherk (1798-1885), Jacobi's former student, who in 1832[41] and 1835[42] published the discovery of no less than five new minimal surfaces. His first paper contained

$$e^z = \frac{\cos y}{\cos x} \tag{8.1}$$

(usually called "*Scherk's minimal surface*") and

$$z = a \log\left[\sqrt{r^2+b^2} + \sqrt{r^2-a^2}\right] - b \arctan \frac{a\sqrt{r^2+b^2}}{b\sqrt{r^2-a^2}} + b\theta + c \tag{8.2}$$

where r, θ are polar coordinates (so that $a = 0$ gives the helicoid and $b = 0$ the catenoid). The second paper contained

$$\sin z = \sinh x \, \sinh y \tag{8.3}$$

[equation 28, p. 198, except for notation] and two other, more complicated minimal surfaces, namely,

$$\cosh\left[2z - \frac{y}{x}\sqrt{2(g+h)} + x\sqrt{2(g-h)}\right] = g + \frac{y^2}{x^2} + \frac{x^2}{4} \tag{8.4}$$

where

$$g = (h^2 + y^2)^{\frac{1}{2}}, \quad h = 1 + \frac{y^2}{x^2} - \frac{x^2}{4},$$

[equation 16, p. 193, with three misprints corrected] and

$$\cosh\left[z + \operatorname{cosec}\frac{x}{2}\sqrt{\frac{1}{2}(g+h)}\right] = \frac{1}{y^2}\left(4\sin^2\frac{x}{2} + h\right) \tag{8.5}$$

where

$$g = \left(h^2 + y^4\sin^2 x\right)^{\frac{1}{2}}, \quad h = 4\sin^2\frac{x}{2} + y^2\cos x$$

[equation 20, p. 195]. Note the gap of sixty years between the discoveries of the catenoid and the right helicoid as minimal surfaces and Scherk's results under discussion. Note further the remarkable accomplishment in deriving *real* surfaces from those Monge-Legendre formulas involving *complex* quantities, as we shall explain in more detail below.

To overcome the obvious standstill in the development on minimal surfaces, the Jablonowski Society of Leipzig (founded in 1774 by Duke Jablonowski) had invited papers on minimal surfaces, and the prize was awarded to Scherk's paper[41]].

In his second paper, read to the Copenhagen Academy in 1833, Scherk admitted [p. 187] that, just as his predecessors, he was unable "to obtain a general construction of the [minimal] surface," and therefore he "proceeded to discover as many particular cases as possible by a uniform method." His method consisted in decomposing the minimal surface equation into two equations, "such that one of them could easily be integrated, and the arbitrary functions occurring in the [general] solution were then determined so that the other equation was satisfied." Thus, (8.1) resulted from this method by setting $s = 0$ (i.e., $z_{xy} = 0$), and so on. Scherk's surfaces attracted repeatedly the interest of other mathematicians; for some details, see [34]].

Scherk's paper of 1835 also contained [p. 190] the famous conjecture that "besides the right helicoid there does not exist a [real] minimal surface which can be generated by the motion of a straight line." This was proved in 1842 by E. Catalan [J. Math. Pures Appl. **7**, 203-211]. It is interesting that this theorem attracted considerable attention – perhaps in reaction to the relative stagnation of the general theory – and several different proofs were subsequently given, for instance, in 1844 by Björling (Sec. 7, l.c.), in 1846 by J.A. Serret [J. Math. Pures Appl. **11**, 451-457], and by M. Roberts [ibid., 300-312], and much later by E. Beltrami (1865), O. Bonnet (1885), and A. Cayley (1888).

Björlings paper seems to be of particular further interest, because in it the author utilized Legendre's method, but with different variables, which led him to the equation

$$w_{ab} + \frac{2a}{a^2 - b^2}w_a - \frac{2b}{a^2 - b^2}w_b = 0. \tag{8.6}$$

This equation is of *class P* in the sense of Bergman's theory; see[20]]. We shall investigate consequences of this at some other occasion.

Björling also solved, for the first time, the problem of determining a minimal surface passing through a given curve C and having along each point of C a given normal direction. This so-called *Björling's problem* attracted further interest, notably in 1860 that of O. Bonnet[5]], who observed that the solution of the problem also entails the determination of those minimal surfaces which contain a given curve as a geodesic, as an asymptotic curve, or as a line of curvature.

Another solution of Björling's problem was given by H.A. Schwarz [J. reine angew. Math. **80**, 280-300] in 1875.

9. Spherical Mapping

With the interest in minimal surfaces renewed by Scherk's and other results, an accelerated further evolution grew on the fertile soil of Gauss's "Disquisitiones", Weierstrassian rigor in general, the contemporary development in the calculus of variations, and the perfection of classical complex analysis in the hands of Riemann and Weierstrass.

In one[30]] of his important papers on geodesy, published in 1852, F. Minding (1806-1885) proposed that it seemed "natural" to use spherical mapping from Gauss's "Disquisitiones" (of which his papers showed a deep understanding) for transplanting the notions of meridians and parallels of a sphere onto other surfaces, and he proved that on a minimal surface these images form an orthogonal net.

That in this paper on geodesy, Minding discussed general surface theory had the reason that he wanted to keep the basic formulas of geodesy independent of special assumptions on the shape of the surface of the earth [p. 68], a very modern viewpoint, nowadays essential in connection with rockets and missiles.

The use of spherical mapping led to further important results on minimal surfaces, by O. Bonnet, announced in four Comptes Rendus notes of 1853-1856, and fully worked out in the great Memoir[5]] of 1860 already mentioned. By employing spherical mapping and the "Bonnet coordinates" $\xi = \phi$, $\eta = \log \tan \frac{1}{2}\theta$, where ϕ is the geographical longitude and θ the complement of the geographical latitude, Bonnet was able to give the general minimal surface formulas a form suitable for deriving *all* real minimal surfaces, for investigating common properties and for giving a new solution of Björling's problem. In particular, he discovered that in those coordinates the first fundamental form of a minimal surface became [with suitable $v = v(\xi, \eta)$ and $w = w(\xi, \eta)$]

$$ds^2 = (v^2 + w^2)\,(d\xi^2 + d\eta^2),$$

whereas the corresponding form for the unit sphere was

$$d\tilde{s}^2 = \frac{d\xi^2 + d\eta^2}{\cosh^2 \eta}.$$

This implied not only the orthogonality already obtained by Minding, but also the isothermality of those "meridians" and "parallels". In turn, it entailed the important result that for a minimal surface, the spherical mapping is *conformal.* This became really basic in 1867 when Christoffel[9] proved that conformality to be characteristic of minimal surfaces:

Theorem 9.1. *A surface is a minimal surface (or a sphere) if and only if its spherical mapping is conformal.*

We finally mention a further result on minimal surfaces based on spherical mapping, obtained in 1863 by J. Weingarten[50] in an investigation of "Weingarten surfaces" (W-surfaces), for which by definition one of the principal radii of curvature is a function of the other, $\rho' = \lambda(\rho)$ in his notation. For these surfaces, Weingarten gave a general representation by line integrals. From it, for $\rho + \rho' = 0$, he obtained another general representation of minimal surfaces; see[14],III,3,306.

10. Culmination Point: Enneper, Weierstrass, Riemann

We are now approaching the culmination point of the extensive efforts to find practically useful general representation formulas for minimal surfaces. This will concern the works of Enneper (1864), Weierstrass (1866), and Riemann (1860/61, posthumously published in 1868).

In his paper[13] of 1864, A. Enneper (1830-1885) assumed a representation with the lines of curvature as the coordinate curves on the surface and then obtained for a minimal surface the representation (using Gauss's notation p, q for the parameters)

$$\begin{aligned} x &= \int \frac{\phi^2(p) - 1}{4\phi'(p)} dp + \int \frac{\psi^2(q) - 1}{4\psi'(q)} dq \\ y &= i \int \frac{\phi^2(p) + 1}{4\phi'(p)} dp - i \int \frac{\psi^2(q) + 1}{4\psi'(q)} dq \\ z &= \int \frac{\phi(p)}{2\phi'(p)} dp + \int \frac{\psi(q)}{2\psi'(q)} dq. \end{aligned} \tag{10.1}$$

Here $\phi(p) = \Phi(p) + i\Psi(p)$, $\psi(q) = \Phi(q) - i\Psi(q)$, where $\Phi(p)$ and $\Psi(p)$ are real for real p, so that the lines of curvature are $p + q =$ *const* and $p - q =$ *const.*

In that paper, Enneper also included the discovery of a minimal surface, parametrically given by

$$\begin{aligned} x &= p + pq^2 - \frac{1}{3}p^3 \\ y &= -q - p^2 q + \frac{1}{3}q^3 \\ z &= p^2 - q^2 \end{aligned} \tag{10.2}$$

and named after him. For a graph and discussion of this surface, see[34],[76]. This surface has several interesting properties; for instance, its lines of curvature are plane.

In a later paper of 1869, Enneper [Zeitschr. Math. u. Phys. **14**, 393-421] investigated properties of minimal surfaces generated by circles, and over a decade later, he discovered another minimal surface, which has the property that a family of parabolas lies on its [Gött. Abh. **29** (1882), 41-50].

The highlight of this period occurred in 1866 when Weierstrass[48] communicated his famous formulas to the Berlin Academy. He began with his motivation, saying that "I have occupied myself with the theory of these ... minimal surfaces in more detail because ... they are intimately related with the theory of analytic functions of a complex argument."

Weierstrass was neither the only nor the first mathematician to consciously look for such a relationship, as we have seen, but at that time, complex analysis had finally been developed far enough that such a relationship could now be exploited most elegantly and completely, and this is precisely what Weierstrass did, before the publication of Riemann's equivalent formulas (see below). The advantage of his formulas [(10.5), below] over others was that his parameter of integration (10.4) had a geometric significance, as we shall see.

Weierstrass chose isothermic parameters p, q on a minimal surface (M), explicitly assumed twice continuous differentiability of x, y, z [p. 40], and set $p + iq = u$. Then

$$x_{pp} + x_{qq} = 0, \quad y_{pp} + y_{qq} = 0, \quad z_{pp} + z_{qq} = 0,$$

that is, x, y, z are harmonic, hence they are the real parts of analytic functions of u, which he denoted by f, g, h. Thus, whereas the theory of *pairs* of conjugate harmonic functions is the theory of *analytic* functions of a complex variable, the theory of *triples* of conjugate harmonic functions (x, y, z) of p, q satisfying $E = G$, $F = 0$ (the condition of p, q to be isothermic) is the theory of *minimal surfaces.* Weierstrass had no notation for E, F, G and denoted the mean curvature by K. From those two conditions and analyticity he concluded that

$$0 = E - 2iF - G = f'^2 + g'^2 + h'^2,$$

where E, F, G are defined as in (3.3), with $u = p$ and $v = q$. To satisfy this condition, he introduced analytic functions G and H of u and set

$$f' = G^2 - H^2, \quad g' = i(G^2 + H^2), \quad h' = 2GH.$$

By integration from a fixed point $(x_0, y_0, z_0) = (x(u^0)), y(u^0), z(u^0))$ he obtained a first triple of formulas (R = real part)

$$\begin{aligned} x &= x_0 + \mathrm{R}\int_{u^0} (G^2(u) - H^2(u))\, du \\ y &= y_0 + \mathrm{R}\int_{u^0} i(G^2(u) + H^2(u))\, du \\ z &= z_0 + \mathrm{R}\int_{u^0} 2G(u)H(u)\, du \end{aligned} \tag{10.3}$$

valid for a simply connected open minimal surface corresponding to the open unit disk. Weierstrass's next idea was the transition from u to

$$s = \frac{H(u)}{G(u)} = \frac{f' + ig'}{-h'}. \tag{10.4}$$

Geometrically, s is the image in the plane through the equator of the unit sphere under the stereographic projection (with the "North pole" as center) of the spherical image of a point of the minimal surface (M). Setting

$$G^2(u)\frac{du}{ds} = F(s),$$

Weierstrass obtained as a second form

$$\begin{aligned} dx &= \mathrm{R}\,(1 - s^2)F(s)\,ds \\ dy &= \mathrm{R}\,i(1 + s^2)F(s)\,ds \\ dz &= \mathrm{R}\,2sF(s)\,ds \end{aligned} \tag{10.5}$$

(usually written in terms of integrals; R = real part). In this way, Weierstrass associated with every analytic function F a unique real minimal surface (M) (except for its position in space), and conversely, with every real minimal surface (M) precisely two analytic functions

$$F(u) \quad \text{and} \quad -u^{-4}\overline{F(-1/\bar{u})},$$

where the bars denote conjugation.

For instance, we get from $F(s) = 1$ Enneper's surface (10.2), from $F(s) = 1/2s^2$ the catenoid $\rho = (x^2 + y^2)^{\frac{1}{2}} = \cosh z$, and from $F(s) = i/2s^2$ the right helicoid $x = y \tan z$.

By setting $\hat{F}''' = F$, Weierstrass obtained the integral-free form [p. 46]

$$\begin{aligned} x &= \mathrm{R}\{(1 - s^2)\hat{F}'' + 2s\hat{F}' - 2\hat{F}\} \\ y &= \mathrm{R}\{i(1 + s^2)\hat{F}'' - 2is\hat{F}' + 2i\hat{F}\} \\ z &= \mathrm{R}\{2s\hat{F}'' - 2\hat{F}'\}. \end{aligned} \tag{10.6}$$

Although these will usually be local representations only (as Weierstrass pointed out clearly), they are useful for general investigations, because due to analyticity many properties in the large can be studied locally.

If $\hat{F}$ is algebraic, so is (M), and Weierstrass also showed [p. 48] that the converse is true. In this way he was the first to give the means for obtaining all real algebraic minimal surfaces.

In conclusion we can state that the theory of minimal surfaces reached a culmination point in the work of Weierstrass, who also brought the basic results of Lagrange, Monge, and Riemann in a final and most satisfactory form, from the viewpoint of both theory and applications.

At that time the task of relating minimal surface theory to the most recent accomplishments of complex analysis was "in the air." This is distinctly shown by the fact that also Riemann concerned himself with this problem, as early as 1860-61, but his work[40] was published later than that of Weierstrass, namely, posthumously in 1868, so that it became less influential. Riemann took the unit sphere and projected it from $(-1,0,0)$ onto the tangent plane at $(1,0,0)$ with axes y', z' parallel to the y and z axes, respectively, and set $\eta = y' + iz'$. He represented the minimal surface by $x = x(y,z)$. Then $-Z\,dy + Y\,dz$ became the differential of a function $\xi = \xi(y,z)$. Here, (X,Y,Z) is the unit surface normal vector. He further showed that $x + i\xi$ depends only on η, and

$$u = \int \left(i\frac{dx}{d\log\eta} \right)^{\frac{1}{2}} d\log\eta$$

is invariant under rotations about the origin in xyz-space. Using this complex u, Riemann gained his representation

$$\begin{aligned} x &= \mathrm{R}\int -2i\left(\frac{du}{d\log\eta}\right)^2 d\log\eta \\ y &= \mathrm{R}\int -i\left(\frac{du}{d\log\eta}\right)^2 \left(\eta - \frac{1}{\eta}\right) d\log\eta \\ z &= \mathrm{R}\int -\left(\frac{du}{d\log\eta}\right)^2 \left(\eta + \frac{1}{\eta}\right) d\log\eta. \end{aligned} \tag{10.7}$$

It is interesting that after a reflection of the surface in the yz-plane these formulas are related to those of Weierstrass by $\eta = s$ and $F(s) = -(u')^2$, except for the notation of the axes.

Riemann's examples included as boundaries a (nonplane) quadrilateral and two polygons in parallel planes. So here we meet with solutions of *Plateau's problem*, to which we turn in the next section.

The inclusion of *complex* minimal surfaces in this whole circle of ideas was strongly and repeatedly advocated by S. Lie (1842-1899), for instance, in 1879 on p. 332 of his basic paper [Math. Ann. **14**, 331-416], in which he also introduced his complex minimal surface

$$2(x - iy)^3 - 6i(x - iy)z - 3(x + iy) = 0.$$

There would be much to say about this idea, which, however, we leave outside the scope of this paper.

11. Plateau's Problem

We understand the problem here in the classical sense, as indicated in Sec. 3, that is, the determination of a minimal surface S in E^3 homeomorphic to a disk and bounded by a given closed Jordan curve C.

Attempts to determine minimal surfaces containing given straight lines or curves reach back to the middle of the previous century (J. Serret [Comptes Rendus Paris **40** (1855), 1078]; O. Bonnet[5]], particularly p. 245, etc.). But Plateau's problem was rightly regarded as much more difficult. For instance, Weierstrass[48],219] fully recognized the inherent complications by saying about his formulas (10.3):

> "The determination of the functions $G(u)$, $H(u)$, appearing in these formulas, for a given boundary of the surface, is generally connected with unsurmountable difficulties. I therefore have restricted myself to investigate in more detail the case in which the boundary consists of straight-line segments..."

In particular, the case of a quadrangle consisting of four edges of a regular tetrahedron was solved and treated in great detail in 1867[44]] by H.A. Schwarz (1843-1921), who discovered the corresponding function $F(s)$ in the Weierstrass formulas (10.5) in the form [l.c., p. 32]

$$F(s) = \frac{2}{\sqrt{1 - 14s^4 + s^8}}. \tag{11.1}$$

With this famous result, which won him a prize of the Berlin Academy, Schwarz was the first to solve a nontrivial case of Plateau's problem, and his solution was *explicit*. (The latter also occurred in Riemann's posthumous paper just discussed.) Short thereafter, A. Schondorff[43]] solved a somewhat more general case. Weierstrass had to evaluate Schwarz's Preisschrift[44]], a duty that gave him the idea to his paper[49]] on periodic minimal surfaces. Moreover, Schwarz and his students, in particular E.R. Neovius, obtained various results that helped to pave the way of the further development.

In 1872, Schwarz[45]] generalized his method to "Schwarzian chains" [l.c., p. 130], in which the boundary consists of portions of polygons and "free boundary portions" in given planes, that is, curves along which the surface must be orthogonal to such a plane. Note that this case also yields a polygon of circular arcs as its spherical image; these are portions of great circles because the free boundary portion must be a geodesic line of curvature.

Weierstrass's paper[48]], from which we have cited, contains [pp. 219, 220] Weierstrass's ideas and program to Plateau's problem for piecewise straight boundaries. He worked this out for a simply connected surface with given polygonal boundary [Math. Werke III, 221-239; posthumously published in 1903], namely, he reduced this problem to that of the determination of two particular solutions of a homogeneous linear differential equation of second order with rational functions of u as coefficients, which are real for real u.

This idea was worked out further by G. Darboux (1842-1917) and included in his standard treatise[10],I,547–601. Darboux used Weierstrass's formulas (10.3) (with t instead of u) and wrote that differential equation in the form [p. 550]

$$\frac{d^2\theta}{dt^2} + P\frac{d\theta}{dt} + Q\theta = 0. \tag{11.2}$$

For given P and Q he called a *family* of minimal surfaces the surfaces obtainable for all pairs of independent particular solutions. For a given Schwarzian chain he determined P and Q (except for some numerical constants) by noting that the adjoint of a surface S belongs to the same family and has as its boundary a Schwarzian chain whose segments correspond to free boundary portions of S and whose free boundary portions correspond to segments of the boundary of S.

Darboux also investigated the inverse problem that a differential equation (11.2) – in particular, the hypergeometric equation [p. 573] – is given and one asks for compatible boundaries of portions of minimal surfaces.

12. Developments Since 1890

Weierstrassian rigor also exerted its effect on minimal surface theory, be it Plateau's problem or the problem of least area (cf. Sec. 3). With respect to area, one had believed that, by analogy with curves, one could define (and obtain) the area of a (portion of a) surface S as the least upper bound of the areas of a sequence of inscribed polyhedra with vertices on S. The naive belief that this bound will always be finite and have a reasonable value was shattered as late as in 1880 by H.A. Schwarz [Ges. Math. Abhandlungen, II, 309-311, 369-370], who gave as a counterexample a sequence of polyhedra consisting of triangles and inscribed in a circular cylinder. Depending on how we shrink the triangles to zero, for a cylinder of radius 1 and height 1 we can obtain the expected value 2π as the area, or larger values or even infinity; see also[8],24 or[22],75. (The same example was found independently in 1882 by Peano.) This surprising result sparked extensive activity, on which Radó[38],7 commented:

> "It is perhaps difficult to understand how mathematicians could ever have been surprised by an example as simple as that devised by H.A. Schwarz. In any case, an overwhelming number of definitions have been subsequently proposed for surface area, and [in 1948] the end is apparently not yet in sight."

The development that began in 1880 showed later that Lebesgue's definition in his thesis[24] of 1902 and Hausdorff's two-dimensional measure of point sets in

E^3 are particularly useful (and, for sufficiently regular surfaces, yield the same values).

In his thesis, Lebesgue [l.c., p. 342] spoke of a "preparation for Plateau's problem." This referred to the Dirichlet problem for the minimal surface equation – usually called the *nonparametric Plateau problem* (cf. Sec. 3). For a boundary curve with a convex orthogonal projection in the xy-plane and no osculating plane perpendicular to the xy-plane, Lebesgue proved the existence of a Lipschitz continuous function $z(x, y)$ on which the given boundary curve bounds a portion of minimal "area" (which in elementary cases agrees with generally accepted values of the area; see Chap. V.2 of Radó[38]). However, it remained open whether this function satisfies the minimal surface equation, equivalent to its being of class C^2.

Now just as Schwarz's above negative result entailed great general activity, so did Weierstrass's condemnation of the Dirichlet principle in 1870 [Math. Werke II, 49-54], resulting in the ingenious existence proofs for the Dirichlet problem for the Laplace equation by C. Neumann (1870), H.A. Schwarz (1870), and H. Poincaré (1890). With this and with Hilbert's Paris talk of 1900, in which Problem 19 addressed elliptic equations, sooner or later one could expect work on the nonparametric Plateau's problem.

In 1909, A. Korn[19] solved Plateau's problem for a curve C differing only slightly from a plane curve, a case that had been proposed as early as 1832 by Poisson [J. reine angew. Math. **8**, 362] (who never published a proof of his solution). In his solution, Korn assumed the projection of C in the xy-plane as well as the function $z = f(x, y)$ giving C in space to be of class $C^{2+\alpha}$ and the $(2 + \alpha)$-norm of f to be sufficiently small. Later work of Radó[36] and others established the same result, assuming only the continuity of f and the convexity of the domain in the xy-plane.

Similar in character was a paper of 1911 by Ch. Müntz[32], in which Müntz combined facts from potential theory, notably, sharp estimates not yet common at that time, with the idea of successive approximation, reminding the reader that the latter had been introduced into analysis in 1885 by H.A. Schwarz and since then popularized by Picard in papers between 1890 and 1902 on the solution of equations.

13. S. Bernstein's Work

At about the same time, between 1910 and 1916 appeared the work of S. Bernstein (1880-1968) on Plateau's problem, as a part of an entirely novel approach to partial differential equations, which gave the whole theory a new direction.

Indeed, Bernstein's famous thesis[1] of 1904 is often considered the beginning of the *modern* theory of partial differential equations. In it, Bernstein proved that C^3-solutions of an elliptic nonlinear analytic equation in two variables must be analytic. This was an answer to Hilbert's Problem 19 of the Paris talk in 1900, in which Hilbert explicitly mentioned the minimal surface equation, which is clearly covered by Bernstein's result.

The thesis also contained the idea and earliest technique of *a-priori estimates*, which was later developed into a powerful general method by J. Leray, J. Schauder, and many others.

In 1910 and 1912, in two deep papers, using his "normal series," Bernstein[2],[3] solved the nonparametric Plateau problem for a curve C satisfying the conditions mentioned in connection with Lebesgue's work in Sec. 12.

Claims by L. Lichtenstein[14],II,3.2,1326 and later by Ch. Müntz [Math. Ann. **94** (1925), 54] that Bernstein's proof contained serious gaps were refuted by Bernstein [ibid., **95** (1926), 585-594]. This interesting polemic simply seems to reveal two facts, the novelty of Bernstein's ideas and the depth and difficult style of his papers. Tibor Radó once told the author that he spent much time in reformulating Bernstein's papers to make them more transparent. In doing so, in 1927 [ibid., **96**, 587-596] he also pointed to several serious gaps in Müntz's above paper of 1925, thereby developing an interesting geometric convergence theory for families of surfaces. In a subsequent answer [ibid., p. 600], Müntz acknowledged the validity of Bernstein's solution of Plateau's problem.

Bernstein[3] also was the first to discuss degrees of nonlinearity of elliptic equations, for which he introduced a notion, which he called *genre*, such that linear and uniformly elliptic equations are of genre 0, and the minimal surface equation is of genre 2; and he proved that with respect to the solvability of the Dirichlet problem, all equations of genre 1 or less behave similar to linear equations, whereas for genre greater than 1, one must expect a different behavior.

As a landmark, in 1916, Bernstein[4] published his famous

Theorem 13.1. *A twice continuously differentiable minimal surface* S : $z = z(x, y)$ *which is defined on the whole* xy*-plane must be a plane.*

Obviously, this is an analog of Liouville's theorem on analytic (hence harmonic) functions, but Bernstein emphasized the basic difference that "whereas the variety of entire (rational or transcendental) harmonic functions is so large, there is only a single entire function which satisfies the equation of minimal surfaces..., namely, the linear function $z = ax + by + c$."

Bernstein's theorem also illustrated that, whereas for harmonic functions the general behavior is typical of solutions of more general *linear* elliptic equations, this no longer holds in the *nonlinear* elliptic case. Furthermore, other differences appear in connection with the following theorem by Bernstein[4] (cf. also [Math. Zeitschr. **26** (1927) 551]), from which Bernstein derived his Theorem 13.1.

Theorem 13.2. *If* $z(x, y)$ *is a bounded twice continuously differentiable solution of*

$$Az_{xx} + 2Bz_{xy} + Cz_{yy} = 0$$

for all values of x, y, *where* A, B, C *are finite functions of* x, y, z, z_x, z_y, z_{xx}, z_{xy}, z_{yy} *with* $AC - B^2 > 0$, *then* $z(x, y)$ *is a constant.*

In contrast to Liouville's theorem, this theorem does not generalize to more than two independent variables. Furthermore, a theorem by A. Harnack states that a harmonic function which is positive on the whole plane must be a constant. In contrast, in Theorem 13.2 we cannot replace "bounded" by "everywhere positive." A counterexample is the everywhere positive solution

$$z(x, y) = \exp(x - e^y)$$

of the equation

$$(1 + e^y)z_{xx} + 2z_{xy} + e^{-y}z_{yy} = 0,$$

given by E. Hopf [Math. Zeitschr. **29** (1929), 744-745].

A simplified proof of Bernstein's Theorem 13.1 was published by Radó in 1927 [ibid. **26**, 559-565].

14. Haar's Variational Approach

The calculus of variations contributed to minimal surface theory, as we have seen. Conversely, minimal surface theory contributed to the calculus of variations in the work by H.A. Schwarz of 1872 [Ges. Math. Abhandlungen, I, 151-167] and its path-breaking extension of 1885 [ibid., I, 223-269], where Schwarz gave the first sufficient condition for a *double* integral to be extremum.

Schwarz's method of 1885 was extended to more general equations in 1917 and 1919 by L. Lichtenstein[27], who, in connection with the minimal surface equation and more general equations, observed the occurrence of "functional branchings", as he called the fact that given boundary values may perhaps correspond to two extremal surfaces in a neighborhood of a surface, instead of just one. For nonlinear *integral* equations, a similar phenomenon had been predicted by E. Schmidt in his classical work of 1908 [Math. Ann. **65**, 370-399], but Lichtenstein was unable to reduce his problems to such integral equations.

On p. 6 of his first paper, Lichtenstein mentioned that his investigations resulted from discussions with A. Haar (1885-1933) in Göttingen in the summer of 1914. Haar told him that he had another method of proving the existence of a field (in the sense of the calculus of variations), thus encouraging Lichtenstein to undertake his work[27].

Haar himself delayed publication of his investigations – perhaps due to his move from Klausenburg to Szeged in 1918 when Siebenbürgen became part of Romania, and due to the hard initial years at the new University of Szeged, where F. Riesz was Haar's colleague.

Haar published his paper[18] on Plateau's problem in 1927, thanking [p. 128] T. Radó for his general support and for a particularly important proof. This work was preceded by two papers on the variation of double integrals [J. reine angew. Math. **149** (1919), 1-18; Acta Szeged **3** (1927), 224-234], which contain [p. 2] the following basic lemma aimed at the relaxation of differentiability assumptions for comparison functions in the calculus of variations, a generalization of the classical Du Bois-Reymond lemma of 1879 (for motivations, see, e.g., R. Courant and D. Hilbert, *Methods of Mathematical Physics*, I, 199-204).

Haar's Lemma. *Let $u(x,y)$ and $v(x,y)$ be continuous in a region B. Suppose that*

$$\iint_B (u\zeta_x + v\zeta_y)\, dx\, dy = 0$$

for all C^1-functions $\zeta(x,y)$ that vanish on the entire boundary of B. Then for any closed curve C in the interior of B,

$$\int_C (u\, dy - v\, dx) = 0,$$

that is, there exists a class C^1-function $\omega(x,y)$ in the interior of B such that $\omega_x = -v$ and $\omega_y = u$.

Whereas the papers on Plateau's problem by Korn, Lichtenstein, and others used methods of successive approximation, Haar, a Ph.D. student of Hilbert, employed the direct method of the calculus of variations. Created by Hilbert in 1900, this method had to be developed further because a minimal sequence $\{z_n(x,y)\}$, needed in the construction of an extremal function $z(x,y)$ for a regular problem

$$\iint F(z_x, z_y)\, dx\, dy = \text{ Min},$$

which must be such that

$$\lim_{n\to\infty} \iint F(z_{n,x}, z_{n,y})\, dx\, dy$$

equals the infimum of the integral, need not converge. The first idea, due to B. Levi and to H. Lebesgue, consisted in the replacement of a minimal sequence by a special one whose *uniform* convergence could be proved. The second idea, due to Fubini, was to abandon uniform convergence and to prove existence by using a minimal sequence which merely converges *almost everywhere.* In his basic work of 1908 [Annali di Mat. (3) **15**, 125], Fubini pointed out that the real difficulty of the method is not the construction of an extremal function, but the proof that it has derivatives. Using this improved direct method and his lemma of 1919 as well as results by H. Rademacher [Math. Ann. **79** (1918-20), 340-359; **81** (1920), 52-63] on the differentiability of functions of two variables, Haar solved Plateau's problem for a curve satisfying the same conditions as in Lebesgue's and Bernstein's work. He finally showed analyticity by a method of Radó [Math. Zeitschr. **24** (1926), 321-327], who, by applying classical theorems of Riemann and Weierstrass, proved that a class C^1 extremal function must be analytic. It is interesting that Radó in turn used results in Haar's paper of 1919, which made Radó's further reasoning elementary, namely, due to the fact that certain theorems in the theory of minimal surfaces could be proved without assuming the existence of second derivatives.

15. On Tibor Radó's Solution of Plateau's Problem

In the previous section we have considered some of Radó's relevant work and influence on the development until 1927, a period of collaboration with Haar, who was ten years older than Radó (1895-1965).

To attack the general Plateau's problem would have been a hopeless task at Schwarz's times. In view of the complicated form a Jordan curve may possess, we have seen that it was necessary to abandon the idea of the explicit determination of solutions and concentrate on the problem of their *existence*. Furthermore, the papers just discussed already reflected the experience gained by the solution of the uniformization problem of complex analysis by Poincaré and Koebe. In addition, progress in functional analysis (by Hadamard, Fréchet, F. Riesz, Hahn, and others) gradually led to a deeper conceptual understanding of functionals in general.

A first *general* solution of Plateau's problem was published in 1928 by R. Garnier in a very extensive paper[16]. Garnier based his method on G.D. Birkhoff's study of singular points of ordinary linear differential equations of 1909 [Trans. Amer. Math. Soc. **10**, 436-470] and on Birkhoff's theory of difference equations [ibid., **12**, (1911), 243-284] as well as on his own previous solution of Riemann's problem for second-order linear differential systems published in an extensive paper of 1926 [Ann. École Norm. Sup. (3) **43**, 177-307]. Garnier succeeded in determining G and H in Weierstrass's formulas (10.3) for a simple closed polygon, where the singular behavior at the points corresponding to corners of the boundary was determined by the geometric nature of the Schwarzian chain. This amounted to a solution by solving the monodromic group problem of Riemann under extremely difficult additional conditions. Garnier then approximated a simple closed curve by inscribed polygons P_n and let $n \to \infty$, accompanied by a complicated investigation of the solution of a monodromic group problem with variable initial conditions [l.c., pp. 116-144]. In this way he was able to extend his solution of Plateau's problem from polygons to unknotted Jordan curves consisting of finitely many arcs of bounded curvature.

Whereas the classical literature contains several *particular* cases, as we have seen, Garnier's result is the first *general* solution, and it is essential that the Jordan curve must satisfy the conditions just stated, but no further conditions – in contrast, for instance, to Bernstein's solution of 1912 or Haar's solution of 1927.

Garnier's method was complicated and left the impression that there should be a simpler way of solving the problem under comparable generality. That this is the case was shown in publications of 1930 and 1931 independently by T. Radó[36] and J. Douglas[11], respectively.

An immediate simplification of Garnier's work was suggested by T. Radó[35] in the following form.

Radó's Approximation Theorem. *Suppose that Plateau's problem is solvable for each simple closed curve C_n of a family $\{C_n\}$ whose lengths form a uniformly bounded sequence, and the sequence $\{C_n\}$ converges in the sense of Fréchet to a simple closed curve C. Then Plateau's problem is solvable for C.*

The proof is straightforward; it uses Helly's selection theorem and a theorem of F. Riesz [Math. Zeitschr. **8** (1923), 87-95] on the boundary values of analytic functions. The paper also contains a uniqueness theorem (whereas no uniqueness statements were possible in Garnier's original method).

This work on Garnier's paper of 1928 provided Radó with one of the main ideas of his own method of solution of Plateau's problem (and the problem of least area; cf. Sec. 3) in Radó's paper[36], namely, that of approximating the given curve by polygons and solving the problem first for the latter. Radó based his method essentially on the classical area integral and on conformal mapping. Inasmuch it was, just like Garnier's work, an extension of the classical approach. However, it differed in the important aspect from Garnier's method that it was independent of the monodromic group problem.

Another main idea was that Radó first gave a solution of a properly chosen "approximate problem," which led to a "polyhedron" bounded by a polygon close to the given boundary curve; for details of this process we must refer to Radó's paper of 1930 [Annals of Math. (2) **31**, 457-469], preceding the actual solution[36] of the same year. In this paper, Radó also observed that his limit process in[35] even works (with proper modification) if instead of solutions for polygons he took *approximate* solutions for polygons. This then gave a complete existence proof, first for the case of a rectifiable Jordan curve and then without difficulty for any Jordan curve which bounded at least one continuous surface (Sec. 3) of finite area.

Radó[36],769 also regarded his method as "a generalization of methods used in the theory of conformal mapping of plane regions, and in particular, of theorems of Carathéodory on the conformal mappings of variable plane regions." He stated his main results in this form[36],791 (see our Sec. 3 for terminology):

> "For any Jordan curve Γ^* in xyz-space, the problem of the least area has infinitely many solutions. Among these solutions, there is always an analytic surface, and if Γ^* bounds at least one continuous surface with finite area, then there is even a minimal surface solving the problem of the least area. For the problem of Plateau itself there follows the existence of the solution for any Jordan curve which bounds at least one continuous surface with a finite area."

The first sentence needs perhaps an explanation. For this, we give an example[24],354 illustrating that to a continuous surface that solves the problem of least area, we can always add a point set without increasing the area or destroying the continuity. Indeed, the unit disk

$$x = u, \quad y = v, \quad z = 0, \quad u^2 + v^2 \leq 1$$

is a first solution of the problem of least area for the unit circle, and a second solution is the continuous surface consisting of that disk and the spine $[0, \frac{1}{2}]$ on the z-axis and represented by

$$x = (2r - 1)\cos\theta, \quad y = (2r - 1)\sin\theta, \quad z = 0$$

if $\frac{1}{2} \leq r \leq 1$, and

$$x = 0, \qquad y = 0, \qquad z = \tfrac{1}{2} - r$$

if $0 \leq r \leq \frac{1}{2}$; here r, θ are the usual polar coordinates defined by $u = r\cos\theta$ and $v = r\sin\theta$.

16. Conclusions

The highlights of the evolution of minimal surface theory considered in this paper show that these surfaces attracted attention for several reasons, first from the viewpoint of the calculus of variations, next for geometric reasons of the theory of curves and surfaces, thereafter from the standpoint of developing complex analysis, and finally, during our century, also in connection with the theory of elliptic partial differential equations. We also have seen that the development in time was far from uniform. Periods of no essential progress alternated with times of successful activity. Weierstrassian rigor, Lebesgue's theory, and new methods in the calculus of variations finally provided the foundations for the culmination on Plateau's problem in the works of Garnier, Radó, Douglas, and McShane within the short time interval of roughly six years.

As indicated at the beginning, this paper was mainly concerned with the roots and development of the ideas of T. Radó, leaving completely aside the works by Douglas and McShane. This task certainly required the investigation of the period from 1860 to 1930, beginning with Riemann's and Weierstrass's works. Before that, we first considered some of the most basic accomplishments prior to that period. We did this in order to show the birth and early growth of the most essential ideas on minimal surfaces, as well as the motivations and origins of much of the work between 1860 and 1930, our central concern.

References

1. S. Bernstein, *Sur la nature analytique des solutions des équations aux dérivées partielles du second ordre*, Math. Ann. **59** (1904) 20-76.
2. S. Bernstein, *Sur les surfaces définies au moyen de leur courbure moyenne ou totale*, Ann. Éc. Norm. Sup. (3) **27** (1910) 233-256.

3. S. **Bernstein**, *Sur les équations du calcul des variations*, Ann. Éc. Norm. Sup. (3) **29** (1912) 431-485.
4. S. Bernstein, *Sur un théorème de géométrie et son application aux équations aux dérivées partielles du type elliptique*, Comm. Soc. Math. Kharkov **15** (1915-17) 431- 485.
5. O. Bonnet, *Mémoire sur l'emploi d'un nouveau système de variables dans l'étude des propriétés des surfaces courbes*, J. Math. Pures Appl. (2) **5** (1860) 153-266.
6. M. Cantor, *Vorlesungen über Geschichte der Mathematik*, 4 vols., Leipzig, Teubner (1907-1908).
7. C. Carathéodory, *Bericht über die Verleihung der Fieldsmedaillen*, Comptes Rendus du Congr. Int. Math. Oslo 1936, vol. I, 308-314, Oslo, Brøggers (1937).
8. L. Cesari, *Surface Area*, Princeton, N.J., Princeton University Press (1956).
9. E.B. Christoffel, *Ueber einige allgemeine Eigenschaften der Minimumsflächen*, J. reine angew. Math. **67** (1867) 218-228.
10. G. Darboux, *Leçons sur la théorie générale des surfaces et les applications géométriques du calcul infinitésimal*, 4 vols., Paris, Gauthier-Villars (1888-1896).
11. J. Douglas, *Solution of the problem of Plateau*, Trans. Amer. Math. Soc. **33** (1931) 263-321.
12. L.P. Eisenhart, *A Treatise on the Differential Geometry of Curves and Surfaces*, New York, Dover (1960; original edition 1909).
13. A. Enneper, *Analytisch-geometrische Untersuchungen*, Zeitschr. Math. u. Phys. **9** (1864) 96-125.
14. E.M.W., *Encyklopädie der mathematischen Wissenschaften*, Leipzig, Teubner (1898-1921).
15. L. Euler, *Methodus inveniendi lineas curvas maximi minimive proprietate gaudentes sive solutio problematis isoperimetrici latissimo sensu accepti*, Opera omnia (1) **24**, Zürich, Orell Füssli (1952).
16. R. Garnier, *Le problème de Plateau*, Ann. Éc. Norm. Sup. (3) **45** (1928) 53-144.
17. C.F. Gauss, *Disquisitiones generales circa superficies curvas*, Comment. Soc. Gott. **6** (1828), Math. C1. 99-146.
18. A. Haar, *Über das Plateausche Problem*, Math. Ann. **97** (1927) 124-158.
19. A. Korn, *Über Minimalflächen, deren Randkurven wenig von ebenen Kurven abweichen*, Abh. Preuss. Akad. Wiss. Berlin, Phys.-Math. C1. II (1909) 1-37.
20. M. Kracht and E. Kreyszig, *Methods of Complex Analysis in Partial Differential Equations with Applications*, New York, Wiley (1988).
21. M. Kracht and E. Kreyszig, *E.W. von Tschirnhaus: His role in early Calculus and his work and impact on Algebra*, in print.
22. E. Kreyszig, *Introduction to Differential Geometry and Riemannian Geometry*, Toronto, University of Toronto Press (1975).

23. J.L. Lagrange, *Essai d'une nouvelle méthode pour déterminer les maxima et les minima des formules intégrales indéfinies*, Miscellanea Taurinensia **2** (1760-1761) 173-195. Oeuvres I, 335-362.
24. H. Lebesgue, *Intégrale, longueur, aire*, Annali di Mat. (3) **7** (1902) 231-359.
25. H. Lebesgue, *Sur le problème de Dirichlet*, Rend. Circ. Math. Palermo **24** (1907) 371-402.
26. A.M. Legendre, *Mémoire sur l'intégration de quelques équations aux différences partielles*, Mém. Acad. Paris (1789) 309-351.
27. L. Lichtenstein, *Untersuchungen über zweidimensionale reguläre Variationsprobleme I, II*, Monatsh. Math. Phys. **28** (1917) 3-51, Math. Zeitschr. **5** (1919) 26-51.
28. E.J. McShane, *Parametrizations of saddle surfaces, with application to the problem of Plateau*, Trans. Amer. Math. Soc. **35** (1933) 716-733.
29. J. Meusnier, *Mémoire sur la courbure des surfaces*, Mém. Acad. Paris (1785) 477-510.
30. F. Minding, *Über einige Grundformeln der Geodäsie*, J. reine angew. Math. **44** (1852) 66-72.
31. G. Monge, *Mémoire sur le calcul intégral des équations aux différences partielles*, Mém. Acad. Paris (1787) 118-192.
32. Ch. Müntz, *Zum Randwertproblem der partiellen Differentialgleichung der Minimalflächen*, J. reine angew. Math. **139** (1911) 52-79.
33. J.C.C. Nitsche, *On new results in the theory of minimal surfaces*, Bull. Amer. Math. Soc. **71** (1965) 195-270.
34. J.C.C. Nitsche, *Vorlesungen über Minimalflächen*, Berlin, Springer (1975).
35. T. Radó, *Some remarks on the problem of Plateau*, Proc. Nat. Acad. Sci. USA **16** (1930) 242-248.
36. T.Radó, *The problem of the least area and the problem of Plateau*, Math. Zeitschr. **32** (1930) 763-796.
37. T.Radó, *On the Problem of Plateau*, Berlin, Springer (1933).
38. T.Radó, *Length and Area*, New York, American Mathematical Society (1948).
39. K. Reich, *Die Geschichte der Differentialgeometrie von Gauss bis Riemann (1828-1868)*, Archive Hist. Exact Sci. **11** (1973) 273-382.
40. B. Riemann, *Ueber die Fläche vom kleinsten Inhalt bei gegebener Begrenzung*, Manuscript worked out in 1866 by K. Hattendorff, Ges. Math. Werke, 2nd ed., 301-337, New York, Dover (1953, original edition 1892 & 1902).
41. H.F. Scherk, *De proprietatibus superficiei, quae hac continetur aequatione: $(1+q^2)r - 2pqs + (1+p^2)t = 0$ disquisitiones analyticae*, Acta Soc. Jab. Nova **4** (1832) 203-280.
42. H.F. Scherk, *Bemerkungen über die kleinste Fläche innerhalb gegebener Grenzen*, J. reine angew. Math. **13** (1835) 185-208.
43. A. Schondorff, *Über die Minimalfläche, die von einem doppeltgleichschenkligen räumlichen Viereck begrenzt wird*, Göttingen, Kaestner (1868).
44. H.A. Schwarz, *Bestimmung einer speciellen Minimalfläche*, Preisschrift der Preuss. Akad. Wiss. Berlin 1867, Ges. Math. Abh. **1**, 6-125, New York,

Dover (1972, original edition 1890).

45. H.A. Schwarz, *Fortgesetzte Untersuchungen über specielle Minimalflächen*, Monatsber. Preuss. Akad. Wiss. Berlin (1872) 3-27. Ges. Math. Abh. **1**, 126-148, New York, Dover (1972).
46. M. Spivak, *A Comprehensive Introduction to Differential Geometry*, 5 vols., Berkeley, Publish or Perish (1979).
47. D.J. Struik, *Outline of a history of differential geometry*, Isis **19** (1933) 92-120, **20** (1933) 161-191.
48. K. Weierstrass, *Untersuchungen über die Flächen, deren mittlere Krümmung überall gleich null ist*, Read 1866, Preuss. Akad. Wiss. Berlin, Math. Werke III, 39-52, 219-220.
49. K. Weierstrass, *Über eine besondere Gattung von Minimalflächen*, Monatsber. Preuss. Akad. Wiss. Berlin (1867) 511-518.
50. J. Weingarten, *Ueber die Oberflächen, für welche einer der beiden Hauptkrümmungshalbmesser eine Function des anderen ist.* J. reine angew. Math. **62** (1863) 160-173.

The Problem of Plateau (pp. 165–180)
ed. Th. M. Rassias

AREA-MINIMIZING m-TUPLES OF k-PLANES

GARY LAWLOR

Department of Mathematics, Princeton University
Fine Hall, Washington Road
Princeton, NJ 08544

ABSTRACT

We prove a sufficient criterion for a collection of k-planes passing through a common point to be area-minimizing. The criterion is similar to, but not as sharp as, the "angle criterion," a necessary and sufficient condition for a pair of oriented k-planes to be area-minimizing. The main results are contained in Theorem 8.3 and Corollaries 8.4 and 8.5.

1. Preliminaries

Can a surface cross through itself and still be area-minimizing?

A compact, k dimensional surface with boundary is called "area-minimizing" if no surface having the same boundary has less area. If a surface S is not compact (such as a k-plane), then we intersect it with an arbitrary closed ball in $\mathbf{R}^n$ and ask whether the resulting compact portion of S is area-minimizing. If all such compact portions of S are minimizing, we say that S itself is area-minimizing.

A "comparison surface" for S is a surface with the same boundary as S. If S is noncompact, a comparison surface is required to have the same boundary as some compact portion of S.

If an immersed k dimensional manifold M crosses through itself at a point p, then there are two (or more) k-planes tangent to M at p. If M is to be area-minimizing, then that union of k-planes must also be minimizing.

Thus, we would like to know exactly which pairs, and which m-tuples, of k-planes through a common vertex are area-minimizing.

If two intersecting k-planes are oriented and we allow only oriented comparison surfaces (requiring *oriented* boundaries to match) then the answer is known. Frank Morgan conjectured that a pair of oriented k-planes is area-minimizing if and only if a simple condition holds for the "characterizing angles" between the planes. The conjecture was proved by complementary work of Dana Mackenzie[1] and the author[2]. That theorem will be stated in Section 3, after we define the characterizing angles.

There are two related questions which are still open. First, if we do not regard orientation, then we are allowing more comparison surfaces, mainly because the orientation is not required to match at the boundary as before. Thus, some minimizing pairs of oriented k-planes are not minimizing when we drop the orientation. A necessary condition for the pair to be minimizing modulo 2 (i.e., without regard to orientation) is that they be area-minimizing with any orientation that is given them. This will also be stated in Section 3 in terms of the characterizing angles. It has been conjectured that this is also a sufficient condition. Frank Morgan[3] proved the conjecture for $k = 2$; see also Morgan[4]. In this paper, we will derive a sufficient condition for all k which is not as sharp, but closely resembles the conjecture.

The second open question is which m-tuples of k-planes through a vertex are area-minimizing, with or without regard to orientation. It is necessary that they be pairwise area-minimizing. I do not believe that this is sufficient, but no counterexamples are known. In this paper, we will give a sufficient condition for m-tuples of k-planes to be minimizing. See Lawlor[2], Section 1, for further references.

2. Overview of method

We will seek to define a retraction Π from a subset $U \subset \mathbf{R}^n$ onto the union of k-planes, which decreases the k-dimensional area of any piece of surface lying in U. Let D be a compact portion of the union of k-planes; then since Π decreases area we can prove that D is area-minimizing as follows:

(1) Let S be any other surface with the same boundary as D. Let S_0 be the intersection of S with the domain of Π.

(2) Because Π is continuous we can show that $\Pi(S_0)$ covers D. Thus,

$$\text{Area}(S) \geq \text{Area}(S_0) \geq \text{Area}(\Pi(S_0)) \geq \text{Area}(D).$$

The main steps involved in finding the retraction are the following:

(1) Decide how to cover a single k-plane P with a family of paths ("base paths") which intersect only at their starting points, and extend to infinity. The base paths determine a family of retractions onto P.

(2) Compute how fast the base paths are diverging from each other (the "spread").

(3) Use the spread as a parameter in a certain P.D.E. in two variables. There is a separate P.D.E. for each base path; to simplify the solution we are able to reduce it to an O.D.E. for each path.

(4) The solution to step (3) determines an area-decreasing retraction onto P, selected from among the family of retractions mentioned in step (1) above. This proves that the k-plane P is area-minimizing (which we already knew). To prove that an m-tuple of k-planes is area-minimizing, we wish to construct a retraction

onto the union. We do this by constructing (if we can) the retractions onto each k-plane separately, in such a way that the domains of the retractions do not intersect.

We will not go into all of the details of computation in this paper, but at this point will give an indication of how the retraction is formed. Think of a retraction from $\mathbf{R}^n$ to P geometrically in terms of the $n-k$ dimensional "surfaces of retraction" which are inverse images of single points of P. We wish to foliate a subset of $\mathbf{R}^n$ with $n-k$ dimensional surfaces S_p, each of which intersects P orthogonally at a single point p, such that the intersection points cover $\mathcal{H}^k$ almost all of P. Then our retraction will be defined by sending the surfaces of retraction down to their intersection points with P. We restrict our attention to retractions whose surfaces of retraction cast one dimensional shadows into the base paths, where the shadow of a point z is defined as the nearest point on P. In contrast, for a general retraction, the shadows would be higher dimensional (if $k < n-1$) and would not line up, but rather overlap and cross each other.

The solution $G(u,v)$ to the differential equation prescribes the rate of growth of the distance of S_p from P as we move out along a base path.

3. Angles between k-planes

In this section we define the space between intersecting k-planes from two different points of view. The first is useful in defining the relationship between a pair of k-planes; the second helps us to generalize to an m-tuple of k-planes through a common point.

For any pair of intersecting k-planes P and Q, we can describe the geometric relationship between them in terms of k "characterizing angles;" see Lawlor[2], Section 4, Morgan[5] Lemma 1, or Harvey and Lawson[6], p. 75. For each i, the characterizing angle β_i is the angle between a vector e_i in P and a vector f_i in Q. The sets $\{e_i\}$ and $\{f_i\}$ are orthonormal bases, chosen in such a way that

(1) β_1 is as small as possible, and

(2) β_2 is as small as possible given that we must satisfy $e_2 \perp e_1$ and $f_2 \perp f_1$, and

(3) β_3 is as small as possible having already fixed e_1, e_2, f_1, and f_2, et cetera. Notice that

$$\beta_1 \le \beta_2 \le \cdots \le \beta_k, \quad \text{and} \quad \beta_{k-1} \le \pi - \beta_k.$$

When P and Q are oriented planes, the resulting orthonormal bases $\{e_i\}$ and $\{f_i\}$ are required to agree with the orientations of P and Q. This will mean that once we have chosen e_1, f_1 through e_{k-1}, f_{k-1}, our choice of e_k and f_k is already determined. For that reason, β_k may be greater than $\pi/2$. If we are not regarding orientation, then we do not make this restriction. In that case, we always have $\beta_k \le \pi/2$, and call the set of angles the "unoriented" characterizing angles (which will differ from the oriented characterizing angles only possibly in the last angle of the set). When necessary for clarity, we will call the unoriented characterizing angles β_i^U.

3.1. Theorem (Mackenzie[1], Lawlor[2]): The pair of oriented k-planes P and Q, with oriented characterizing angles β_i between them, is area-minimizing if and only if

$$\beta_1 + \cdots + \beta_{k-1} \geq \beta_k.$$

As mentioned in Section 1, if we are not regarding orientation of the k-planes and the comparison surfaces, then in order for the pair to be minimizing, it must at least be (oriented) area-minimizing with respect to any orientations we could put on P and Q. Let us translate this requirement into a statement about the unoriented characterizing angles. If

$$\beta_1^U + \cdots + \beta_k^U \geq \pi$$

then since

$$\beta_k^U \leq \pi - \beta_k^U,$$

we also have

$$\beta_1^U + \cdots + \beta_{k-1}^U \geq \beta_k^U.$$

Then whether $\beta_k^U = \beta_k$ or $\beta_k^U = \pi - \beta_k$, the above inequalities imply the angle criterion stated in the theorem. Thus, the following conjecture is at least a necessary condition.

3.2. Conjecture: The pair of k-planes P, Q is area-minimizing mod 2 if and only if

$$\beta_1^U + \cdots + \beta_k^U \geq \pi.$$

In this paper, we will prove a weaker form of the unproved half of the conjecture, namely that the pair is minimizing modulo 2 as long as

$$\beta_1^U + \cdots + \beta_k^U \geq 2\pi.$$

We now look at the angles between planes from a different point of view. Instead of describing the separation between two planes, we will define a cushion about a single plane, and see how this relates to the characterizing angles. The cushion will allow us to generalize to a theorem about more than two planes intersecting at a point. We stated in Section 2 that we want to construct an area-decreasing retraction onto each of the planes, in such a way that the domains of the retractions do not intersect. The purpose of the cushion about each plane is that it will contain the domain of the retraction onto that plane. We will determine how wide a cushion must be to contain the domain of one of our area-decreasing retractions, then state the theorems in Section 8 in terms of k-planes surrounded by nonintersecting, sufficiently wide cushions.

A "cushion" is a cone-shaped, open n-dimensional set around a k-plane P, defined as follows. Choose a k-plane T intersecting P. The associated cushion about P is the union of open $(n-k)$-disks normal to P, whose radius is as large as possible so that the disks do not intersect T.

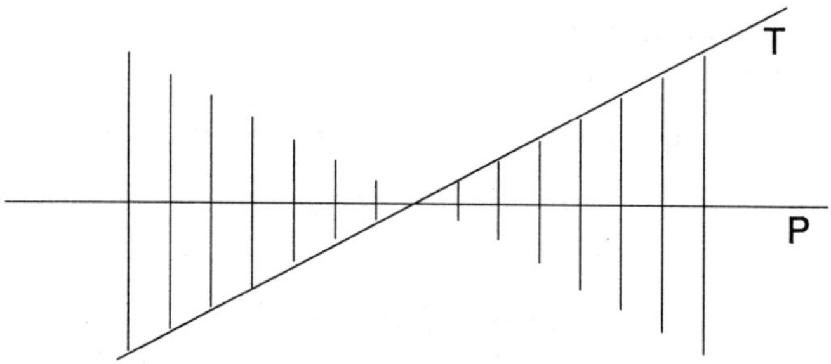

Figure 1. A cushion about P.

3.3. Definition: Let P and T be k-planes in $\mathbf{R}^n$, intersecting at the origin, with nonzero characterizing angles $\theta_1, \ldots, \theta_k$ between them. Since the angles are nonzero, the planes only intersect at the origin, implying that $n \geq 2k$. Let $\Theta = (\theta_1, \ldots, \theta_k)$. We define the "$\Theta$-cushion" about p (with respect to $\{e_i\}$) to be the union of open $n-k$ disks centered on P and normal to P, which are as large as possible without intersecting T. The origin is the only point of P not contained in one of the disks.

3.4. Lemma: In Definition 3.3, the disk centered at the point $p = x_1 e_1 + \cdots + x_k e_k$ has radius

$$r(p) = \left(\sum x_i^2 \tan^2 \theta_i\right)^{1/2}.$$

Proof: Let y_i be the coordinate functions in $\mathbf{R}^n$. It is convenient to assume that P is the horizontal k-plane given by $y_{k+1} = \cdots = y_n = 0$, and that T is spanned by the vectors

$$\cos(\theta_i)\frac{\partial}{\partial y_i} + \sin(\theta_i)\frac{\partial}{\partial y_{k+i}}.$$

Then the nearest (in fact, only) point of intersection of T with the normal space to P at p is

$$(x_1, \ldots, x_k, x_1 \tan(\theta_1), \ldots, x_k \tan(\theta_k), 0, \ldots, 0),$$

which is a distance $\left(\sum x_i^2 \tan^2 \theta_i\right)^{1/2}$ from p.

3.5. Lemma: Let P and Q be two k-planes separated by characterizing angles $2\theta_1, \ldots, 2\theta_k$ with $0 < \theta_i \leq \pi/4$. Then there is a Θ-cushion about P and a Θ-cushion about Q which are disjoint.

Proof: Let $\{e_i\}$ and $\{f_i\}$ be the orthonormal bases for P and Q prescribed by the characterizing angles construction, so that the angle between e_i and f_i is $2\theta_i$. We want to show that the Θ-cushions about P and Q with respect to these bases are disjoint.

Since $\theta_i > 0$, it must be that $n \geq 2k$. By projecting onto the $2k$-plane spanned by the bases of P and Q, we may as well assume that $n = 2k$.

Let $z \in \mathbf{R}^{2k}$. Suppose that the nearest point to z in P is $x = \sum x_i e_i$, and the nearest point to z in Q is $y = \sum y_i f_i$. We want to show that it cannot be true that both

$$\operatorname{dist}(z, x) < \left(\sum x_i^2 \tan^2 \theta_i\right)^{1/2}$$

and

$$\operatorname{dist}(z, y) < \left(\sum y_i^2 \tan^2 \theta_i\right)^{1/2}.$$

It suffices to show that

$$\operatorname{dist}^2(z, x) + \operatorname{dist}^2(z, y) \geq \sum (x_i^2 + y_i^2) \tan^2 \theta_i.$$

Now for each i let E_i and F_i be unit vectors in the plane spanned by e_i and f_i, such that

$$E_i \perp e_i \qquad \text{and} \qquad F_i \perp f_i.$$

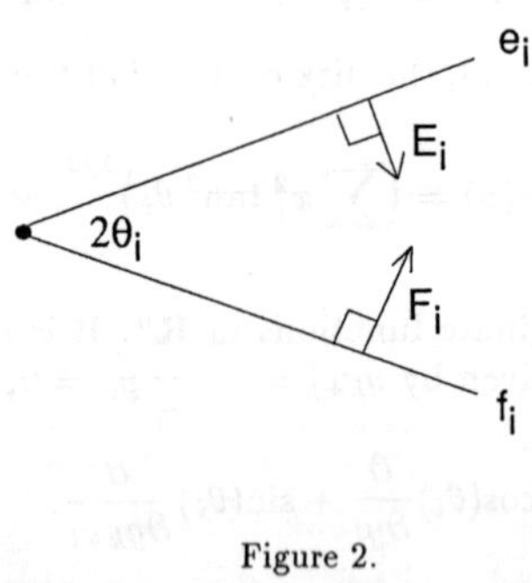

Figure 2.

We can write

$$z = \sum x_i e_i + a_i E_i = \sum y_i f_i + b_i F_i.$$

Then

$$\operatorname{dist}^2(z, x) + \operatorname{dist}^2(z, y) = \sum a_i^2 + b_i^2.$$

It suffices to prove that for each i,

$$a_i^2 + b_i^2 \geq (x_i^2 + y_i^2) \tan^2 \theta_i.$$

For convenience, we will drop the subscript i.

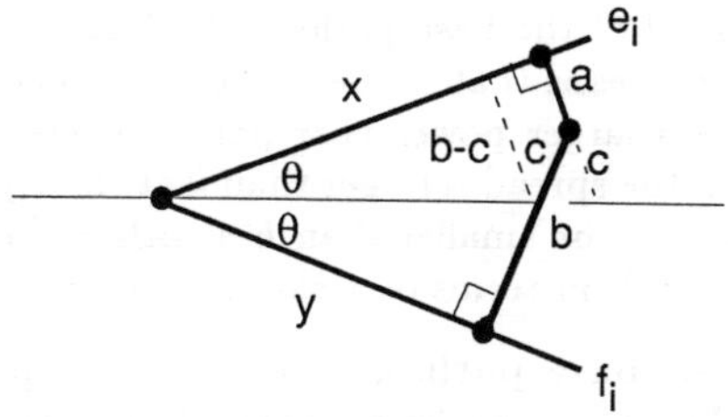

Figure 3. Prove: $a^2+b^2 \geq (x^2+y^2)\tan^2\theta$ if $0<\theta<\pi/4$.

By comparison with the dotted segment of length $b-c$, we can see that since $2\theta \leq \pi/2$,

$$a \leq b - c.$$

Now

$$a + c = x \tan\theta$$
$$b - c = y \tan\theta$$
$$a^2 + 2ac + c^2 = x^2 \tan^2\theta$$
$$b^2 - 2bc + c^2 = y^2 \tan^2\theta$$
$$a^2 + b^2 + 2c(c + a - b) = (x^2 + y^2)\tan^2\theta$$

and $c + a - b \leq 0$, so that

$$a^2 + b^2 \geq (x^2 + y^2)\tan^2\theta,$$

as desired.

4. The base paths

As stated in Section 2, to construct our area-decreasing retraction we begin by covering the k-plane P with a family of base paths. This is defined locally by a continuous map from $[0,\infty) \times U \subset \mathbf{R} \times \mathbf{R}^{k-1} = \mathbf{R}^k$ to P, which is one to one for $x_1 > 0$. Each base path is then the image of $[0,\infty) \times \{q\}$ for some $q \in U$. We will let $t \in [0,\infty)$ be a non-unit speed parameter on the base path, and define $u = u(t) = \int_0^t |\gamma'(u(s))|ds$ to be the length of γ from 0 to t.

A valid choice of base paths would be simply the rays extending from the origin. In this case, the spread of the base paths (see Section 5) would equal k everywhere. However, this would yield a retraction supported in a cushion with equal angles around the plane; we also want to allow cushions with some angles larger and some smaller. We need the spread of the base paths to be larger wherever the cushion angle is to be smaller. This means that we need the cross section of a little bundle of base paths to grow like a larger power near points where the cushion angle is smaller. It turns out that if the spread is larger than k at some points of the k-plane, automatically the spread must be smaller than k at other points. This is why we might expect to obtain a result in terms of a *sum* of angles in a cushion.

4.1 Definition (choice of base paths): Let P be a k-plane in $\mathbf{R}^n$, with an orthonormal basis $\{e_i\}$. Let $\theta_1, \ldots, \theta_k$ be a choice of positive angles for which we hope that the Θ-cushion about P contains a retraction onto P. For each i, let $\alpha_i = \tan(\theta_i) > 0$. For each k-tuple of real numbers $(t, a_2, \ldots, a_k)$ with $t \geq 0$, define maps f_1 and f_{-1} from $\mathbf{R}^k$ to P by

$$f_1(t, a_2, \ldots, a_k) = (t^{\alpha_1}, \ldots, a_k t^{\alpha_k}),$$

$$f_{-1}(t, a_2, \ldots, a_k) = (-t^{\alpha_1}, \ldots, a_k t^{\alpha_k}).$$

Here the image point in P is written with respect to the basis $\{e_i\}$. For each $A = (a_1, a_2, \ldots, a_k)$ with $a_1 = 1$ or -1, define the base path

$$\gamma_A(t) = f_{a_1}(t, a_2, \ldots, a_k) = (a_1 t^{\alpha_1}, \ldots, a_k t^{\alpha_k}).$$

These paths cover $\mathcal{H}^k$-almost every point of P, and no point except the origin is covered more than once, as required.

5. The spread

The next step is to compute a certain measure of how fast the paths are diverging from each other, which we call the "spread."

In order to compute the spread of these base paths, we need to know the Jacobian of f.

$$\frac{\partial f}{\partial t} = (a_1 \alpha_1 t^{\alpha_1 - 1}, \ldots, a_k \alpha_k t^{\alpha_k - 1})$$

$$\frac{\partial f}{\partial a_2} = (0, t^{\alpha_2}, 0, \ldots, 0)$$

$$\vdots$$

$$\frac{\partial f}{\partial a_k} = (0, \ldots, 0, t^{\alpha_k}).$$

The length of the wedge product of these vectors is

$$J_f(t) = \alpha_1 t^{\alpha_1+\cdots+\alpha_k-1}.$$

Using the formula for spread given in Lawlor[7], we compute

$$\mathcal{S}(\gamma(T)) = \frac{\int_0^T |\gamma'(t)|dt\ J_f(T)}{|\gamma'(T)|\ \int_0^T J_f(t,0)dt}$$

$$= \frac{u\alpha_1 T^{\alpha_1+\cdots+\alpha_k-1}}{\sqrt{\alpha_1^2(\frac{x_1}{T})^2+\cdots+\alpha_k^2(\frac{x_k}{T})^2}\ \int_0^T \alpha_1 t^{\alpha_1+\cdots+\alpha_k-1}dt}$$

$$= \left(\sum \alpha_i\right)\frac{u}{\sqrt{\sum \alpha_i^2 x_i^2}},$$

where (as noted before) u is the length of the path segment $\gamma[0,T]$. Here we have written the spread as a real-valued function on P. By abuse of notation we will also write $\mathcal{S}(u) \equiv \mathcal{S}(\gamma(T))$, where $u = u(T)$ is the length of $\gamma[0,T]$.

For example, along the jth coordinate axis (which itself is a straight base path), $u = x_j$ and the spread is $\frac{1}{\alpha_j}\sum \alpha_i$. As another example, if all of the α_i are equal, then every path is straight, so that $u = \sqrt{\sum x_i^2}$, and the spread is k everywhere.

6. The partial differential equation

As in Lawlor[7], we now let Π be the retraction onto the k-plane P induced by these base paths and by some choice of function $G_\gamma(u,v)$ for each base path γ. As noted at the end of Section 2, if we take the $n-k$ dimensional inverse image of a point, $\Pi^{-1}(p)$, and project it orthogonally onto P, the "shadow" always lies along a base path. The function G tells us (in a way described in Lawlor[7]) how fast the radius of $\Pi^{-1}(p)$ increases as we move outward along the base path.

Since the normal space to P is constant along a base path (in fact, everywhere), and the curvature is zero, we can use the main theorem of Lawlor[7] which states that the retraction Π decreases area if $G_\gamma(u,v)$ satisfies the boundary value problem

$$(G + \frac{u}{\mathcal{S}}\frac{\partial G}{\partial u})^2 + (\frac{u}{\mathcal{S}}\frac{\partial G}{\partial v})^2 \leq 1 \tag{1}$$

$$G(u,0) = 1.$$

For each base path we need to solve this for G_γ and see where $G_\gamma = 0$. Then for each u, as pointed out in Section 4, the value of v for which $G(u,v)$ vanishes is the radius of a disk $\xi(p)$ normal to P and centered at $p = \gamma(T)$. Here, T is the value chosen so that the length of the path segment $\gamma[0,T]$ is u. The union of

$\xi(p)$ is the domain of the retraction Π. Then to prove that a union of two or more k-planes is area-minimizing, we need only arrange (if possible) that the domains of the respective retractions onto the k-planes not intersect.

7. Reducing to an ordinary differential equation

We now have, for each base path γ, a partial differential inequality, Eq. 1, with a given boundary value. How can we solve all of them, or estimate their solutions? How can we solve any one of them?

The answer to the second question is that we must solve with a numerical approximation method, being especially careful near $v = 0$ where the P.D.E. is singular. The we must prove the existence of a solution for each path γ.

In order to avoid doing this from scratch, we will reduce the two variables u, v to a single variable and make use of information in a table in the minimizing cones paper[8]. The inequality in one variable is not equivalent; it will imply Eq. 1 but the result will not be as sharp.

In the minimizing cones paper[8], Section 2.3.8, the differential inequality

$$\left(g(\tau) - \frac{\tau}{k}g'\right)^2 + \left(\frac{g'}{k}\right)^2 \leq 1 \tag{2}$$
$$g(0) = 1$$

is studied. For each value of $k \geq 3$ there is a solution g which reaches zero; the larger k is, the sooner g reaches zero.

Given a solution $g(\tau)$ of Eq. 2, we will see that

$$G(u,v) = g\left(\frac{\mathcal{S}}{\inf(\mathcal{S})} \cdot \frac{v}{u}\right)$$

is a solution of Eq. 1.

7.1. Proposition:

Let $\alpha_1 \geq \alpha_2, \ldots, \alpha_m > 0$.

Choose a base path

$$\gamma(t) = (x_1(t), \ldots, x_m(t)) = (a_1 t_1^\alpha, \ldots, a_m t_m^\alpha).$$

Let $u = u(t)$ be the length of the path from 0 to t.

Let $\mathcal{S}(u)$ be the spread function calculated in Section 5, for the choice of base paths from Section 4.

Let $L = \inf_\gamma \mathcal{S}(u) = \frac{1}{\alpha_1}\sum \alpha_i$.

Suppose that for this value of L there is a function $g(\tau)$ which satisfies

$$\left(g(\tau) - \frac{\tau}{L}g'\right)^2 + \left(\frac{g'}{L}\right)^2 \leq 1$$

$$g(0) = 1$$

$$g(\tau_0) = 0 \quad \text{for} \quad \text{some} \quad \tau_0 > 0$$

and

$$g'(\tau) \leq 0 \quad \text{on} \quad [0, \tau_0].$$

Then there is a solution of the differential inequality from Section 6

$$(G + \frac{u}{\mathcal{S}}\frac{\partial G}{\partial u})^2 + \left(\frac{u}{\mathcal{S}}\frac{\partial G}{\partial v}\right)^2 \leq 1$$

$$G(u, 0) = 1,$$

namely the function

$$G(u, v) = g\left(\frac{v\mathcal{S}}{uL}\right).$$

For each u, this solution G exists for $v \in [0, uL\tau_0/\mathcal{S}]$, and equals zero at $v = uL\tau_0/\mathcal{S}$.

Proof. By Lemma 7.2,

$$\frac{u}{\mathcal{S}}\frac{\partial G}{\partial u} = \frac{-\alpha_1 v \sum \alpha_i^3 x_i^2}{(\sum \alpha_i)(\sum \alpha_i^2 x_i^2)^{3/2}} g' \geq 0,$$

and

$$\frac{u}{\mathcal{S}}\frac{\partial G}{\partial v} = \frac{g'}{L}.$$

Since $G + \frac{u}{\mathcal{S}}\frac{\partial G}{\partial u}$ is a sum of nonnegative numbers, it suffices to show that

$$\frac{u}{\mathcal{S}}\frac{\partial G}{\partial u} \leq \frac{-\tau g'}{L},$$

which will be true if

$$\frac{\alpha_1 v \sum \alpha_i^3 x_i^2}{(\sum \alpha_i)(\sum \alpha_i^2 x_i^2)^{3/2}} \leq \frac{\tau}{L} = \frac{v\mathcal{S}}{uL^2} =$$

$$\frac{v \sum \alpha_i}{\sqrt{\sum \alpha_i^2 x_i^2}} \cdot \frac{\alpha_1^2}{(\sum \alpha_i)^2} = \frac{\alpha_1^2 v}{(\sum \alpha_i)\sqrt{\sum \alpha_i^2 x_i^2}},$$

i.e.,

$$\sum \alpha_i^3 x_i^2 \leq \alpha_1 \sum \alpha_i^2 x_i^2$$

which is true because $\alpha_1 \geq \alpha_i$ for all i.

7.2. Lemma. With notation as in Proposition 7.1, if

$$G(u,v) = g\left(\frac{v\mathcal{S}}{uL}\right)$$

then

$$\frac{u}{\mathcal{S}}\frac{\partial G}{\partial u} = \frac{-\alpha_1 v \sum \alpha_i^3 x_i^2}{(\sum \alpha_i)(\sum \alpha_i^2 x_i^2)^{3/2}} g'$$

and

$$\frac{u}{\mathcal{S}}\frac{\partial G}{\partial v} = \frac{g'}{L}.$$

Proof. The second statement follows immediately since

$$\frac{\partial G}{\partial v} = \frac{\mathcal{S}}{uL} g'.$$

To prove the first assertion, we differentiate by t; since u is a function of t but v does not depend on t,

$$\frac{\partial G}{\partial u} = \frac{\frac{dg}{dt}}{\frac{du}{dt}}.$$

Now

$$\frac{dg}{dt} = \frac{d}{dt}\left(g\left(\frac{v\mathcal{S}}{uL}\right)\right) = \frac{d}{dt}\left(\frac{v\mathcal{S}}{uL}\right)\cdot g' = \frac{vg'}{L}\frac{d}{dt}\left(\frac{\mathcal{S}}{u}\right);$$

$$\frac{\mathcal{S}}{u} = \frac{\sum \alpha_i}{\sqrt{\sum \alpha_i^2 x_i^2}}$$

$$\begin{aligned}
\frac{d}{dt}\left(\frac{\mathcal{S}}{u}\right) &= \left(\sum \alpha_i\right)\left(-\frac{1}{2}\left(\sum \alpha_i^2 x_i^2\right)^{-3/2}\left(2\sum \alpha_i^2 x_i \frac{dx_i}{dt}\right)\right) \\
&= -\left(\sum \alpha_i\right)\left(\sum \alpha_i^2 x_i^2\right)^{-3/2} \sum \alpha_i^2 x_i a_i t^{\alpha_i - 1} \\
&= -\frac{1}{t}\left(\sum \alpha_i\right)\left(\sum \alpha_i^2 x_i^2\right)^{-3/2} \sum \alpha_i^3 x_i^2
\end{aligned}$$

so that

$$\begin{aligned}
\frac{dg}{dt} &= -\frac{vg'}{Lt}\left(\sum \alpha_i\right)\left(\sum \alpha_i^2 x_i^2\right)^{-3/2} \sum \alpha_i^3 x_i^2 \\
&= -\frac{\alpha_1 v g'}{t}\left(\sum \alpha_i^2 x_i^2\right)^{-3/2} \sum \alpha_i^3 x_i^2
\end{aligned}$$

and

$$\frac{\partial G}{\partial u} = -\left(\frac{du}{dt}\right)^{-1} \cdot \frac{\alpha_1 v g'}{t}\left(\sum \alpha_i^2 x_i^2\right)^{-3/2} \sum \alpha_i^3 x_i^2$$

$$= \frac{-\alpha_1 v(\sum \alpha_i^3 x_i^2) g'}{\left(\sum \alpha_i^2 x_i^2\right)^2}.$$

Multiplying this by u/S gives the result as desired.

8. Theorems

We are now ready to state and prove the main theorems of this paper.

8.1 Proposition: Let P be a k-plane in $\mathbf{R}^n$. Choose an orthonormal basis for P, and a Θ-cushion about P, where $\Theta = (\theta_1, \ldots, \theta_k)$ is some choice of k angles such that $0 < \theta_i < \pi/2$, and $\theta_1 \geq \theta_i$ for all i. Let $\alpha_i = \tan\theta_i$, and let

$$L = \frac{\sum \alpha_i}{\alpha_1}.$$

Let τ_0 be the smallest value of τ for which $g(\tau)$ can equal zero, where g satisfies the differential inequality

$$(g - \frac{\tau g'}{L})^2 + (\frac{g'}{L})^2 \leq 1$$

$$g(0) = 1.$$

If $\tau_0 \leq \alpha_1$ then there is an area-decreasing retraction onto P which is supported inside the Θ-cushion about P.

If L is an integer from 3 to 12, τ_0 can be looked up in Table 1.4.1 of Lawlor[8], using the first row of the table (corresponding to $\alpha^2 = 0$) and using $k = L$ to tell us which column we want. The table gives an angle in degrees, whose tangent is τ_0. If $L > 3$ is not an integer, one can use the next integer smaller than L (or 12 if $L > 12$) in Table 1.4.1, and then apply Proposition 1.4.2. (It is important to note that in Proposition 1.4.2 the implied restriction that $k \geq 12$ does not apply in the case that $\alpha^2 = 0$.)

Proof of proposition: Let $G(u,v) = g(\tau) = g(\frac{vS(u)}{uL})$. By the calculations in the previous section, this function $G(u,v)$ satisfies the differential inequality

$$(G + \frac{u}{S}\frac{\partial G}{\partial u})^2 + \left(\frac{u}{S}\frac{\partial G}{\partial v}\right)^2 \leq 1.$$

Now this means that for each u, $G(u,v)$ vanishes at

$$v = \tau_0 h(u) = \tau_0 \frac{uL}{S(u)} = \frac{\tau_0}{\alpha_1}\sqrt{\sum \alpha_i^2 x_i^2} \leq \sqrt{\sum \alpha_i^2 x_i^2},$$

which is the radius of the theta-cushion at $\gamma(u) = (x_1, \ldots, x_k)$. By Section 6, there is an area-decreasing retraction onto P whose domain lies within the Θ-cushion about P.

8.2 Definition: If $\theta_1, \ldots, \theta_k$ are such as to satisfy the hypotheses of the previous proposition, then we say that a Θ-cushion about a k-plane is a "sufficient cushion."

8.3 Theorem: Let $P_1, \ldots, P_m$ be a set of k-planes through a common vertex in $\mathbf{R}^n$. For each $j \in [1, m]$, let $\Theta_j = (\theta_{1j}, \ldots, \theta_{kj})$ be a k-tuple of angles. Suppose that we can find a (single, fixed) orthonormal basis for each of the P_i with respect to which the Θ_j cushions about P_j are all disjoint in $\mathbf{R}^n$. If the Θ_j cushions are all sufficient, then the union of the k-planes is area-minimizing modulo 2 (and thus also area-minimizing with respect to any set of orientations on P_i.)

Proof: Let S be a surface whose boundary equals the boundary of a compact portion B of the union of k-planes. Let Π_i be area-decreasing retractions onto the planes P_i whose domains are contained, respectively, in the nonintersecting Θ_i cushions. Let Π be the retraction defined piecewise by the Π_i. Then Π is well-defined and area-decreasing, and maps the part of S intersecting its domain onto B, so that

$$\text{Area}(S) \geq \text{Area}(\Pi(S)) \geq \text{Area}(B).$$

8.4 Corollary: Let P_i be k-planes as above, with angles $\theta_{1j}, \ldots, \theta_{kj}$ defining nonintersecting Θ-cushions. Suppose that for all i, j, $0 < \theta_{ij} \leq \pi/4$. If for each j,

$$\sum_{i=1}^{k} \theta_{ij} \geq \pi$$

then the union of k-planes is area-minimizing mod 2.

Proof: We will show that the Θ-cushions are sufficient. In the proof, we drop the extra subscript; thus, let $\theta_1, \ldots, \theta_k$ be positive angles less than or equal to $\pi/4$. For notation's sake, let θ_1 be the largest of the angles. Let $\alpha_i = \tan\theta_i$. Then

$$\sum \alpha_i > \sum \theta_i \geq \pi > 3.$$

Suppose for a moment that $\theta_1 = \pi/4$. To apply Proposition 8.1, note that

$$L = \frac{\sum \alpha_i}{\alpha_1} > 3.$$

We look up τ_0 in Table 1.4.1 of Lawlor[8], and find that $\tau_0 < \tan(40°) < 1 = \alpha_1$ so that in this case the Θ-cushion is sufficient. Now if $\theta_1 < \pi/4$, then L is larger and so τ_0 is smaller; in fact, Proposition 1.4.2 of Lawlor[8] tells us directly that

$$\tau_0 < \alpha_1$$

still holds. Thus by Proposition 8.1 there are area-decreasing retractions onto each P_i. Their domains are the nonintersecting Θ-cushions, so together they define a well-defined area-decreasing retraction onto the union of the planes P_i.

8.5 Corollary: Let P and Q be two intersecting k-planes in $\mathbf{R}^n$. If the (unoriented) characterizing angles between P and Q sum to at least 2π, then $P \cup Q$ is area-minimizing mod 2.

Proof: This is a corollary of 8.4 and Lemma 3.5. Let $\beta_1, \ldots, \beta_k$ be the unoriented characterizing angles between P and Q, and let $\theta_i = \frac{1}{2}\beta_i$. Each $\theta_i \leq \pi/4$ because unoriented characterizing angles are always at most $\pi/2$. By Lemma 3.5, there is a Θ-cushion about P and a Θ-cushion about Q which are disjoint. Since $\sum \theta_i \geq \pi$, by Corollary 8.4, $P \cup Q$ is area-minimizing modulo 2.

Note that if one or more of the characterizing angles is zero, we reduce dimension and eliminate the zero angle(s) by using the fact that a surface T is area-minimizing if and only if its Cartesian product with a line is area-minimizing.

Acknowledgment: The author would like to thank the National Science Foundation for supporting the research and preparation of this paper through a mathematics postdoctoral fellowship.

REFERENCES

1. D. Nance (now Mackenzie), *Sufficient conditions for a pair of n-planes to be area-minimizing*, Math. Ann. **279** (1987), 161–164.

2. G. R. Lawlor, *The angle criterion*, Invent. Math. **95** (1989), 437–446.

3. F. Morgan, *On the singular structure of two-dimensional area minimizing surfaces in* $\mathbf{R}^n$, Math. Ann. **261** (1982), 101–110.

4. F. Morgan, *Calibrations modulo* ν, Adv. in Math. **64** (1987), 32–50.

5. F. Morgan, *On the singular structure of three-dimensional area minimizing surfaces*, Trans. Amer. Math. Soc. **276** (1983), 137–143.

6. R. Harvey and H. B. Lawson, Jr., *Calibrated geometries*, Acta Mathematica **148** (1982), 47–157.

7. G. R. Lawlor, *A path covering method for proving a minimal surface to be area-minimizing*, in preparation.

8. G. R. Lawlor, *A sufficient criterion for a cone to be area-minimizing*, Memoirs Amer. Math. Soc., to appear (1991).

The Problem of Plateau (pp. 181–189)
ed. Th. M. Rassias

MORSE INDEX OF COMPLETE MINIMAL SURFACES

SHIN NAYATANI
Mathematical Institute
Tôhoku University
Sendai 980, Japan

Abstract

After a general discussion of the subject, we give an exposition of some quantitative results which have been obtained in the last several years. It includes an account of the author's recent results. We also determine the index of Hoffman-Meeks' minimal surfaces with three ends when the genus is not very large.

1. Introduction and preliminaries

Let M be a complete oriented minimal surface in $\mathbf{R}^3$. The Jacobi operator of M is given by

$$L_M = -\Delta + 2K,$$

where Δ and K are the Laplace-Beltrami operator and the Gauss curvature of M respectively. If Ω is a bounded domain in M, its index is defined as the number of negative eigenvalues of the Dirichlet eigenvalue problem

$$L_M u = \lambda u \quad \text{in } \Omega, \qquad u = 0 \quad \text{on } \partial\Omega.$$

Choosing an exhaustion $\{\Omega_j\}_{j=1}^{\infty}$ of M, we define $\mathrm{Ind}(M)$, the index of M, as the limit of the indices of Ω_j, which may be infinite. Fischer-Colbrie [6], Gulliver and Lawson [7], [8] independently proved that the index of M is finite if and only if the total Gauss curvature of M is finite, thus gave a geometric characterization of a surface with finite index. A naturally arising problem here is to study the index quantitatively, that is, to compute or to estimate it in terms of geometric invariants of the surface such as the total curvature. In view of the result just mentioned, we may assume that the

*Partially supported by the Kawai Foundation and the Yukawa Foundation.

total curvature of M is finite. By a theorem of Osserman, such M is conformally equivalent to a compact Riemann surface with finitely many punctures and the Gauss map $G : M \to S^2$ extends to the compactified surface holomorphically. Let $\overline{M}$ and $\overline{G} : \overline{M} \to S^2$ denote the compactified surface and the extended Gauss map respectively. Notice that the total curvature of M is equal to (-4π) times the degree of $\overline{G}$ and so they are essentially the same thing.

We now let $G : \Sigma \to S^2$ be an arbitrary nonconstant holomorphic map defined on a compact Riemann surface Σ. We fix a conformal metric on Σ and consider the operator $L = -\Delta - | dG |^2$, acting on functions on Σ. The quadratic form associated to L is given by

$$Q(u, u) = \int_\Sigma (| du |^2 - | dG |^2 u^2)\, dA, \quad u \in C^\infty(\Sigma).$$

Notice that Q does not depend on the chosen metric. We now define $\mathrm{Ind}(G)$, the index of G, as the number of negative eigenvalues of L. It can also be defined as the dimension of a maximal subspace of $C^\infty(\Sigma)$ on which Q is negative definite. By this latter definition, $\mathrm{Ind}(G)$ is well-defined, that is, independent of the particular choice of metric. The kernel of L, $N(G) = \{u \in C^\infty(\Sigma) \mid Lu = 0\}$, is also an invariant of G. We define $\mathrm{Nul}(G)$, the nullity of G, as the dimension of $N(G)$. We note that $LN(G) = \{a \cdot G \mid a \in \mathbf{R}^3\}$ is a three-dimensional subspace of $N(G)$ and so $\mathrm{Nul}(G) \geq 3$.

We now let M be a complete oriented minimal surface with finite total curvature. As was observed by Fischer-Colbrie [6], the index of M coincides with the index of $\overline{G}$, the extended Gauss map of M. Moreover, the space $N(\overline{G})$ (restricted to M) exactly consists of bounded solutions of the Jacobi operator. These facts motivate the study of the index and the nullity of a holomorphic map defined on a compact Riemann surface. For other related geometric situations, see [11].

At this point, we should mention that, although the index and the nullity are invariants of G, the eigenvalues of L themselves are not; they actually depend on the metric which we chose to define them. It turns out that a good choice of metric is the one induced by G from S^2. This metric degenerates at the ramification points of G. However, with this choice of metric, the operator L takes a simple form of $-\Delta - 2$ ($\equiv L_G$). We denote the k-th eigenvalue of this operator by $\lambda_k(G)$, which is now an invariant of G.

2. Some quantitative results

The following result due to Tysk [15] is the first quantitative result on the index of a holomorphic map.

Theorem 1 *Let $G : \Sigma \to S^2$ be a nonconstant holomorphic map of degree d. Then*

$$\mathrm{Ind}(G) + \mathrm{Nul}(G) \leq 7.7d.$$

The proof is based on the comparison of the heat kernel of the metric ds_G^2 on Σ with that of the standard metric on S^2.

The above result concerns the upper bound for the index. As for the lower bound, the following observation is useful. Let u be a nonzero element of $N(G)$. Suppose that the set $u^{-1}(0)$ divides the surface Σ into N components. Let $\mathrm{Ind}(G) = i$. Then u is an $(i+1)$−th eigenfunction of L. Courant's nodal domain theorem [2] then implies that $N \leq i+1$, or $i \geq N-1$. We have proved:

Proposition 2 ([1], [12]) *Let u and N be as above. Then*

$$\mathrm{Ind}(G) \geq N - 1.$$

Let M be an oriented minimal surface in $\mathbf{R}^3$. If l is a line in $\mathbf{R}^3$, then the infinitesimal rotation around l is a Killing vector field on $\mathbf{R}^3$. Hence the inner product with the unit normal vector of M satisfies the Jacobi equation. We denote this function by u_l. Also it is well-known that the support function of M (that is, the inner product of the position vector and the unit normal vector) is a solution of the Jacobi operator. We now assume that M is complete and of finite total curvature. The following theorem due to Choe [1] gives sufficient conditions under which these functions are bounded and thus lie in $N(\overline{G})$, where $\overline{G}$ is the extended Gauss map of M.

Theorem 3 *Let M be a complete oriented minimal surface in $\mathbf{R}^3$ with finite total curvature. Suppose that each end of M is embedded.*
(a) *If the normal vectors at infinity are all parallel to a line l, then the function u_l is bounded.*
(b) *If each end of M is planer, then the support function of M is bounded.*

The proof of (a) is given in [1] and uses Schoen's characterization of embedded ends (see [14]). The proof of (b) is similar. Notice that the assumption in (a) is satisfied if M is embedded.

We now give examples. Let $G_d : \mathbf{C} \cup \{\infty\} \to S^2$, $d = 1, 2, \cdots$, be the holomorphic map defined by $\Pi \circ G_d(z) = z^d$, where $\Pi : S^2 \to \mathbf{C} \cup \{\infty\}$ is the stereographic projection from the north pole. Let $a = (0,0,1)$ and $u = a \cdot G_d$. Then $u^{-1}(0)$ divides the surface into $2d$ components. By Proposition 2, we obtain $\mathrm{Ind}(G_d) \geq 2d - 1$. Actually we can solve the eigenvalue problem associated with the operator L_{G_d} by the separation of variable method (see [12]). As a result,

$$\mathrm{Ind}(G_d) = 2d - 1 \quad \text{and} \quad \mathrm{Nul}(G_d) = 3. \tag{1}$$

This determines the index of Jorge-Meeks' $(d+1)$-ends catenoid [10] since G_d is exactly the extended Gauss map of this surface. On the other hand, Choe [1]showed that the index of Hoffman-Meeks' surface with three ends and genus γ [9] is not less than $2\gamma+1$. In fact, he observed that $u_l^{-1}(0)$ divides the surface into at least $2\gamma+2$ components, where u_l is the function as in Theorem 3 (a).

Ejiri-Kotani [4] and Montiel-Ros [11] generalized the notion of a planer end to a non-embedded end and observed that Theorem 3 (b) continues to hold if the ends are not necessarily embedded. They also proved:

Theorem 4 *Let $G : \Sigma \to S^2$ be a nonconstant holomorphic map. Then each non-linear element of $N(G)$ (that is, an element of $N(G) \setminus LN(G)$) is expressed as the support function of a complete branched minimal surface with planer ends whose extended Gauss map is G.*

Theorem 4 gives, via Weierstrass representation formula, a holomorphic description of the space $N(G)$. Using this, these authors proved the following:

Theorem 5 *Let $G : \mathbf{C} \cup \{\infty\} \to S^2$ be a nonconstant holomorphic map of degree d. Then*

$$\mathrm{Ind}(G) \leq 2d - 1,$$

and for generic choice of G

$$\mathrm{Ind}(G) = 2d - 1 \quad \textit{and} \quad \mathrm{Nul}(G) = 3.$$

Sketch of proof. Let $\mathcal{M}_d$ denote the space of all the rational functions of degree d. It is identified with the complement of an algebraic hypersurface in the complex projective space $P^{2d+2}(\mathbf{C})$ and as such it is a complex manifold. Let $\mathcal{N}_d$ denote the subset of $\mathcal{M}_d$ where Nul > 3. Then it can be shown, using the description of $N(G)$, that $\mathcal{N}_d$ is an analytic subvariety and hence does not disconnect $\mathcal{M}_d$. It follows from this, the continuity of the eigenvalues and (1) that Ind is identically equal to $2d-1$ on $\mathcal{M}_d \backslash \mathcal{N}_d$. By the continuity of the eigenvalues again, we obtain Ind $\leq 2d - 1$ everywhere on $\mathcal{M}_d$.

We now define a one-parameter family of conformal diffeomorphisms $\mathcal{A}_t$, $0 < t < \infty$, of S^2 by

$$\Pi \circ \mathcal{A}_t \circ \Pi^{-1}(w) = tw, \quad w \in \mathbf{C} \cup \{\infty\}.$$

In [13] the author has proved the following theorem:

Theorem 6 *Let $G : \Sigma \to S^2$ be a nonconstant holomorphic map of degree d and $G_t = \mathcal{A}_t \circ G$, $0 < t < \infty$. Let ν be the number of distinct poles of the meromorphic function $g = \Pi \circ G$. Then the following estimates hold for all sufficiently small t:*

$$\mathrm{Ind}(G_t) \geq 2d - \nu; \tag{2}$$

$$\mathrm{Ind}(G_t) + \mathrm{Nul}(G_t) \leq 2d + \nu + 1; \tag{3}$$

$$\mathrm{Nul}(G_t) \leq 2\nu + 1. \tag{4}$$

In particular, if $\nu = 1$, then we have

$$\mathrm{Ind}(G_t) = 2d - 1 \quad \textit{and} \quad \mathrm{Nul}(G_t) = 3$$

for all sufficiently small t.

Sketch of proof. As t tends to zero, G_t converges smoothly, away from the poles of g, to the constant map which maps the whole surface onto the south pole. In this process, the energy of G_t becomes more and more concentrated near the poles, say p_i's. At the limit, a holomorphic map $\tilde{G}_i : \overline{\mathbf{C}}_i \to S^2$ bubbles off at each p_i, where $\overline{\mathbf{C}}_i$ is a copy of the Riemann sphere. With appropriate choice of parameter on $\overline{\mathbf{C}}_i$, the map $\tilde{G}_i$ is given by $\Pi \circ \tilde{G}_i(z) = z^{d_i}$, where d_i is the order of p_i. We now take the disjoint union

$$\tilde{G} \equiv \amalg \tilde{G}_i : \amalg \overline{\mathbf{C}}_i \to S^2.$$

Since $\mathrm{Ind}(\tilde{G}_i) = 2d_i - 1$, $\mathrm{Nul}(\tilde{G}_i) = 3$ by (1) and $\Sigma_{i=1}^{\nu} d_i = d$, we have

$$\mathrm{Ind}(\tilde{G}) = 2d - \nu \quad \text{and} \quad \mathrm{Nul}(\tilde{G}) = 3\nu.$$

On the other hand, it can be shown that $\lambda_k(G_t)$ converges to $\lambda_k(\tilde{G})$ as t tends to zero. Moreover, when t moves away from zero, at least $\nu - 1$ of the null eigenvalues of $L_{\tilde{G}}$ initially go up. Thus we obtain (2) and (3), from which (4) follows.

Using their description of $N(G)$, Montiel-Ros [11] observed that, if the genus of Σ is zero, $\mathrm{Nul}(G_t)$ is constant, where $G_t = \mathcal{A}_t \circ G$. This fact together with Theorem 6 gives the following:

Theorem 7 (**[13]**) *Let $G : \mathbf{C} \cup \{\infty\} \to S^2$ be a nonconstant holomorphic map of degree d and $\nu = \nu(G)$ the minimal number of distinct points in $G^{-1}(q)$ when q runs over S^2. Then we have*

$$\mathrm{Ind}(G) \geq 2d - \nu,$$
$$\mathrm{Ind}(G) + \mathrm{Nul}(G) \leq 2d + \nu + 1,$$

and

$$\mathrm{Nul}(G) \leq 2\nu + 1.$$

In particular, if $\nu(G) = 1$, then

$$\mathrm{Ind}(G) = 2d - 1 \quad \textit{and} \quad \mathrm{Nul}(G) = 3.$$

Using this theorem, we can determine the index and the nullity of the rational function $z^m + \frac{1}{z^n}$ to be $2d - 2$ and 5 respectiovly if $d \equiv m + n \geq 3$ (see [13]). Ejiri [3] has also used Theorem 7 to compute the index of certain rational functions.

In the case of nonzero genus, Montiel-Ros' description still implies that $\mathrm{Nul}(G_t)$ is constant except for finite number of values of t. Moreover, this number is bounded from above by an explicit constant times the genus (see [13]). We thus obtain the following:

Theorem 8 (**[13]**) *Let $G : \Sigma \to S^2$ be a nonconstant holomorphic map of degree $d \geq 2$. Then*

$$\mathrm{Ind}(G) \geq d - 3\gamma + 1,$$

and

$$\mathrm{Ind}(G) + \mathrm{Nul}(G) \leq 3d + 3\gamma,$$

where γ is the genus of Σ.

It should be mentioned that Ejiri-Micallef [5] gave a similar upper bound by a quite different method.

3. Index of Hoffman-Meeks' surfaces

In this section we try to compute the index of Hoffman-Meeks' surfaces with three ends [9]. The computation is only outlined since in [13] the details are supplied for the genus one case.

Let Σ_γ, $\gamma \geq 1$, be the compact Riemann surface

$$\Sigma_\gamma = \{(z, w) \in (\mathbf{C} \cup \{\infty\})^2 \mid w^{\gamma+1} = z^\gamma(z^2 - 1)\},$$

and let $p_0 = (0, 0)$, $p_\infty = (\infty, \infty)$. Let $G : \Sigma_\gamma \to S^2$ be the holomorphic map defined by

$$\Pi \circ G(z, w) = w.$$

The degree of G is $\gamma + 2$ and the point p_∞ is the only pole of the meromorphic function w. By Theorem 6, $\mathrm{Ind}(G_t) = 2\gamma + 3$ and $\mathrm{Nul}(G_t) = 3$ for sufficiently small t. Moreover, computation based on Montiel-Ros' description shows that, if $2 \leq \gamma \leq 37$, there are exactly three values of t, say $t_1 < t_2 < t_3$, such that $\mathrm{Nul}(G_t) > 3$. In fact, $\mathrm{Nul}(G_t) = 4$ if $t = t_1, t_2$ and $= 5$ if $t = t_3$. The values t_i are given as follows:

$$t_1 = (\int_0^1 t^{\frac{1}{\gamma+1}}(1 - t^2)^{-\frac{1}{\gamma+1}} dt)^{\frac{1}{2}} (\int_0^1 t^{-\frac{1}{\gamma+1}}(1 - t^2)^{\frac{1}{\gamma+1}} dt)^{-\frac{1}{2}};$$

$$t_2 = (\int_0^1 t^{-\frac{\gamma}{\gamma+1}}(1 - t^2)^{-\frac{1}{\gamma+1}} dt)^{\frac{1}{2}} ((2\gamma + 3) \int_0^1 t^{\frac{\gamma}{\gamma+1}}(1 - t^2)^{\frac{1}{\gamma+1}} dt)^{-\frac{1}{2}};$$

$$t_3 = (\int_0^1 t^{-\frac{1}{\gamma+1}}(1 - t^2)^{\frac{1}{\gamma+1}} dt \int_0^1 t^{-\frac{\gamma-1}{\gamma+1}}(1 - t^2)^{-\frac{2}{\gamma+1}} dt)^{\frac{1}{2}}$$
$$\times (\gamma \int_0^1 t^{\frac{1}{\gamma+1}}(1 - t^2)^{-\frac{1}{\gamma+1}} dt \int_0^1 t^{\frac{\gamma-1}{\gamma+1}}(1 - t^2)^{\frac{2}{\gamma+1}} dt)^{-\frac{1}{2}}.$$

We can also write down nonlinear elements of $N(G_t)$ at these values explicitely and will do it. Let

$$\eta_1 = \frac{z}{w((\gamma + 2)z^2 - \gamma)^2} dz, \quad \eta_2 = i \frac{(\gamma + 2)z^2 + \gamma}{w((\gamma + 2)z^2 - \gamma)^2} dz,$$

$$\eta_3 = \frac{(a_\gamma - b_\gamma w^2)z}{w^2((\gamma + 2)z^2 - \gamma)^2} dz, \quad \eta_4 = i \frac{(a_\gamma + b_\gamma w^2)z}{w^2((\gamma + 2)z^2 - \gamma)^2} dz,$$

where

$$a_\gamma = \int_0^1 t^{\frac{1}{\gamma+1}}(1-t^2)^{-\frac{1}{\gamma+1}}dt, \quad b_\gamma = \int_0^1 t^{-\frac{1}{\gamma+1}}(1-t^2)^{\frac{1}{\gamma+1}}dt.$$

Denoting the ramification locus of G by $\mathcal{R}$, let $X_i : \Sigma_\gamma \setminus \mathcal{R} \to \mathbf{R}^3$ be the branched minimal immersion defined by

$$X_i(p) = \mathrm{Re} \int_{p_0}^{p} (1 - t_i{}^2 w^2, i(1 + t_i{}^2 w^2), 2t_i{}^2 w)\, \eta_i.$$

Then $u_i = X_i \cdot G_{t_i}$ extends over to $\mathcal{R}$ smoothly and gives an element of $N(G_{t_i}) \backslash LN(G_{t_i})$ $(t_4 = t_3)$.

We now examine how the extra null eigenvalues at $t = t_1, t_2$ and t_3 behave when t is perturbed. Let $V_\lambda(t)$ denote the eigenspace of L_{G_t} corresponding to an eigenvalue λ. Consider the conformal diffeomorphisms κ, λ and τ of Σ_γ defined by

$$\kappa(z,w) = (\bar{z}, \bar{w}), \quad \lambda(z,w) = (-z, \rho w),$$

$$\tau = \lambda^{-1}\kappa\lambda,$$

where $\rho = e^{\frac{\gamma\pi}{\gamma+1}i}$. It is easy to see that these are isometries with respect to the metric $ds^2_{G_t}$ for all t. Hence they act L^2-orthogonally on $V_\lambda(t)$. We define mutually orthogonal subspaces $V_\lambda(t)^+$ and $V_\lambda(t)^-$ of $V_\lambda(t)$ by

$$V_\lambda(t)^\pm = \{u \in V_\lambda(t) \mid u \circ \kappa = \pm u, \ u \circ \tau = \pm u\}.$$

Let Ω be a domain in Σ_γ given by

$$\Omega = \{(z,w) \in \Sigma_\gamma \mid \mathrm{Im} z > 0, \ w = |\, z^\gamma (z^2-1)\,|^{\frac{1}{\gamma+1}} \times \exp \frac{\gamma \arg z + \arg(z+1) + \arg(z-1)}{\gamma+1} i\}.$$

We consider the eigenvalue problem

$$L_{G_t} u = \lambda u \quad in\ \Omega \tag{5}$$

under Neumann and Dirichlet boundary conditions. Let $V^N{}_\lambda(t)$ (resp. $V^D{}_\lambda(t)$) denote the eigenspace of the Neumann (resp. Dirichlet) eigenvalue problem corresponding to an eigenvalue λ. Let $u \in V_\lambda(t)^+$. Then it is easily verified that $u|_\Omega$ satisfies (5) together with Neumann boundary condition and thus belongs to $V^N{}_\lambda(t)$. Conversely, if $v \in V^N{}_\lambda(t)$, then extending v so that the resulting function is invariant by κ and τ, we get a function which belongs to $V_\lambda(t)^+$. Thus we have a natural bijective correspondence between $V_\lambda(t)^+$ and $V^N{}_\lambda(t)$. We can establish the similar correspondence between $V_\lambda(t)^-$ and $V^D{}_\lambda(t)$. Let $\lambda^N{}_k(t)$ (resp. $\lambda^D{}_k(t)$) denote the k-th eigenvalue of the Neumann (resp. Dirichlet) eigenvalue problem. These are continuous in t. Moreover, $\lambda^N{}_k(t)$ (resp. $\lambda^D{}_k(t)$) is monotonically non-increasing (resp. non-decreasing) in t. We now outline the proof of this assertion. It is easily verified that the meromorphic function w maps Ω bijectively onto the region $\{\zeta \in \mathbf{C} - \{0\} \mid 0 < \arg\zeta < \frac{\gamma+2}{\gamma+1}\pi\} \setminus (\{se^{\frac{\pi}{\gamma+1}i} \mid 0 < s \leq c_\gamma\} \cup \{se^{\pi i} \mid 0 < s \leq c_\gamma\})$, where $c_\gamma = ((\frac{\gamma}{\gamma+2})^{\frac{\gamma}{2}} \frac{2}{\gamma+2})^{\frac{1}{\gamma+1}}$. Thus G maps (Ω, ds_G^2) isometrically onto an open $\frac{\gamma+2}{2(\gamma+1)}$-sphere from which two geodesic segments, emanating from the south pole and running toward the north pole, are deleted. Clearly this statement continues to hold with G replaced by G_t, and as t increases, the deleted geodesic segments become longer. Then the space of functions with vanishing boundary values becomes smaller and the standard max-mini characterization of eigenvalues implies that the Dirichlet eigenvalues of the spherical Laplacian increase with t. Since G_t is an isometry, the same is true of $\lambda^D{}_k(t)$. The proof for $\lambda^N{}_k(t)$ is similar. Our problem now amounts to the study of symmetry of the functions u_i, $i = 1, ..., 4$. In fact, it is not difficult to check that $u_1 \in V_0(t_1)^+$, $u_2 \in V_0(t_2)^-$, $u_3 \in V_0(t_3)^+$ and $u_4 \in V_0(t_3)^-$. Therefore the extra null eigenvalue at t_1 is decreasing there and the one at t_2 is increasing. At t_3, one extra null eigenvalue is decreasing and the other increasing. We can now conclude:

Theorem 9 *Let $G : \Sigma_\gamma \to S^2$ be as above. If $2 \leq \gamma \leq 37$, then*

$$\mathrm{Ind}(G_t) = \begin{cases} 2\gamma+3 & \textit{if } t \leq t_1,\ t_2 \leq t < t_3,\ t_3 < t, \\ 2\gamma+4 & \textit{if } t_1 < t < t_2, \\ 2\gamma+2 & \textit{if } t = t_3. \end{cases}$$

If $\gamma \geq 38$, there are more values of t other than t_i, $i = 1, 2, 3$, such that $\mathrm{Nul}(G_t) > 3$. It is plausible that all of them are greater than t_3, but we have not been able to check this.

Since the extended Gauss map of Hoffman-Meeks' surface with three ends and genus γ is, up to rotation, nothing but G_{t_2} (see [9]), we obtain:

Corollary 10 *The index of Hoffman-Meeks' surface with three ends and genus γ, $\gamma \leq 37$, is $2\gamma + 3$.*

References

[1] J. Choe, *Index, vision number, and stability of complete minimal surfaces*, Arch. Rat. Mech. Anal. 109 (1990), 195-212.

[2] R. Courant and D. Hilbert, Methods of mathematical physics, Vol. 1, Interscience, New York, 1953.

[3] N. Ejiri, *Minimal deformations from a holomorphic map of S^2 onto $S^2(1)$ to a minimal surface in $S^4(1)$*, preprint.

[4] N. Ejiri and M. Kotani, *Index and flat ends of minimal surfaces*, preprint.

[5] N. Ejiri and M. J. Micallef, in preparation.

[6] D. Fischer-Colbrie, *On complete minimal surfaces with finite Morse index in three manifolds,* Invent. Math. 82 (1985), 121-132.

[7] R. Gulliver, *Index and total curvature of complete minimal surfaces,* Proc. Symp. Pure Math. 44 (1986), 207-211.

[8] R. Gulliver and H. B. Lawson, Jr., *The structure of stable minimal hypersurfaces near a singularity,* Proc. Symp. Pure Math. 44 (1986), 213-237.

[9] D. A. Hoffman and W. H. Meeks, *Embedded minimal surfaces of finite topology,* Ann. Math. 131 (1990), 1-34.

[10] L. P. Jorge and W. H. Meeks, *The topology of complete minimal surfaces of finite total Gaussian curvature,* Topology 22 (1983), 203-221.

[11] S. Montiel and A. Ros, *Schrödinger operators associated to a holomorphic map,* Global Differential Geometry and Global Analysis (Lecture Notes in Math. 1481), ed. by B. Wegner et al., Springer, Berlin, 1991, 147-174.

[12] S. Nayatani, *Lower bounds for the Morse index of complete minimal surfaces in Euclidean 3-space,* Osaka J. Math. 27 (1990), 453-464.

[13] S. Nayatani, *Morse index and Gauss maps of complete minimal surfaces in Euclidean 3-space,* preprint.

[14] R. M. Schoen, *Uniqueness, symmetry, and embeddedness of minimal surfaces,* J. Differ. Geom. 18 (1983), 791-809.

[15] J. Tysk, *Eigenvalue estimates with applications to minimal surfaces,* Pacific J. Math. 128 (1987), 361-366.

This paper is in final form and no version will appear elsewhere.

The Problem of Plateau (pp. 190–220)
ed. Th. M. Rassias

REMOVABLE SINGULARITIES OF STATIONARY FIELDS

THOMAS H. OTWAY
Department of Mathematics, Yeshiva University
New York, New York 10033 USA

ABSTRACT

The following problems in higher-dimensional field theory are considered: the removability of isolated and accumulation-point singularities from low-energy gauge fields; L^p conditions for removing certain classes of submanifold singularities from higher-dimensional gauge fields; extensions to harmonic maps (nonlinear sigma models) and to Yang-Mills fields coupled to matter fields; an *a priori* inequality for a relevant class of elliptic subsolutions.

0. Introduction

A crucial element in the solution of the Plateau problem was the replacement of the area functional in the minimization problem by the Dirichlet (or *energy*) functional. A natural generalization of this problem involves the study of maps which are critical points of their energy functional with respect to a homotopy class of maps between Riemannian manifolds.[18] This is called the *harmonic map problem.* The higher-dimensional theory of such maps, introduced by Eells and Sampson,[6] has found numerous applications to physics and continuum mechanics; some of the more important are the nonlinear sigma and Skyrme models of quantum field theory,[8,36] and the analysis of nematic liquid crystals.[12]

A map u between an n-dimensional Riemannian manifold M and an m-dimensional Riemannian manifold N is said to be *harmonic* if it is twice continuously differentiable and if it is a critical point of the quadratic functional

$$E_M(u) = \int_M |\nabla_M u|^2 \, dM \tag{0.1}$$

with respect to finite-energy maps from M to N. (Here ∇_M denotes the gradient on TM.) We assume throughout that N is smoothly embedded in a higher-dimensional euclidean space $\mathbb{R}^k$. (If N is compact this embedding exists and is an isometry by the Nash-Moser theorem.) Then if $\gamma_{\alpha\beta}(x)$ is the metric tensor on M we can write

$$E_M(u) = \int_M \delta_{ij}\gamma^{\alpha\beta}(x)\partial_\alpha u^i(x)\partial_\beta u^j(x)\sqrt{\det\gamma}\,dx^1 \wedge \cdots \wedge dx^n\,, \tag{0.2}$$

where $1 \leq i,j \leq k$, $1 \leq \alpha, \beta \leq n$, δ_{ij} is the kronecker symbol, and (here and below) we use the Einstein convention for repeated indices.

What physicists call a *field* corresponds geometrically to a section of a vector bundle. The map $u : M \to N$ automatically defines a bundle u^*TM. The energy density in (0.2) can be considered the squared norm of a section-valued 1-form du, and the functional $E_M(u)$ can be considered the energy of the *field du*.

Similarly, if we define over M a principal bundle P with compact structure group G, we can associate to P a vector bundle X, the fiber of which is the lie algebra $\mathfrak{G}$ of G. The curvature Ω of a connection ω on X can be characterized as a $\mathfrak{G}$-valued section $\Omega_\omega \in \Gamma(\Lambda^2 \otimes \mathrm{End}(X))$; Ω_ω is a *Yang-Mills field* (*gauge field*) if it is a critical point of the energy functional

$$E_M(\Omega) = \int_M |\Omega_\omega|^2\, dM\,.$$

We denote a Yang-Mills field by the 2-form F_A, where A is a connection 1-form on X. Physically A represents the potential of the field F_A, and they are related by the formula [c.f. (0.8)]

$$F_A = dA + \tfrac{1}{2}[A, A]\,.$$

Here d is the exterior derivative on lie-algebra-valued sections and [,] is the lie bracket on $\mathfrak{G}$.

Yang-Mills fields over $\mathbb{R}^2$ and $\mathbb{R}^3$ occur in models of superconductivity and magnetic monopoles, respectively. If $M = \mathbb{R}^4$, Yang-Mills fields over M are the sourceless fields of quantum field theory. If M is a Riemannian manifold of even dimension $n > 4$, Yang-Mills connections have applications to the geometry of holomorphic vector bundles and figure in several well-known conjectures of algebraic geometry.[42] There is also some potential physical interest in higher-dimensional gauge fields in connection with membrane theory interactions, although in this case one expects the energy functional to be somewhat different from the one considered here.

Many of the analytic results for harmonic maps have analogues for Yang-Mills fields, with the important exceptions that in the latter case the notions of weak solution and ellipticity are problematic. Both difficulties are due to the invariance of critical points of $E(F_A)$ under the action of the group of automorphisms of X. (Physically this group corresponds to the *internal symmetries* of the field.)

We define a *stationary field* as one which is a critical point of its energy with respect to an appropriate admissible class. In the harmonic map case this class is the set of maps in $L^\infty(M,N) \cap H^{1,2}(M,N)$; in the Yang-Mills case it is the set of connections ω for which $\Omega_\omega \in L^2(M)$. There are at least two ways to take variations and we expect a stationary field to be critical with respect to both types. The first way is to take variations in the tangent space of the image of the map (or in the space of infinitesimal deformations of the connection, in the Yang-Mills case). This method has been described in detail for harmonic maps[30] and for Yang-Mills fields.[4] The second way is to take variations by reparametrizing the domain. This method is described in detail (in Ref. 28) for both harmonic maps and Yang-Mills fields; but since it is by far the less familiar of the two types of variations we recall its essential features here.

Define the set of *r-variations* of a finite-energy map u on a manifold M to be the variations generated by 1-parameter reparametrizations of M. This is the set

$$S(u) = \left\{ w \in H_0^{1,2}(M, \mathbb{R}^k) \mid w = u \circ \varphi_t \right\}$$

for φ_t a 1-parameter family of compactly supported C^1 diffeomorphisms of M with $\varphi \in (-\varepsilon, \varepsilon)$, $\varphi_0 =$ identity and $\varphi_{s+t} = \varphi_t \circ \varphi_s$.

The initial velocity field for the flow generated by these reparametrizations is the *variation vector field*

$$\zeta(x) = \left.\frac{d\varphi(x,t)}{dt}\right|_{t=0}, \qquad x \in M .$$

The *first r-variation* is the derivative

$$\delta E_M^r(u) = \left.\frac{d}{dt}\right|_{t=0} \int_M |\varphi_t^* \, du|^2 \, dM . \tag{0.3}$$

A map is called *r-stationary* if $\delta E_M^r(u) = 0$ for all families φ_t defined by $S(u)$.

By construction, $\varphi_t^* \, du$ is a section-valued 1-form. However, if ω is a $\mathfrak{G}$-valued connection it is *not* true that $\varphi_t^*\omega$ is necessarily a $\mathfrak{G}$-valued connection. Thus the above argument must be modified slightly in the Yang-Mills case. In particular, φ_t must be lifted to the principal bundle by parallel transport along $x_t = \varphi_t(x)$. This construction is described in Ref. 28.

One can compute expression (0.3) explicitly. Since our applications will be local, we can assume that M is an open domain of Euclidean space and introduce some simplifications over the arguments in Ref. 28.

Write

$$\varphi_t(x) = \tilde{x} = x + t\zeta(x) + O(t^2) \ ,$$

where

$$\zeta(x) = \frac{\partial \tilde{x}}{\partial t}\bigg|_{t=0}$$

is the variation vector field. We compute

$$\begin{aligned}
\frac{d}{dt}\bigg|_{t=0} (\varphi_t^* \, du) &= \frac{d}{dt}\bigg|_{t=0} \partial_i u(\tilde{x}) \, d\tilde{x}^i \\
&= \frac{d}{dt}\bigg|_{t=0} \partial_i u(\tilde{x}) \frac{\partial \tilde{x}^i}{\partial x^j} \, dx^j \\
&= \frac{d}{dt}\bigg|_{t=0} \partial_i u(\tilde{x}) \left(\delta_j^i + t \frac{\partial \zeta^i}{\partial x^j} \right) dx^j \ .
\end{aligned}$$

Let $y = \varphi_t^{-1}(x)$. Then in terms of y,

$$\frac{d}{dt}\bigg|_{t=0} \partial_i u(\tilde{x}) \, d\tilde{x}^i = \partial_i u(x) \frac{\partial \zeta^i}{\partial x^j} \, dx^j \ . \tag{0.4}$$

Since $\tilde{x}$ is only defined to first order in t, the Jacobian J of the transformation $x \to y$ satisfies

$$\frac{d}{dt}\bigg|_{t=0} J(\varphi_t^{-1}) = \frac{d}{dt}\bigg|_{t=0} \left| \frac{\partial x}{\partial \tilde{x}} \right| = -\operatorname{div} \zeta \ . \tag{0.5}$$

Using the chain rule, (0.4), and (0.5),

$$\begin{aligned}
\delta E^r(u) &= \frac{d}{dt}\bigg|_{t=0} \int \langle \varphi_t^* \, du, \varphi_t^* \, du \rangle \, dy \\
&= \int \langle du(x), du(x) \rangle \frac{d}{dt}\bigg|_{t=0} J(\varphi_t^{-1}) \, dx \\
&\quad + 2 \int \left\langle \frac{d}{dt}\bigg|_{t=0} [\partial_i u(\tilde{x}) \, d\tilde{x}^i], \partial_i u(\tilde{x}) \, d\tilde{x}^i \right\rangle dx
\end{aligned}$$

$$= -\int |du|^2 \operatorname{div}\zeta \, dx + 2\int \left\langle \partial_i u(x) \frac{\partial \zeta^i}{\partial x^j}\, dx^j, du \right\rangle dx$$

$$= -\int |du|^2 \operatorname{div}\zeta \, dx + 2\int \langle du(\nabla_{e_i}\zeta), du(e_i)\rangle \, dx \,, \tag{0.6}$$

where $\{e_i\}$, $i = 1, \ldots, n$ is an orthonormal frame field.

An essential feature of both harmonic maps and Yang-Mills fields is the nonlinearity of their Euler-Lagrange equations. Thus in general we can expect stationary points to have singularities. But since solutions to the Yang-Mills equations are invariant under the action of G-valued maps, we can often "remove" the singularity by applying such a map (called a *gauge transformation*). If the map is continuous, it will preserve the topology of X, *i.e.*, no new singularities will be introduced. In the examples that we shall consider, the purpose of the gauge transformation will always be to transform the Yang-Mills equations into an elliptic system, for which a strong regularity theory exists. The Yang-Mills equations in such a gauge bear a marked resemblance to the equations for a harmonic map into the Lie group G. Thus many of the removable singularities theorems for gauge fields imply analogous statements about harmonic maps.

The relation between the Yang-Mills and harmonic map equations can be seen by writing the two systems in coordinate form. The Euler-Lagrange equations of the Yang-Mills energy functional can be written

$$D_A^* F_A = 0 \,, \tag{0.7}$$

where D_A^* is the formal adjoint of the exterior covariant derivative D_A on X. In local coordinates, (0.7) becomes

$$\partial_i F_{ij} + [A_i, F_{ij}] = 0 \,.$$

If we add the *Lorentz* (*Hodge*, *Coulomb*) condition

$$\partial_i A_i = 0 \,,$$

we obtain a quasilinear elliptic system in the potential A, using the relation

$$F_{ij} = \partial_j A_i - \partial_i A_j + A_i A_j - A_j A_i \,. \tag{0.8}$$

Since

$$(dA)_{ij} = \partial_j A_i - \partial_i A_j$$

and

$$[A_i, A_j] = A_iA_j - A_jA_i = (A \wedge A)_{ij} ,$$

the Yang-Mills equations in an elliptic gauge can be written in the compact notation

$$(dd^* + d^*d)A + \tfrac{1}{2}d^*[A,A] + *[A,*dA] + *\big[A,*[A,A]\big] = 0 , \qquad (0.9)$$

where $dd^* + d^*d$ is the Hodge Laplacian on 1-forms and $* : \Lambda^p \to \Lambda^{n-p}$ is the Hodge involution.

The harmonic map equation has a nonlinear structure similar to (0.9). In local coordinates,

$$\Delta_M u^i + \gamma^{\alpha\beta}\Gamma^i_{j\ell}\big(u(x)\big)\partial_\alpha u^j \partial_\beta u^\ell = 0 , \qquad (0.10)$$

where

$$\Delta_M = \frac{1}{\sqrt{\gamma}}\partial_\alpha\big(\sqrt{\gamma}\,\gamma^{\alpha\beta}\partial_\beta\big)$$

is the Laplace-Beltrami operator on M and

$$\Gamma^i_{j\ell} = \tfrac{1}{2}g^{ik}(\partial_j g_{k\ell} - \partial_k g_{j\ell} + \partial_\ell g_{jk}) .$$

Due to the existence of a smooth embedding of N into $\mathbb{R}^k$ we can rewrite (0.10) in the form

$$(d^*d + dd^*)u = a(du, du) ,$$

where a is the second fundamental form of N in $\mathbb{R}^k$. [Compare also inequalities (1.3) and (3.9), below.]

In Section 3 we discuss the removability of singularities in harmonic maps. However, the problem is less important in the harmonic map case than in the case of gauge fields. Since the notion of weak solution is well-defined for harmonic maps, boundary-value problems for harmonic maps can be reduced to the question of the regularity of certain weak solutions. The boundary-value problem for the Yang-Mills equations is harder. Gauge-equivalent connections need not lie in the same function space, a fact which precludes a good notion of weak solution. If one replaces the idea of a weak solution with that of a limit point of a sequence of approximations, one encounters the problem that a sequence of C^∞ approximations with uniformly bounded energy will converge to a Yang-Mills connection only on the complement of a singular set of codimension 4. In dimension 4 this singular set is removable, and the Dirichlet and

Neumann problems have been solved (only) in this dimension.[16] In higher dimensions the singularity does not seem to be removable under a finite-energy hypothesis. Thus the regularity problem in gauge theory is closely connected with the removability of singularities.

Geometrically, the removability of singularities in the Yang-Mills equations is connected to Donaldson's Theorem[7] on the existence of exotic differentiable structures on $\mathbb{R}^4$. In physical theories based on the Yang-Mills equations one needs to know the existence and location of singularities (*e.g.*, Ref. 11). Thus the emphasis in this paper will be on singularities of Yang-Mills fields. In physical theories these fields are generally coupled to matter fields (sources); but we shall find in Section 3 that the interesting mathematical problems are often already present in the "sourceless" case.

Perhaps the main goal in the study of any nonlinear partial differential equation is an understanding of the way in which a geometric object inherits a certain smoothness by being a solution of the equation. This goal is impossible to reach in the absence of an n-dimensional theory. For example, the Yang-Mills energy is a conformally invariant functional in dimension 4, so the energy of any Yang-Mills field can be made locally small in this dimension; in lower dimensions any finite-energy Yang-Mills connection is continuous by the Sobolev theorem. (If $F_A \in L^p$ for $p > n/2$, then the scalar $\|F_A\|_{L^\infty}$ satisfies the hypotheses of Theorem 5.3.1 of Ref. 18 and is thus bounded. The Sobolev theorem then gives a Hölder estimate for A.) For these reasons the critical dimensions for the *analysis* of the Yang-Mills equations are $n > 4$; in these dimensions any smoothness that the connection attains must illuminate some property of the equations themselves. (An exactly analogous argument holds for harmonic maps, in which the critical domain dimension is $n = 2$.) In this paper we are exclusively concerned with smoothness properties possessed by higher-dimensional fields.

The remainder of the paper is organized into four sections. Section 1 is an introduction to the removable singularities problem for Yang-Mills fields and includes some preliminary results. Section 2 contains theorems for low-energy gauge fields. Section 3 concerns the coupling of Yang-Mills fields to matter and the extension of the results in Sections 1 and 2 to harmonic maps. In Section 4 we prove a theorem for a class of elliptic subsolutions of which the fields considered here are a special case.

In the sections that follow we assume that the euclidean unit ball B^n is centered at the origin of coordinates in $\mathbb{R}^n$. Unless otherwise stated, C [or $C(n)$]

denotes a generic, positive, dimensional constant the value of which may change from line to line.

1. Removability of singularities in Yang-Mills fields

In dimensions 2 and 3, all finite-energy Yang-Mills fields are regular by the Sobolev theorem. In dimensions 5 and higher, singular fields with finite energy can be explicitly constructed.[39,42,21] The critical case is dimension 4. In this dimension Uhlenbeck showed that isolated singularities in finite-energy Yang-Mills fields can be removed by a continuous gauge transformation.[39]

It is natural to ask whether singularities in dimensions $n > 4$ can be removed if the energy is assumed to be bounded above by a prescribed (small) constant. Physically this condition is suggested by the hope that any singularity must be accompanied by a detectable increase of energy. Variationally the condition is natural since the admissible class of connections is determined by specifying the energy functional rather than, for instance, by trying to define an abstract $H^{p,r}$ space of gauge-invariant connections.

In the next section we show the existence of a positive dimensional constant κ for which any gauge field with energy smaller than κ cannot have point singularities. This result was proven independently in Refs. 19 and 25. The proof given here follows Ref. 25 but with some modifications (essentially introduced in Ref. 19). In addition we show that in sufficiently high dimensions fields with energy below κ cannot even have an infinite sequence of singularities, provided the sequence lies in a $C^{1,\alpha}$ curve and contains a single limit point.

First, however, we state some of the essential lemmas from gauge theory and classical analysis. We also prove some removable singularities theorems for higher-order singularities in which the low-energy condition is replaced by an L^p condition on the field at the singularity.

A Yang-Mills connection satisfies an elliptic partial differential equation if and only if it is d-coclosed. As we mentioned in the preceding section, this is a gauge-dependent condition. However, we have the following fundamental result by K. Uhlenbeck:

Lemma 1.1. (Uhlenbeck[40]). *Let* $X = B^n \times \mathbb{R}^\ell$, $G \subset SO(\ell)$, $2p \geq n$, $D = d + \widetilde{A}$ *for* $\widetilde{A} \in H^{1,p}(B^n, \mathbb{R}^\ell \times \mathfrak{G})$. *Then there exists* $\kappa(n) > 0$ *and* $c(n) < \infty$ *such that if* $\|F_A\|^{n/2}_{L^{n/2}} \leq \kappa(n)$, *then* D *is gauge equivalent by an element* $s \in H^{2,p}(B^n, G)$ *to a connection* $d + A$, *where* A *satisfies*

i) $d^*A = 0$

ii) $\|A\|_{H^{1,p}} \leq c(n)\|F\|_{L^p}$.

(See also Ref. 41.)

If $F_A \in L^{n/2}$, then this norm can be made small by a conformal dilation which leaves the Yang-Mills equations invariant (Ref. 34, Lemma 1.1). Thus it might seem that the goal of a proof of the removability of singularities in low-energy gauge fields is to show that the Yang-Mills field is in the space $L^{n/2}$ on a domain which includes the singularity. Indeed, this is the first step; but notice that the map s in Lemma 1.1 is not guaranteed continuous unless $F_A \in L^p$ for some $p > n/2$ (in which case the Sobolev embedding theorem applies to the map s). Thus before we apply Lemma 1.1 we must show that F_A lies in a higher L^p space than $L^{n/2}$. This step requires some technical lemmas from classical elliptic theory:

Lemma 1.2. (Serrin[31]). *Denote by $U(\Sigma)$ the class of smooth functions $\bar{\eta}(x)$ which satisfy $0 \leq \bar{\eta} \leq 1$ and vanish in some neighborhood of the compact set Σ. Let Σ have zero s-capacity for $1 \leq s \leq n$. Then there exists a sequence of functions $\bar{\eta}^{(\nu)}$ contained in $U(\Sigma)$ such that $\bar{\eta}^{(\nu)} \to 1$ a.e. and $\|\nabla\bar{\eta}^{(\nu)}\|_{L^s} \to 0$.*

Lemma 1.3. (Carlson[43]). *Let the compact set Σ have Hausdorff dimension m, $0 < m < n-1$. Then the s-capacity of Σ is zero, where $s = n - m - \varepsilon$ for ε in the interval $(0, n-m-1)$. Moreover, if Σ is a compact Lipschitz manifold of codimension s, then the s-capacity of Σ is zero.*

Although the lemma is well-known, the assertion proven in Ref. 43 is not precisely Lemma 1.3 but rather a closely related result. Thus we include here a brief sketch of Carlson's proof:

Idea of the proof of Lemma 1.3. First we recall that the *s-dimensional outer measure* of a set Q is the number

$$\mathcal{H}^s(Q) = \lim_{\delta \to 0} \mathcal{H}^s_\delta(Q) ,$$

where

$$\mathcal{H}^s_\delta(Q) = \inf \sum_i r_i^s , \qquad s \geq 0 ;$$

the infimum is taken over all countable coverings of Q by families of open balls of radius $\{r_i\} < \delta$. The *s-dimensional Hausdorff measure* H^s is the restriction of $\mathcal{H}^s$ to *measurable sets, i.e.*, to sets Q^* such that for all sets $A \subset \mathbb{R}^n$,

$$\mathcal{H}^s(A) = \mathcal{H}^s(A \cap Q^*) + \mathcal{H}^s(A/Q^*) .$$

By construction

$$\mathcal{H}^s_\delta(Q) \geq \delta^{s-t}\mathcal{H}^t_\delta(Q) ,$$

so if $\mathcal{H}^t(Q) > 0$ and $s < t$, then $\mathcal{H}^s(Q) = \infty$. Similarly, if $s < t$ and $\mathcal{H}^s(Q) < \infty$, then $\mathcal{H}^t(Q) = 0$. A consequence of these sort of arguments is the existence of a unique number $\dim_H Q$, the *Hausdorff dimension* of Q, such that

$$H^s(Q) = \begin{cases} \infty & \text{if and only if } 0 \leq s < \dim_H Q , \\ 0 & \text{if and only if } \dim_H Q < s < \infty . \end{cases}$$

Now the *s-capacity* of a compact set $Q \subset \mathbb{R}^n$ is the number

$$\text{cap}_s(Q) = \inf \int_Q |\nabla\psi|^s * 1 , \qquad 1 \leq s < n ,$$

where the infimum is taken over all functions $\psi \in H^{1,s}(\mathbb{R}^n) = H_0^{1,s}(Q)$ which are ≥ 1 in the interior of Q.

For simplicity assume that Q is a Lipschitz manifold. We intend to show that if $H^s(Q) = 0$ for $0 < s \leq m-1$, then $\text{cap}_{n-s}(Q) = 0$. This is the heart of the proof of Lemma 1.3 since if $H^s(Q) = 0$ then $\dim_H Q < s$.

Consider a sequence $\{x_i\}_{i=1}^m$ of points and choose $\{r_i\}_{i=1}^m$ so that the sequence $\{B_{r_i}(x_i)\}_{i=1}^m$ is an open covering of Q. Define

$$\ell_i(r) = \frac{2r_i - \rho}{r_i}$$

for $r_i \leq \rho \leq 2r_i$. Set

$$\varphi_i(x) = \begin{cases} 1 , & \text{for } x \in B_{r_i}(x_i) \\ \ell_i(|x - x_i|) , & \text{for } x \in B_{2r_i}(x_i) - B_{r_i}(x_i) \\ 0 , & \text{for } x \in \mathbb{R}^n - B_{2r_i}(x_i) . \end{cases}$$

Then

$$|\nabla\varphi_i| = r_i^{-1}$$

inside the annulus $B_{2r_i}(x_i) - B_{r_i}(x_i)$. Define

$$\psi(x) = \sup_{1 \leq i \leq m} \varphi_i(x) ;$$

then $\psi(x) \geq 1$ on Q (since $\ell_i > 1$), $\psi \in H^{1,n-s}(\mathbb{R}^n)$, and

$$\int_Q |\nabla\psi|^{n-s} * 1 \leq \sum_{i=1}^m \int_{B_{2r_i}(x_i)} r_i^{s-n} * 1 \leq C(n) \sum_{i=1}^m r_i^s .$$

This implies that if $H^s(Q)$ vanishes, then necessarily $\text{cap}_{n-s}(Q)$ vanishes, which essentially proves Lemma 1.3.

Since we consider singular sets which are not points, we require the following result, which uses Lemma 1.2 and Lemma 1.3 to slightly extend a lemma by L.M. Sibner[33]:

Lemma 1.4. *Let $u \in C^\infty(B^n/\Sigma)$, $u \geq 0$ satisfy*

$$\Delta u + f(x)u \geq 0 \tag{1.1}$$

in B^n/Σ, where B^n is the unit ball in $\mathbb{R}^n$, Σ is a compact set of Hausdorff dimension $m < n-2$, and $f \in L^{n/2}(B^n)$. Suppose $u \in L^{2q_0\alpha} \cap L^{2q}(B^n)$ where $\alpha = (n-m-\varepsilon)/(n-m-2-\varepsilon)$ for some small $\varepsilon > 0$ and for $(q_0, q) \mid \frac{1}{2} < q_0 < q$. Then in a sufficiently small ball $\widetilde{B}^n$ strictly contained in B^n, $\forall \eta \in C_0^\infty(\widetilde{B}^n)$, $\eta \geq 0$, u satisfies

$$\int_{\widetilde{B}^n} \eta^2 |\nabla u^q|^2 \, dx \leq C \int_{\widetilde{B}^n} |\nabla \eta|^2 u^{2q} \, dx \, .$$

Proof. Integrating ∇u by parts using (1.1) gives

$$\int_{B^n} \nabla u \cdot \nabla \xi \, dx \leq \int_{B^n} |f| u \xi \, dx$$

$\forall \xi \in C_0^\infty(B^n)$ such that $\xi \geq 0$. Now let

$$\xi = \eta \bar{\eta} G(u)$$

for $\eta \geq 0$, $\bar{\eta} \geq 0$, $\bar{\eta}(x) = 0$ for x in a neighborhood of Σ, $G(u) = S(u)S'(u)$, where S is the *Serrin test function*[31]

$$S(u) = \begin{cases} u^q & \text{for } 0 \leq u \leq \ell \\ \dfrac{1}{q_0}\left[q\ell^{q-q_0}u^{q_0} + (q_0 - q)\ell^q\right] & \text{for } \ell \leq u \, . \end{cases}$$

Then if $\widetilde{B}^n$ is chosen so that $\|f\|_{L^{n/2}(\widetilde{B}^n)}$ is sufficiently small relative to the Sobolev constant and η is chosen so that $\eta \in C_0^\infty(\widetilde{B}^n)$, we have

$$\begin{aligned} &\int_{\widetilde{B}^n} (\eta\bar{\eta})^2 |\nabla S|^2 \, dx \\ &\leq C\left(\int_{B^n} |\nabla \eta|^2 \bar{\eta}^2 S^2 \, dx + \int_{B^n} \eta |\nabla \bar{\eta}|^2 S^2 \, dx\right) \\ &\leq C\left(\int_{B^n} |\nabla \eta|^2 \bar{\eta}^2 S^2 \, dx + \|\nabla \bar{\eta}\|^2_{L^{n-m-\varepsilon}(B^n)} \|u^{2q_0}\|_{L^\alpha(B^n)}\right) . \end{aligned} \tag{1.2}$$

Choosing $\bar{\eta} = \bar{\eta}^{(\nu)}$ to be the sequence of Lemma 1.2, we conclude that the last term in the sum on the right-hand side of (1.2) tends to zero as $\nu \to \infty$. The proof is completed by letting ℓ tend to infinity and invoking classical convergence theorems.

Now we can state and prove our first removable singularities theorem for the Yang-Mills equations.

Theorem 1.5. *a) Let F_A be a C^∞ solution of the Yang-Mills equations* (0.7) *in B^n/Σ, where B^n is the unit ball in $\mathbb{R}^n$, $n > 4$, and Σ is a smooth compact submanifold of codimension k, $k > 4$. If $F_A \in L^{n/2}(B^n)$, then F_A is equivalent via a continuous gauge transformation to a solution $F_{\widetilde{A}} \in C^\infty(B^n)$.*

b) The conclusion of a) still holds if Σ has codimension $k \geq 3$ provided $n \geq 6$.

Theorem 1.5 was proven for the case $\Sigma = \{0\}$ by L.M. Sibner.[33] The arguments for the more general case are similar. Although the low-energy condition which will be considered in the next section can be considered natural from a physical or variational perspective, the natural conditions for removing singularities in elliptic theory seem to be L^p conditions or related notions such as s-capacity.[31]

To prove the theorem we first observe that the hypotheses imply $F_A \in L^p(B^n)$ for some $p > n/2$. In case a) this follows from Lemma 1.4 and the inequality[39]

$$\Delta|F| + C(n)|F|^2 \geq 0\ . \tag{1.3}$$

In case b) it follows from Lemma 1.4 and the inequality[2]

$$\Delta|F|^{1-\varepsilon} + C(n)|F|^{2-\varepsilon} \geq 0$$

for some $\varepsilon = \varepsilon(n) > 0$. Applying a covering argument analogous to Section 5 of Ref. 35 (but simpler), we find that Lemma 1.1 can still be applied if B^n is replaced by B^n/Σ. Lemma 1.1 guarantees the existence of a continuous gauge transformation g taking A to a connection $\widetilde{A} = g(A)$ which satisfies an elliptic system of the form (0.9) in B^n/Σ. This system can be attacked by the techniques of classical elliptic theory. In particular, applying Theorem 1 of Ref. 17 we conclude that $\widetilde{A}$ is Hölder continuous and applying Theorem 6.6 iii) of Ref. 13 we find that $\widetilde{A}$ is continuously differentiable. Since the system (1.4) is diagonal, it is now an easy matter to successively take derivatives and complete the proof of Theorem 1.5 by elementary elliptic bootstrapping.

The case of codimension-2 singularities in 4-dimensional gauge fields is considered in Ref. 35. In this case one encounters holonomy phenomena also found in point singularities of 2-dimensional fields.[37]

2. Low energy fields

A basic technical lemma in removable singularities theorems for low-energy fields is the monotonicity formula of Price[28] for r-stationary fields on a manifold

M. (See also Allard.[1]) This formula generalizes to the case of fields which are r-stationary on a manifold M/Σ, where Σ is a singular set of positive Hausdorff dimension, provided that the dimension of M is sufficiently large.

Lemma 2.1. *Let F_A be a smooth r-stationary Yang-Mills field on B^n/Σ, $n > 5$, where Σ is an infinite sequence of point singularities contained in a $C^{1,\alpha}$ curve Γ. Let F_A satisfy the growth condition*

$$|F_A(x)| \leq C\rho^{-2} , \tag{2.1}$$

where ρ is the distance from $x \in B^n/\Sigma$ to Γ. Then $\forall\, x_0 \in B_{1/4}$ and all $0 < r \leq R \leq 1/2$, F_A satisfies the **monotonicity inequality**

$$r^{4-n} \int_{B_r(x_0)} |F_A|^2 * 1 \leq R^{4-n} \int_{B_R(x_0)} |F_A|^2 * 1 . \tag{2.2}$$

Proof. The analogue for euclidean Yang-Mills fields of Equation (0.6) is the expression[28]

$$\int_M \Big\{ |F_A|^2 \operatorname{div} \zeta - 4\big\langle F_A(\nabla_{e_i}\zeta, e_j), F_A(e_i, e_j)\big\rangle \Big\} * 1 = 0 , \tag{2.3}$$

where ζ is the initial velocity field for the lifted reparametrization of M and $\{e_i\}$, $i = 1, \ldots, n$ is an orthonormal basis. (If M is euclidean, then ∇_{e_i} is just the directional derivative along e_i.)

Choose $r = |x|$, $e_1 = \partial/\partial r$, $e_j = \vartheta_j$, $j = 2, \ldots, n$, and

$$\zeta = \bar{\eta}^{(\nu)} \eta(r) r \cdot \frac{\partial}{\partial r} ,$$

where the function $\bar{\eta}^{(\nu)}$, vanishing in a neighborhood of Σ, is the sequence of Lemma 1.2; $\eta(r) \in C_0^\infty[0,1]$ is a function chosen so that $\eta'(r) \leq 0\ \forall\, r$ and

$$\eta(r) = \begin{cases} 1 & \text{for } r \leq \tau,\ \tau \in (0,1) \\ 0 & \text{for } r > \tau + \delta,\ \delta > 0 . \end{cases}$$

Evaluating (2.3) for this choice of ζ we obtain, after some algebra,

$$- \int_B [\bar{\eta} r \eta' + \bar{\eta}\eta(n-4)] |F_A|^2 * 1 \geq$$
$$- C \Big\{ \int_B |\nabla \bar{\eta}^{(\nu)}| r\eta |F_A|^2 * 1 + \int_B \bar{\eta} r \eta' |(F_A)_{r\vartheta_j}\, dr \wedge d\vartheta^j|^2 * 1 \Big\} .$$

Here $(F_A)_{r\vartheta_j}\, dr \wedge d\vartheta^j$ is the contraction of the 2-form F_A by the radial basis vector (and we have of course summed over repeated indices).

If s is the codimension of Σ, then

$$\int_B |\nabla \bar{\eta}^{(\nu)}| r |F_A|^2 * 1 \leq \|\nabla \bar{\eta}^{(\nu)}\|_{L^{s-\varepsilon}(B)} \|F_A\|_{L^{2(s-\varepsilon)/(s-\varepsilon-1)}(B)} \,, \tag{2.4}$$

where in this case ε is the constant of Lemma 1.3. Now (2.1) implies that

$$\int_B |F_A|^p * 1 = \int_{B/\Gamma} |F_A|^p * 1 \leq C \int_0^1 \rho^{-2p+n-2}\, d\rho < \infty$$

for $p < (n-1)/2$. Thus if $n > 6$ we use the fact that $\Sigma \subset \Gamma$ and take $s = n-1$, $\varepsilon = 0$. If $n = 6$, take $s = n - 1 + \tilde{\delta}$, where $\tilde{\delta} > 0$ and ε is chosen so small that

$$2(s-\varepsilon)/(s-\varepsilon-1) < (n-1)/2 = 5/2 \,.$$

In either case the right-hand side of (2.4) tends to zero as $\nu \to \infty$.

Define

$$\eta(r) = \eta_\tau(r) = \varphi(r/\tau) \,.$$

Then

$$\tau \frac{\partial}{\partial \tau} \eta_\tau(r) = -r\eta'(r) \geq 0 \,.$$

Thus we obtain

$$\int_B \left[\tau \frac{\partial}{\partial \tau}(\eta_\tau) + \eta(4-n) \right] |F_A|^2 * 1 \geq$$
$$C \int_B \tau \left(\frac{\partial}{\partial \tau} \eta_\tau \right) |(F_A)_{r\vartheta_j}\, dr \wedge d\vartheta^j|^2 * 1 \,. \tag{2.5}$$

Multiplying (2.5) by the integrating factor τ^{3-n} and letting δ tend to zero, we obtain

$$\frac{d}{d\tau} \left(\int_{B_\tau} |F_A|^2 * 1 \right) \geq 0 \,.$$

This completes the proof of Lemma 2.1.

If $n \geq 7$, then the monotonicity inequality holds inside B^n for larger singular sets, without hypothesis (2.1). We can prove this via a technical lemma due to D. Costa and G. Liao:

Lemma 2.2. (Costa-Liao[5]). *Let F_A be a smooth r-stationary Yang-Mills field on B^n/Σ, where Σ is a smooth, compact submanifold of B^n and $d = \dim \Sigma < n-4$. Then $\forall\, x_0 \in B_{1/4} \cap \Sigma$, $0 < r \leq R \leq 1/2$, F_A satisfies the inequality*

$$r^{d+4-n} \int_{D_r} |F_A|^2 * 1 \leq R^{d+4-n} \int_{C_R} |F_A|^2 * 1 + C\big(f(R) - f(r)\big) \int_{C_R} |F_A|^2 * 1 \,, \qquad (2.6)$$

where

$$f(t) = \begin{cases} \ln t & \text{if } n = d+6 \\ t^{d+6-n}/(d+6-n) & \text{if } n \neq d+6 \,, \end{cases}$$

$$C_t = \big\{x \mid \operatorname{dist}(x, \Sigma) \leq t\big\} \cap B \,,$$

$$D_t = C_t \cap B_{3/4} \,.$$

The proof of Lemma 2.2 is given in Ref. 5. Using ideas by Costa and Liao, we can prove the following corollary:

Corollary 2.3. *Assume the conditions of Lemma 2.2 but take $d = 1$ and $n \geq 7$. Then the Yang-Mills field F_A satisfies inequality (2.2) for $x_0 \in B_{1/4}$, $0 < r \leq R \leq 1/2$.*

Proof. Evaluate (2.3) choosing

$$\zeta(x) = \xi_\sigma(\lambda)\eta_\tau(r) r \partial_r(x) \,,$$

where r is the radial coordinate, $x \in B_{1/4}(x_0)$, $\lambda = \operatorname{dist}(x, \Sigma)$, $\eta_\tau(r)$ is defined as in the proof of Lemma 2.1,

$$\xi_\sigma = \begin{cases} 0 & \text{if } 0 \leq \lambda \leq \sigma \\ 1 & \text{if } \lambda \geq 2\sigma \,, \end{cases}$$

and $0 \leq \xi_\sigma \leq 1$ if $\sigma \leq \lambda \leq 2\sigma$. Notice that the "bad" term in (2.3) satisfies

$$\int_{B_{1/2}} |d\xi_\sigma| |F_A|^2 * 1 \leq C(2\sigma^{-1}) \int_{C_{2\sigma}} |F_A|^2 * 1 \,. \qquad (2.7)$$

Taking $r = 2\sigma$ and $R = 1/2$ in (2.3) gives

$$(2\sigma)^{-1} \int_{C_{2\sigma}} |F_A|^2 * 1 \leq C\sigma^{-1}\sigma^{n-d-4}(1 + \sigma^{d+6-n}) \int_{B_1} |F_A|^2 * 1 \qquad (2.8)$$

for $n > d+6$ and a similar expression involving $\sigma^{-1}\sigma^{n-d-4}\ln(\sigma^{-1})$ if $n = d+6$. In either case the right-hand side of (2.8) vanishes as σ tends to zero.

The remainder of the argument is exactly analogous to the proof of Lemma 2.1.

Remark. *a) If Σ is a line singularity, then Lemma 2.2 holds for $n \geq 7$.*

b) If Σ is an isolated point singularity, then the arguments of Ref. 19 (and Ref. 15) show that the monotonicity inequality holds $\forall\ n > 4$.

An important consequence of the above lemmas is an *a priori* estimate for Yang-Mills fields with singularities. The case $\Sigma = \{0\}$ was proven by Nakajima,[19] adapting an argument for C^∞ harmonic maps by Schoen and Uhlenbeck.[30] (See also Ref. 15.) Since the following lemma is a straightforward generalization of these results using Lemma 2.1 and Corollary 2.3, we omit the proof.

Lemma 2.4. *Let F_A be an r-stationary C^∞ Yang-Mills field in B^n/Σ. If Σ is a smooth, compact submanifold of B^n of codimension > 6, let $n > 6$. If Σ is an infinite sequence of point singularities contained in a $C^{1,\alpha}$ curve $\Gamma \subset\subset B^n$, let $n \geq 6$ and let (2.1) hold. If $\Sigma = \{0\}$, let $n \geq 5$. Then there is a constant $\kappa(n) > 0$ such that*

$$\int_{B^n} |F_A|^2 * 1 \leq \kappa$$

implies $\forall\ x \in B^n/\Sigma$,

$$|F(x)| \leq C(n)\kappa\rho^{-2}\ , \tag{2.9}$$

where $\rho = \mathrm{dist}(x, \Sigma)$.

We are now ready to state and prove our two removable singularities theorems for low-energy gauge fields.

Theorem 2.5. [19,25] *Let F_A be an r-stationary C^∞ solution of the Yang-Mills Equations (0.7) in $B^n/\{0\}$, $n > 4$. Then there is a constant $\kappa(n) > 0$ such that*

$$\int_{B^n} |F_A|^2 * 1 \leq \kappa$$

implies that F_A is equivalent via a continuous gauge transformation to a C^∞ Yang-Mills field $F_{\tilde{A}}$ in B.

If $n = 4$ a small-energy condition can be obtained from a finite-energy condition, since solutions to (0.7) are conformally invariant and the energy functional is conformally invariant (only) in dimension 4. Thus in dimension 4 the results of this section follow from known results[39,35] for finite-energy fields. Finite-energy fields are regular for $n \leq 3$ by the Sobolev theorem. This leads us to conclude that low-energy conditions are natural in dimensions exceeding 4.

The idea of the proof is to try to show that $F \in L^{n/2}(B)$, and then to apply Theorem 1.5, also in the special case $\Sigma = \{0\}$. (Notice that in our case Lemma 2.4 gives immediately, by integration in spherical coordinates, the result $F \in L^p(B)\ \forall\, p < n/2$. However there are known singular solutions of (0.7) that satisfy this L^p condition.[39]) Thus Theorem 2.5 is an immediate corollary of the following lemma:

Lemma 2.6. *Under the hypotheses of Theorem 2.5, $F_A \in L^{n/2}(B)$.*

Proof. We want to compare two estimates. The first is a trivial consequence of Lemma 2.4:

$$\int_{B^n} |F_A|^{n/2} * 1 = \int_{B^n} |F_A|^{(n-2)/2} |F_A| * 1 \le C(\kappa) \int_{B^n} |x|^{2-n} |F_A| * 1 . \quad (2.10)$$

The idea is to bound the right-hand side of (2.10) by a finite number. This requires a second estimate which is more subtle. Integrating $u = |F_A|$ by parts using (1.3), we obtain

$$\int_{T_\varepsilon} \nabla\xi \cdot \nabla u * 1 \le C \int_{T_\varepsilon} \xi u^2 * 1 \qquad (2.11)$$

for $\xi \in C_0^\infty(T_\varepsilon)$, $\xi \ge 0$, $T_\varepsilon = \{x \mid \varepsilon < |x| < R\}$, $R \le 1/2$. Choose

$$\xi(x) = \zeta(x)\psi(x) ,$$

where ζ is the trapezoidal cut-off function

$$\zeta(x) = \begin{cases} 1 & \text{for } 2\varepsilon \le |x| \le R/2 , \\ 0 & \text{for } |x| \le \varepsilon,\ |x| \ge R , \end{cases}$$

$$\left.\begin{aligned} \|\nabla\zeta\|_{L^\infty} &\le C_1 \varepsilon^{-1} , \\ \|\nabla\zeta\|_{L^\infty} &\le C_2 \varepsilon^{-2} ; \end{aligned}\right\} \qquad (2.12)$$

ψ is the weighted Green's function

$$\psi(x) = |x|^{4-n} .$$

After some algebra we can write (2.11) in the form

$$\begin{aligned} &\int_{T_\varepsilon} \zeta u \psi \Big[\frac{-\Delta\psi}{\psi} - C(n) u\Big] * 1 \\ &\quad \le \int_{T_\varepsilon} (\psi|\Delta\zeta| + 2|\nabla\zeta|\,|\nabla\psi|)\, u * 1 \\ &\quad \le \int_{\varepsilon<|x|<2\varepsilon} (\psi|\Delta\zeta| + 2|\nabla\zeta|\,|\nabla\psi|)\, u * 1 + C(R) , \end{aligned} \qquad (2.13)$$

where we have used the fact that the derivatives of ζ vanish on a portion of T_ε. Estimates (2.12) and Lemma 2.4 imply that the right-hand side of (2.13) is bounded above by a finite constant independent of ε. We can calculate a lower bound for the left-hand side of (2.13) using

$$\frac{-\Delta\psi}{\psi} - C(n)u \geq \frac{2(n-4) - \widetilde{C}(n)\kappa}{|x|^2} . \tag{2.14}$$

The right-hand side of (2.14) is positive provided $\kappa < 2(n-4)\widetilde{C}^{-1}(n)$, where $\widetilde{C}(n)$ is in this case the product of the constants in (2.9) and (1.3). Substituting (2.14) into the left-hand side of (2.13) and using the definition of ψ, we find that as ε tends to zero the left-hand side of (2.13) is bounded below by a finite, positive, dimensional multiple of the right-hand side of (2.10). This completes the proof of Lemma 2.6 and thus of Theorem 2.5.

In proving the next result we use a scaling law for the Yang-Mills energy. We can derive this law from the properties of differential forms.

Since a p-form is a geometric object, it must be invariant under scale transformations $x \to e^\lambda x$. That is, if $u(x) = u_{i_1 \ldots i_p}(x)\, dx^{i_1} \wedge \cdots \wedge dx^{i_p}$, then necessarily $u(e^\lambda x) = u(x)$. But since

$$\begin{aligned} u(e^\lambda x) &= \tilde{u}_{i_1 \ldots i_p}(e^\lambda x) e^{p\lambda}\, dx^{i_1} \cdots dx^{i_p} \\ &= u(x) = u_{i_1 \ldots i_p}(x)\, dx^{i_1} \cdots dx^{i_p} , \end{aligned}$$

this invariance forces

$$\tilde{u}_{i_1 \ldots i_p}(e^\lambda x) = e^{-p\lambda} u_{i_1 \ldots i_p}(x) .$$

The Yang-Mills field F_A is a 2-form with L^2-norm

$$\int_\Omega F_A \wedge *F_A = \int_\Omega *(F_A \wedge *F_A)\, d(\text{vol}) = \int_\Omega |F_A|^2 * 1 ,$$

where $* : \Lambda^p \to \Lambda^{n-p}$ is the Hodge involution. Thus under local scale transformations $x \to rx$, $r > 0$, the Yang-Mills energy satisfies

$$\int_{B^n} |F_A(x)|^2 * 1 = r^{4-n} \int_{B_r^n} |F_A(rx)|^2 * 1 . \tag{2.15}$$

Here we have used the fact that the n-dimensional volume form satisfies $dy = r^n\, dx$ for $y = rx$. The significance for us of identity (2.15) is that the hypothesis of small energy is not a conformally invariant condition. This is a crucial problem in proving the following result:

Theorem 2.7. *Let F_A be an r-stationary C^∞ Yang-Mills field in B^n/Σ, where $\Sigma = \{\xi_j\}_{j=1}^\infty$ is a sequence of apparent singular points of F_A converging to a single limit point at the origin. Assume that $\Sigma \subset \Gamma$, where Γ is a $C^{1,\alpha}$ curve. Let $n \geq 6$, and if $n = 6$ assume in addition that $\forall\, x \in B^n/\Sigma$ F_A satisfies*

$$|F_A(x)| \leq C\rho^{-2}\ , \tag{2.16}$$

where $\rho = \mathrm{dist}(x, \Gamma)$. Then there is a positive dimensional constant κ such that

$$\int_{B^n} |F_A|^2 * 1 \leq \kappa$$

implies that F_A extends via a continuous gauge transformation in B^n/Σ to a C^∞ Yang-Mills field $F_{\tilde{A}}$ over B.

Examples of convergent sequences of point singularities can be explicitly constructed, *e.g.*, by methods analogous to those of Ref. 9. All *known* examples of singular gauge fields satisfy (2.16); the constant C in this inequality is not assumed to be small. Notice that although the proof of Theorem 2.5 extends immediately to the case of any finite number of point singularities, the proof fails if there is an infinite sequence of point singularities tending to a limit. This is because the smoothness of the radial test functions used in proving Lemma 2.6 depend on their distance from the singularity. The test functions become singular in a neighborhood of the limit point and cannot be used in that region. However, we can prove Theorem 2.7 as an almost immediate consequence of the other results of this section.

Proof of Theorem 2.7. Decompose $B^n/\{0\}$ into a countable union of n-dimensional annuli $\bigcup_{i=1}^\infty V_i$, where $V_i = \{x \mid 2^{-i-1} \leq |x| \leq 2^{-i}\}$. We want to take an arbitrary annulus V_k into the "standard" annulus V_0 by a conformal dilation f. This map leaves the field Equations (0.7) invariant. The limit point ξ_∞ is fixed under this transformation. We assume that each V_i contains only a finite number of singularities. Otherwise the smallest i for which V_i contained an infinite number of singularities would contain a second limit point in B; but we know that the only limit point lies at the origin. The idea of the construction is to apply the low-energy arguments of Theorem 2.5 at each singular point x_α in V_0. Of course we do not know *a priori* that the energy of f^*F_A is small. At this point we invoke Lemma 2.1 and Corollary 2.3. If $n \geq 7$, then the monotonicity inequality holds without assumption (2.16) by the remark following the proof of

Corollary 2.3. If $n = 6$ we assume (2.16) and obtain the monotonicity inequality from Lemma 2.1. In either case the monotonicity inequality implies

$$\rho^{4-n} \int_{B_\rho(x)} |F_A|^2 * 1 \leq 2^{n-4} \int_{B_{1/2}(x)} |F_A|^2 * 1 \leq 2^{n-4}\kappa \,, \tag{2.17}$$

where κ is the constant of Theorem 2.7. Inequality (2.17) is valid for any $B_\rho(x) \subset V_i$ (in particular we want to apply it for large i) and becomes a low-energy condition under the dilation f via (2.15).

We now conclude that f^*F_A has small energy in V_0, is Yang-Mills there, and has at worst a finite number of isolated point singularities in V_0. By the arguments of Theorem 2.5, f^*F_A is thus equivalent via a continuous gauge transformation to a C^∞ solution in V_0. The pull-back map will introduce no new singularities, and the gauge transformations are continuous across the boundaries of V_i by the arguments on page 26 of Ref. 39. (Note that we can show $f^*F_A \in L^{n/2}(V_0)$ by the arguments of Lemma 2.6 without introducing any gauge transformations; this allows us to apply the arguments of Refs. 39 or 32 to show that any gauge transformations we introduce will be continuous across the boundary.) Pulling back, we conclude that F_A is gauge-equivalent to a C^∞ field in $B^n/\{0\}$. Now Theorem 2.5 implies that F_A is gauge-equivalent to a C^∞ field in B^n.

3. Coupled fields and harmonic maps

The field equations studied by physicists generally involve *coupled fields*: an externally imposed gauge field is assumed to interact with the field induced by a particle. If the particle is a boson (*e.g.*, has integer spin), the *Higgs model* leads to the coupled *Yang-Mills-Higgs* equations on an open domain $\mathcal{O} \subset \mathbb{R}^n$:

$$\left.\begin{aligned} D_A^* F_A &= [D_A\varphi_{\mathcal{H}}, \varphi_{\mathcal{H}}] \\ D_A^* D_A \varphi_{\mathcal{H}} &= \left(\frac{\lambda}{2}\right)\left(|\varphi_{\mathcal{H}}|^2 - \mu^2\right)\varphi_{\mathcal{H}} \,. \end{aligned}\right\} \tag{3.1}$$

Here the constant λ is a physical parameter (the *interaction*, or *coupling* constant), generally — but not exclusively — taken to be nonnegative; our results are independent of the sign of λ. The constant μ represents the mass of the particle; the physical requirement that μ have dimensions of inverse length corresponds to the geometric requirement that μ be a smooth section of a trivial real line bundle over $\mathcal{O}$: under coordinate transformations $y = e^k x$, $\mu_y = e^{-k}\mu_x$.

There is no consensus on a geometric representation of $\varphi_{\mathcal{H}}$ beyond the requirement that $\varphi_{\mathcal{H}}$ be a smooth section of ad X. One interpretation takes $\varphi_{\mathcal{H}} \in \Gamma$ (ad $X \otimes \ell$), where ℓ is the determinant of the volume bundle on $\mathcal{O}$ raised to the power $1/n$. This interpretation has the geometric advantage that the coupled field energy transforms homogeneously under rescalings.[26,27,34] The results of this section require only that $\varphi_{\mathcal{H}}$ be a smooth section of ad X satisfying (3.1).

If the Yang-Mills field interacts with the field of a fermion (a particle with half-integer spin), the relevant field equations are the *Yang-Mills-Dirac* equations

$$\left.\begin{aligned} D_A^* F_A &= (\varphi_{\mathfrak{D}}, \varphi_{\mathfrak{D}}) \\ \not{\partial}_A \varphi_{\mathfrak{D}} &= m\varphi_{\mathfrak{D}} \, . \end{aligned}\right\} \tag{3.2}$$

Here $\not{\partial}_A$ is the covariant Dirac operator on bundle-valued spinors; the *mass* m of the fermion is an eigenvalue of $\not{\partial}_A$. The product $(\varphi_{\mathfrak{D}}, \varphi_{\mathfrak{D}})$ in the first of Equations (3.2) is the expression

$$\langle \varphi_{\mathfrak{D}}, \varphi_{\mathfrak{D}} \rangle = -\tfrac{1}{2}\langle \varphi_{\mathfrak{D}}, e^i \cdot \rho(\sigma^\alpha)\varphi_{\mathfrak{D}} \rangle \sigma_\alpha \otimes e_i \, , \tag{3.3}$$

where $\{\sigma_\alpha\}$ is an orthogonal basis for the lie algebra $\mathfrak{G}$ and ρ is a unitary representation of the gauge group G. By analogy to the physical model, expression (3.3) is called the *current* of $\varphi_{\mathfrak{D}}$.

A mathematically analogous set of equations results if the operator $\not{\partial}$ on spinors is replaced by the operator $d + d^*$ on forms (see, *e.g.*, Ref. 23).

Removable point singularities theorems for pure Yang-Mills fields have been extended to solutions of (3.1) and (3.2) in every dimension.[26,34,32,33,37,23,22] Analytically, the two systems have somewhat different properties. The spinor field has a different conformal weight under rescalings than does the Higgs field, a global property which affects some of the local *a priori* estimates. More importantly, the Weitzenböck formula for the covariant Dirac operator contains a lower-order term which is a product of F_A and $\varphi_{\mathfrak{D}}$. Thus the second-order form of the second of Equations (3.2) is

$$\nabla_A^* \nabla_A \varphi_{\mathfrak{D}} + \tfrac{1}{2} e^i \cdot e^j \cdot F_{ij}(\varphi_{\mathfrak{D}}) = m^2 \varphi_{\mathfrak{D}} \, , \tag{3.4}$$

where ∇_A denotes the total covariant derivative on sections of the spin bundle. This is the form of the Dirac equation useful in analysis, and in this equation $\varphi_{\mathfrak{D}}$ is coupled to A firstly via the covariant derivative and secondly via the term involving the curvature F_A. By comparison, in the Higgs equation [the second of

Equations (3.1)], $\varphi_{\mathcal{H}}$ is coupled to A only through the covariant derivative (this is what physicists call *minimal coupling*). The analytic difference is that in the Higgs model an improvement in the smoothness of $\varphi_{\mathcal{H}}$ results in a simultaneous improvement in the smoothness of its coefficients; this does not occur in the Dirac model.

On the other hand, the removability of singularities in the Dirac model is trivial under certain conditions for which the problem is nontrivial in the Higgs model. These conditions involve the coupled fermion energy

$$E_{\text{fermion}} = \int_\Omega \left\{ |F_A|^2 + \langle \not{\partial}_A \varphi_{\mathfrak{D}}, \varphi_{\mathfrak{D}} \rangle \right\} * 1 \, .$$

This functional is indefinite due to the symmetry of the spectrum for $\not{\partial}_A$ about the origin. Thus point singularities in $\varphi_{\mathfrak{D}}$ tend to "integrate out" of the energy functional.

This phenomenon is not relevant if the equations are considered apart from their variational context and if E_{fermion} is replaced by a positive-definite functional. Such problems are solved in Refs. 26, 23 and 22.

Corresponding to inequality (1.3) for pure Yang-Mills fields are the following subelliptic inequalities satisfied by solutions of the coupled equations[34,22,14]:

$$\begin{aligned}
&\Delta|f| + \left(|\varphi_{\mathcal{H}}|^2 + 2|f|\right)|f| + 2|g|^2 \geq 0 \, , \\
&\Delta|g| + \left[\left(1+|\lambda|\right)|\varphi_{\mathcal{H}}|^2 + |\lambda|\,|w| + 4|f|\right]|g| \geq 0 \, , \\
&\tfrac{1}{2}\Delta\left(|\varphi_{\mathcal{H}}|^2\right) + \frac{|\lambda|}{2}\left(|\varphi|^2 + \mu^2\right)|\varphi_{\mathcal{H}}|^2 \geq 0 \, , \\
&\Delta|F_A| + C\left(|F_A|^2 + |\varphi_{\mathfrak{D}}|\,|D_A\varphi_{\mathfrak{D}}|\right) \geq 0 \, , \\
&\Delta|\varphi_{\mathfrak{D}}| + C|F_A|\,|\varphi_{\mathfrak{D}}| \geq 0 \, ,
\end{aligned} \tag{3.5}$$

where $f = *F_A$, $g = D_A\varphi_{\mathcal{H}}$, $w = \frac{1}{2}(\mu^2 - |\varphi_{\mathcal{H}}|^2)$. From inequalities (3.5) one can derive inequalities satisfied by a so-called *total field* with desirable conformal properties. For example,[34] if we define a scalar density

$$h = |F_A| + |D_A\varphi_{\mathcal{H}}| + |\varphi_{\mathcal{H}}|^2 \, ,$$

then (3.5) implies

$$\Delta h + \left[\left(10 + 2|\lambda|\right)h + |\lambda|\mu^2\right]h \geq 0 \, , \tag{3.6}$$

an inequality similar to (1.3). Using (3.6) one can extend Theorem 1.5 to the Higgs model (the extension to the Dirac model is similar):

Theorem 3.1. [24] *a) Let $(F_A, \varphi_{\mathcal{H}})$ be a C^∞ solution of (3.1) in B/Σ, where B is an n-dimensional ball in $\mathbb{R}^n$, $n > 4$ and Σ is a smooth compact submanifold of codimension k, $k > 4$. Suppose $F_A \in L^{n/2}(B)$ and $\varphi_{\mathcal{H}} \in L^n(B)$. Then $(F, \varphi_{\mathcal{H}})$ is equivalent via a continuous gauge transformation in B/Σ to a C^∞ solution in B.*

b) The conclusion of a) holds if $k \geq 3$ provided $n > 6$ and $D_A\varphi_{\mathcal{H}} \in L^p(B)$ for some $p > 3$.

In order to prove Theorem 3.1 a) using inequality (3.6) we need an L^p estimate for $D_A\varphi_{\mathcal{H}}$:

Lemma 3.2. *Under the hypotheses of Theorem 3.1 a), $|D_A\varphi_{\mathcal{H}}| \in L^r(B)$ for any finite r.*

Proof. We first show that $D_A\varphi_{\mathcal{H}} \in L^2(B)$. Define $\xi = (\eta\bar{\eta})^2\varphi_{\mathcal{H}}$, where $\eta \in C_0^\infty(B, [0,1])$ and $\bar{\eta} = \bar{\eta}^{(\nu)}$ is the sequence of Lemma 1.2. Integrating $D_A\varphi_{\mathcal{H}}$ by parts against $D_A\xi$ we find by a standard calculation that

$$\begin{aligned}\int_B (\eta\bar{\eta})^2|D_A\varphi_{\mathcal{H}}|^2 * 1 &\leq C\int_B [(\eta\bar{\eta})^2 + |\nabla(\eta\bar{\eta})|^2]|\varphi_{\mathcal{H}}|^2 * 1 \\ &\quad + \int_B (\eta\bar{\eta})^2|\varphi_{\mathcal{H}}|^4 * 1 \\ &\leq C'\left(1 + \int_B \eta^2|\nabla\bar{\eta}|^2|\varphi_{\mathcal{H}}|^2 * 1\right). \qquad (3.7)\end{aligned}$$

The last integral on the right in (3.7) is finite by Lemma 1.2, the hypotheses on $\varphi_{\mathcal{H}}$, and the Hölder inequality.

Inequalities (3.5) immediately yield

$$\Delta|D_A\varphi_{\mathcal{H}}| + \left[(1 + \tfrac{3}{2}|\lambda|)(\mu^2 + |\varphi_{\mathcal{H}}|^2) + 4|F_A|\right]|D_A\varphi_{\mathcal{H}}| \geq 0\,. \qquad (3.8)$$

Inequalities (3.7) and (3.8) allow us to complete the proof of Lemma 3.2 by iterating Lemma 1.4 and the Sobolev inequality.

Minkowski's inequality now implies that the total field h is in the space $L^{n/2}(B)$. Apply Lemma 1.4 again, this time using (3.6), to conclude that $F_A \in L^p(B)$ for some $p > n/2$. A generalized form of Lemma 1.1 analogous to Ref. 35 allows us to fix a gauge in which the pair $(A, \varphi_{\mathcal{H}})$ satisfies the diagonal elliptic system

$$\left.\begin{aligned}(dd^* + d^*d)A + \tfrac{1}{2}d^*[A, A] + *[A, *F] &= [D_A\varphi_{\mathcal{H}}, \varphi_{\mathcal{H}}] \\ d^*d\varphi_{\mathcal{H}} + d^*[A, \varphi_{\mathcal{H}}] + *[A, *[A, \varphi_{\mathcal{H}}]] &= \frac{\lambda}{2}(|\varphi_{\mathcal{H}}|^2 - \mu^2)\varphi_{\mathcal{H}}\,.\end{aligned}\right\}$$

The proof of Theorem 3.1 a) can be completed by applying known methods.[17] The proof of Theorem 3.1 b) is an easy modification of the above argument.

All the results of the preceding sections have analogues for harmonic maps. Although some of these analogues are already known, even in such cases the Yang-Mills arguments suggest new proofs and, in some cases, technical extensions. The feature of the Yang-Mills case that tends to generate new proofs of the harmonic map case is the necessity in gauge theory to replace arguments for elliptic equations by arguments for subelliptic inequalities in order to obtain initial estimates. (This aspect of the theory is pursued in the next section.)

If u is a harmonic map, then the inequality satisfied by the 1-form du is,[3,38]

$$\Delta|du| + k|du|^3 \geq 0\,, \qquad k \geq 0\,. \tag{3.9}$$

The constant k depends on the sectional curvature of the target manifold N.

S. Takakuwa[38] proved that if a harmonic map is stationary on a Riemannian manifold M and C^∞ in $M - \{p\}$, where p is a point on M, then $u \in C^\infty(M)$ provided $du \in L^n$ in a neighborhood of p, where $n = \dim M \geq 3$.

Since the problem is local, the curvature of M plays an insignificant role. Thus we can directly apply Lemma 1.4: Using (3.9) and choosing $f(x) = |du(x)|^2$ we obtain (via the Sobolev embedding theorem) a version of Takakuwa's result valid for higher-dimensional singular sets (in higher dimensions) and lacking the hypothesis of stationarity. In the case $\Sigma = \{p\}$ the lemma gives an elementary proof of Takakuwa's theorem for dimensions $n > 3$, again without the hypothesis of stationarity.

A low-energy theorem also exists for harmonic maps with point singularities in dimensions $n \geq 3$. It is due to G. Liao.[15] (The 2-dimensional version, due to Sacks and Uhlenbeck,[29] resembles Uhlenbeck's removable singularities theorem for 4-dimensional gauge fields.[39]) Following the proof of Theorem 2.5 we can construct a new, elementary proof of Liao's low-energy theorem for dimensions $n > 3$. The idea is to show that $|du| \in L^n(B)$ and then use Takakuwa's theorem. Argue as in the proof of Lemma 2.6, but replace inequality (1.3) by inequality (3.9) and choose $\psi(x) = |x|^{3-n}$. (Replace Lemma 2.4 by the original version for harmonic maps.[30])

Using Liao's theorem, we can extend Theorem 2.7 to the harmonic map case by replacing the monotonicity formula for Yang-Mills fields by one for harmonic maps and adjusting the dimension range accordingly.

Our precise result is:

Theorem 3.3. *Let $u : M \to N$ be an r-stationary map from M to N, where M and N are smooth, compact Riemannian manifolds and $n = \dim M \geq 5$. Suppose that u is harmonic in B^n/Σ, where $B^n \subset\subset M$ is a unit ball and $\Sigma = \{\xi_i\}_{j=1}^{\infty}$ is a sequence of apparent singular points of u converging to a single limit point at the center of B^n. Assume that $\Sigma \subset \Gamma$, where Γ is a $C^{1,\alpha}$ curve. If $n = 5$ assume in addition that $\forall x \in B^n/\Sigma$, $|du|$ satisfies*

$$|du(x)| \leq C\rho^{-1} \tag{3.10}$$

where $\rho = \operatorname{dist}(x, \Gamma)$. Then there is a positive constant $\kappa(N, n, |\gamma|_{C^2})$ such that

$$\int_{B^n} |du|^2 \, dM \leq \kappa$$

implies that u is harmonic in B^n, where γ is the metric tensor on M.

Sketch of the proof. The relevant monotonicity inequality is

$$r^{2-n} \int_{B_r} |du|^2 \, dx \leq R^{2-n} \int_{B_R} |du|^2 \, dx \tag{3.11}$$

$\forall\, 0 < r \leq R \leq \frac{1}{2}$. (We approximate the metric locally by a euclidean one and center B^n at the origin.) We obtain this inequality for $n \geq 6$ by arguments exactly analogous to the proof of Corollary 2.3, replacing (2.1) by (0.6) as in Ref. 5. For the case $n = 5$ we use arguments exactly analogous to the proof of Lemma 2.1. The domain is then partitioned as in the proof of Theorem 2.7 and Liao's low-energy theorem for harmonic maps is applied in a reference annulus. Analogously to the Yang-Mills case, inequality (3.10) is used to show that the low-energy hypothesis is preserved under conformal dilations.

Remark. An obvious conjecture is that Theorem 2.7 is true $\forall\, n > 4$ without hypothesis (2.16), Theorem 3.3 is true $\forall\, n > 2$ without hypothesis (3.10), and that in both cases Σ need not be contained in a $C^{1,\alpha}$ curve.

4. An a *priori* inequality for elliptic subsolutions

In this section we investigate the extent to which the local properties of gauge fields are artifacts of their global (geometric) properties. That is, we assume that a nonnegative scalar function u smoothly satisfies an inequality of the form (1.3) on an open domain of $\mathbb{R}^n$. We do not assume that u possesses any variational structure, nor do we assume that it is derived from a geometric object. What can one say analytically about the function u? In particular, can one prove an *a priori* estimates of the form (2.9)? [Notice that since u does not live on a bundle it does not obey a scaling law, and that since the constant in (2.9) does not depend on the $L^{n/2}$-norm of the solution, we do not have recourse to Theorem 5.3.1 of Ref. 18 even in the stronger form due to Uhlenbeck (Ref. 39, Theorem 3.5).] To what extent, if any, is the L^∞ norm of u controlled locally by the L^2 norm?

Suppose that $u \geq 0$ smoothly satisfies an inequality of the form (1.3) and that $\max_{x\in\Omega} u(x)$ occurs at a point $x_0 \in \text{int } \Omega$. By a choice of coordinates we can assume that x_0 lies at the origin and that $B_1^n(x_0) = B_1^n(0) = B_1 \subset \Omega$. We find that there is a positive constant κ independent of u such that $\|u\|_{L^2(B_1)} \leq \kappa$ implies that $\|u\|_{L^\infty(B_{1/2})} \leq 1$.

Note that inequalities of the form (1.3) are satisfied by many objects of variational interest. For example, Einstein metrics satisfy such an inequality although the variational equations satisfied by such metrics have a more complex nonlinear structure than the systems studied here (see, *e.g.*, Ref. 21).

In this section we prove

Theorem 4.1. *Let $u \geq 0$, $u \in C^\infty(B_1)$ satisfy*

$$\Delta u(x) + u^2(x) \geq 0 \tag{4.1}$$

and

$$\|u\|_{L^\infty(B_{1/2})} = \|u\|_{L^\infty(B_1)} \tag{4.2}$$

for $x \in B_1 = B_1^n(0)$, $n > 4$. Then there is a positive constant κ independent of u such that

$$\|u\|_{L^2(B_1)} \leq \kappa \tag{4.3}$$

implies

$$\|u\|_{L^\infty(B_{1/2})} \leq 1\,.$$

Remark. Identity (4.2) would be unnatural if (4.1) belonged to the class of expressions for which a maximum principle holds, but this is not the case. In fact, (4.1) includes certain expressions[10] for which the radial derivative of the solution is negative.

Proof. Suppose no such κ exists. Then there is a sequence $\{u_i\}_{i=1}^\infty$ of C^∞ functions satisfying (4.1) and (4.2) for which

$$\begin{aligned} &u_i(x) \geq 0 \quad \forall\, x \in B_1\,, \\ &\|u_i\|_{L^2(B_1)} \leq 1/i\,, \end{aligned} \tag{4.4}$$

but

$$\|u_i\|_{L^\infty(B_{1/2})} > 1 \ . \tag{4.5}$$

Define

$$v_i(x) := u_i(x)/u_{i,0}^\alpha \ ,$$

where

$$u_{i,0}^\alpha := \|u_i\|_{L^\infty(B_{1/2})}^\alpha \tag{4.6}$$

for some small $\alpha > 0$. We have for $\delta > 0$ and all i

$$\begin{aligned} \|v_i\|_{L^{2+\delta}(B_1)} &= \frac{1}{u_{i,0}^\alpha}\left(\int_{B_1} u_i^{2+\delta}\, dx\right)^{1/(2+\delta)} \\ &\le \left(\frac{\|u_i^{\delta/(2+\delta)}\|_{L^\infty(B_1)}}{u_{i,0}^\alpha}\right)\left(\int_{B_1} u_i^2\, dx\right)^{1/(2+\delta)} \\ &\le (1/i)^{2/(2+\delta)} \end{aligned} \tag{4.7}$$

by (4.2) and (4.4)–(4.6), provided we choose δ so small that $\delta/(2+\delta) \le \alpha$.

Thus if the conclusion of the theorem is false there must exist a sequence $\{v_i\}_{i=1}^\infty$ of C^∞ functions satisfying

i) $\Delta v_i + u_{i,0}^\alpha v_i^2 \ge 0$;

ii) $v_i \in L^{2+\delta}(B_1) \quad \forall\, \delta \in [0, \alpha]$;

iii) $\|v_i\|_{L^\infty(B_{1/2})} = \|u_i/u_{i,0}^\alpha\|_{L^\infty(B_{1/2})} = \|u_i\|_{L^\infty(B_{1/2})}^{1-\alpha} > 1$

by (4.5) for $\alpha \in (0,1)$.

Inequality i) implies that

$$\int_{B_1} \nabla\zeta\nabla v_i\, dx \le u_{i,0}^\alpha \int_{B_1} \zeta v_i^2\, dx \tag{4.8}$$

$\forall\, \zeta \in C_0^\infty(B_1)$, $\zeta \ge 0$. Choose

$$\zeta = \eta^2 G(v_i) \tag{4.9}$$

for $\eta \in C_0^\infty(B_1)$, $0 \le \eta \le 1$, and $G(v_i) = H(v_i)H'(v_i)$, where

$$H(v_i) = H_i = \begin{cases} v_i \ \text{ if } \ 0 \le v_i \le \ell \ , \\ \left(\dfrac{4}{2+\delta}\right)\left[\ell^{(2-\delta)/4} v_i^{(2+\delta)/4} + \left(\dfrac{\delta-2}{4}\right)\ell\right] \ \text{ if } \ \ell \le v_i \ . \end{cases}$$

(The function H_i is a special case of the test function $S(u)$ in the proof of Lemma 1.4.) Substituting (4.9) into (4.8) yields

$$C\int_{B_1}\eta^2|\nabla H_i|^2\,dx \le 2\int_{B_1}|\eta\nabla H_i|\;|\nabla\eta H_i|\,dx$$

$$+\,u_{i,0}^{\alpha}\int_{B_1}\eta^2 v_i H_i^2\,dx = 2I_1 + I_2\ . \tag{4.10}$$

By Young's inequality

$$I_1 \le \varepsilon\int_{B_1}\eta^2|\nabla H_i|^2\,dx + \frac{1}{\varepsilon}\int_{B_1}|\nabla\eta|^2 H_i^2\,dx\ ,$$

where we choose ε to be a small positive constant. We can choose η so that $\eta \equiv 1$ on $B_{1/2}$ and

$$\|\nabla\eta\|_{L^\infty(B_1)} \le C\ .$$

Then we have

$$\int_{B_1}|\nabla\eta|^2 H_i^2\,dx \le C\int_{B_1}\left(\frac{16\ell^{(2-\delta)/2}}{[2+\delta]^2}\right)v_i^{(2+\delta)/2}\,dx$$

$$\le C'(\delta,\ell)\int_{B_1}v_i^{(2+\delta)/2}\,dx \le C'(\delta,\ell)\|v_i\|_{L^{2+\delta}(B_1)}^{(2+\delta)/2}\ , \tag{4.11}$$

which tends to zero as $i\to\infty$ by (4.7).

$$I_2 \le u_{i,0}^{\alpha}\int_{B_1}v_i H_i^2\,dx$$

$$\le u_{i,0}^{\alpha}\left(\int_{B_1}v_i^2\,dx\right)^{1/2}\left(\int_{B_1}H_i^4\,dx\right)^{1/2}$$

$$=\left(\int_{B_1}u_i^2\,dx\right)^{1/2}\left(\int_{B_1}H_i^4\,dx\right)^{1/2}$$

$$\le C(\delta,\ell)\|u_i\|_{L^2(B_1)}\;\|v_i\|_{L^{2+\delta}(B_1)}^{(2+\delta)/2}\ , \tag{4.12}$$

which also tends to zero as $i\to\infty$. (Let ℓ be some convenient root of i.)

Estimates (4.10)-(4.12) together imply that the sequence $\{H_i\}_{i=1}^{\infty}$ is bounded in $H^{1,2}(B_{1/2})$ independently of i. Standard arguments imply that a subsequence, which we also denote $\{H_i\}$, tends to zero weakly in $H^{1,2}(B_{1/2})$, strongly in $L^2(B_{1/2})$, and pointwise almost everywhere in $B_{1/2}$. But then the limit H_∞ has an essential supremum of zero in $B_{1/2}$.

By definition, $H_\infty = H(v_\infty)$. If $v_\infty \in [0, \ell]$, then $H_\infty \equiv 0$ a.e. if and only if $v_\infty \equiv 0$ a.e. If $v_\infty \geq \ell$, then $H_\infty = 0$ a.e. if and only if $v_\infty = [(2-\delta)/4]^{4/(2+\delta)}\ell$ a.e.; this is a contradiction unless $\ell = 0$, in which case $v_\infty \equiv 0$ a.e. So $\|v_\infty\|_{L^\infty(B_{1/2})} = 0$, which contradicts iii).

I am grateful to L.M. Sibner and K. Uhlenbeck for many discussions about gauge theory.

References

1. W. Allard, On the first variation of a varifold, *Ann. Math.* **95** (1972) 417–491.
2. S. Bando, A. Kasue, and H. Nakajima, On a construction of coordinates at infinity on manifolds with fast curvature decay and maximal volume growth, *Invent. Math.* **97** (1989) 313–349.
3. S. Bochner, Harmonic surfaces in Riemannian metric, *Trans. Amer. Math. Soc.* **47** (1940) 146–154.
4. J-P. Bourguignon and H.B. Lawson, Jr., Stability and isolation phenomenon for Yang-Mills fields, *Commun. Math. Phys.* **79** (1981) 189–203.
5. D. Costa and G. Liao, On removability of a singular submanifold for weakly harmonic maps, *J. Fac. Sci. Univ. Tokyo Sect.* IA **35** (1988) 321–344.
6. J. Eells and J.H. Sampson, Harmonic mappings of Riemannian manifolds, *Amer. J. Math.* **86** (1964) 109–160.
7. D. Freed and K. Uhlenbeck, *Instantons and Four-Manifolds*, (Springer, Berlin-Heidelberg-New York-Tokyo, 1984).
8. W-D. Garber, S.N.M. Ruijsenaars, E. Seiler, and D. Burns, On finite-action solutions of the nonlinear sigma model, *Ann. Phys.* **199** (1979) 305–325.
9. M. Giaquinta and E. Guisti, The singular set of the minima of certain quadratic functionals, preprint.
10. B. Gidas, W-M. Ni, and L. Nirenberg, Symmetry and related properties via the maximum principle, *Commun. Math. Phys.* **68** (1979) 209–243.
11. J. Glimm and A. Jaffe, Droplet model for quark confinement, *Phys. Rev. D* **18**, (1978) 463–467.
12. R. Hardt, D. Kinderlehrer, and F-H. Lin, Existence and partial regularity of static liquid crystal configurations, *Commun. Math. Phys.* **105**, (1986) 547–570.
13. S. Hildebrandt, Nonlinear elliptic systems and harmonic mappings, in *Proceedings of the 1980 Beijing Symposium on Differential Geometry and*

Differential Equations, (Science Press–Gordon and Breach, Beijing-New York, 1982).

14. A. Jaffe and C.H. Taubes, *Vortices and Monopoles*, (Birkhauser, Boston, 1980).
15. G. Liao, A regularity theorem for harmonic maps with small energy, *J. Differential Geometry* **22** (1985) 233–241.
16. A. Marini, Ph.D. Dissertation, The University of Chicago, 1990.
17. M. Meier, Removable singularities for weak solutions of quasilinear elliptic systems, *J. reine angew. Math.* **344** (1983) 87–101.
18. C.B. Morrey, Jr., "Multiple Integrals in the Calculus of Variations," Springer, Berlin-Heidelberg-New York-Tokyo, 1966.
19. H. Nakajima, Removable singularities for Yang-Mills connections in higher dimensions, *J. Fac. Sci. Univ. Tokyo* **34** (1987) 299–307.
20. H. Nakajima, Compactness of the moduli space of the Yang-Mills connections in higher dimensions, *J. Math. Soc. Jpn.* **40** (1988) 383–392.
21. H. Nakajima, Hausdorff convergence of Einstein 4-manifolds, *J. Fac. Sci. Univ. Tokyo, Sect.IA* **35** (2) (1988) 411–424.
22. T.H. Otway, Removable singularities in coupled Yang-Mills-Dirac fields, *Commun. Partial Differential Equations* **12** (9), (1987) 1029–1070.
23. T.H. Otway, The coupled Yang-Mills-Dirac equations for differential forms, *Pacific J. Math.* **146** (1), (1990) 103–113.
24. T.H. Otway, Higher-order singularities in coupled Yang-Mills-Higgs fields, *Nonlinear Analysis T.M.A.* **15** (3), (1990) 239–244.
25. T.H. Otway and L.M. Sibner, Point singularities of coupled gauge fields with low energy, *Commun. Math. Phys.* **111** (1987) 275–279.
26. T.H. Parker, Gauge theories on four dimensional Riemannian manifolds, *Commun. Math. Phys.* **85** (1982) 563–602.
27. T.H. Parker, Conformal fields and stability, *Math. Z.* **185** (1984) 305–319.
28. P. Price, A monotonicity formula for Yang-Mills fields, *Manuscripta Math.* **43** (1983) 131–166.
29. J. Sacks and K. Uhlenbeck, The existence of minimal immersions of 2-spheres, *Ann. of Math.* (2) **113** (1981) 1–24.
30. R. Schoen, Analytic aspects of the harmonic map problem, in *Seminar on Nonlinear Partial Differential Equations*, ed. S.S. Chern, (Springer, Berlin-Heidelberg-New York-Tokyo, 1985).
31. J. Serrin, Local behavior of solutions of quasilinear equations, *Acta Math.* **111** (1964) 247–302.
32. L.M. Sibner, Removable singularities of Yang-Mills fields in $\mathbb{R}^3$, *Compositio Math.* **53** (1984) 91–104.
33. L.M. Sibner, The isolated point singularity problem for the coupled Yang-Mills equations in higher dimensions, *Math. Ann.* **271** (1985) 125–131.

34. L.M. Sibner and R.J. Sibner, Removable singularities of coupled Yang-Mills fields in $\mathbb{R}^3$, *Commun. Math. Phys.* **93** (1984) 1–17.
35. L.M. Sibner and R.J. Sibner, Classification of singular Sobolev connections by their holonomy, preprint.
36. T.H.R. Skyrme, A nonlinear field theory, *Proc. R. Soc.* **A260** (1961) 127–138.
37. P.D. Smith, Removable singularities for Yang-Mills-Higgs equations in two dimensions, preprint.
38. S. Takakuwa, On removable singularities of stationary harmonic maps, *J. Fac. Sci. Univ. Tokyo Sect. IA*, **32** (1985) 373–395.
39. K. Uhlenbeck, Removable singularities in Yang-Mills fields, *Commun. Math. Phys.* **83** (1982) 11–29.
40. K. Uhlenbeck, Connections with L^p bounds on curvature, *Commun. Math. Phys.* **83** (1982) 31–42.
41. K. Uhlenbeck, The Chern classes of Sobolev connections, *Commun. Math. Phys.* **101** (1985) 449–457.
42. K. Uhlenbeck and S.T. Yau, On the existence of Hermitian-Yang-Mills connections in stable vector bundles, *Commun. Pure Appl. Math.* **39**S (1986) 257–293.
43. H. Wallin, A connection between α-capacity and L^p-classes of differentiable functions, *Ark. Math.* **5** (24) (1964) 331–341.

The Problem of Plateau (pp. 221–228)
ed. Th. M. Rassias

SECOND VARIATION FORMULAS FOR WILLMORE SURFACES

BENNETT PALMER*

Fachbereich 3, Technische Universität Berlin
Strasse des 17. Juni 136
1000 Berlin 12

ABSTRACT

We study the second variation of the Willmore functional for surfaces with umbilics using the second variation of area for the conformal Gauss map.A type of Morse Smale index theorem for Willmore surfaces is proved.

1.Introduction

Let $x{:}M \longrightarrow S^3$ be a smooth, oriented, immersed surface.The conformal Gauss map assigns to each $p \in M$ the oriented two sphere tangent to M at p,whose mean curvature and normal agree with the mean curvature and normal of M at x(p).This sphere is called the central sphere.The totality of all oriented two spheres of S^3 may be identitied with the 4 dimensional deSitter space $S^4_1 := \{Y \in \mathbb{E}^5_1 \mid \langle Y,Y \rangle = 1\}$,which is a Lorentzian manifold with constant +1 sectional curvature.

The mapping $Y{:}M \longrightarrow S^4_1$ is weakly conformal with degeneracies precicely at the umbilics of M.In fact,the conformal factor is such that the Willmore integral W(x) of x is transformed into the area integral A(Y) of Y ,that is

$$W(x) = \int_M (H^2+1-G)\, dA_x = \int_{Y(M)} dA_Y = A(Y)$$

where H and G are respectively the mean and Gauss curvatures of M.It is then not difficult to see that M is a Willmore surface precisely when $\Sigma := Y(M)$ is a surface whose mean curvature,in S^4_1, is weakly equal to zero.

Assume that M is not part of a two sphere.At a generic point of M the map Y may be ``inverted´´ as follows.The normal space of Σ at Y contains two isotropic (null) lines.If the light cone in $\mathbb{E}^5_1$ is projectivized to a 3-sphere,one of these lines,which we denote by X, corresponds to $x(p) \in S^3$.A remarkable theorem of G.Thomsen states that if M is a Willmore surface,then the second isotrpic line represents again a ,possibly branched,

* Partially supported by SFB 288

Willmore surface in S^3,which is called the dual surface.It was shown by R.Bryant that this dual surface can be smoothly extended across the umbilic set to all of M.

For a spacelike surface in a Lorentzian manifold of dimension ≥ 4,the concept of stability does not ,in general,make sense,as the following shows.Let $Y:\Sigma \longrightarrow S^4_1$ be a spacelike surface of zero mean curvature and let N(Y) denote its normal bundle.The induced metric in a fibre of N(Y) is Lorentzian of signature (1,1).In the usual way,one can define the Hodge operator,denoted by $*$,which reverses one isotropic direction and fixes the other.For a smooth section $\xi \in C^\infty(N(Y))$,we denote the second variation of area in the ξ direction by $\delta^2_\xi A(Y)$.Then it is not hard to show that $\delta^2_\xi A(Y) + \delta^2_{*\xi} A(Y) < 0$ holds for every $\xi \in C^\infty(N(Y))$, $\xi \not\equiv 0$.Thus on an arbitrarily small subdomain the second variation operator is not positive semidefinite.On the other hand ,it follows from section 4 below that this operator is neither locally negative semidefinite.

The second variation of the Willmore functional ,$\delta^2 W(x)$,can, however ,be related to the second variation of area $\delta^2 A(Y)$.In a previous paper [1] we showed ,if M contains no umbilics,then those normal variations of Y arising as variations through conformal Gauss maps are those satisfying $\langle J(\xi) , X \rangle \equiv 0$ .Consequently M as above is a stable Willmore surface if and only if

$$\inf_{\langle J(\xi) ,X \rangle \equiv 0} - \int \langle \xi, J(\xi) \rangle \, dA_Y \geq 0 \qquad (1.1)$$

where J denotes the Jacobi operator of Σ.

In this note we extend these ideas to surfaces with umbilics by again identifying the set of allowable variations. We then discuss a spectral formulation for the stability condition (1.1).This leads to a type of Morse-Smale index theorem for Willmore surfaces .

2.A second variation formula for surfaces with umbilics

In this section we show how the section variation of the Willmore functional W(x) can be related to the second variation of area for Σ.This was done previously by the author[1] under the assumptian that M contained no umbilics.We assume that M is compact.

Let $x:M \longrightarrow S^3$ be an oriented Willmore surface. Let $\Sigma := Y(M)$ and let U denote the set of umbilics in M.Let $M^* := M \setminus U$ and $\Sigma^* := Y(M^*)$.Define

$\mathcal{A} := \{ s \in C^\infty(Y^* T S^4_1) \mid \langle J((s|_{M^*})^\perp), X \rangle \equiv 0 \}$.

Here $\perp$ denotes projection onto the normal bundle of Σ^* in S^4_1.

Theorem 2.1 Let M be an oriented Willmore surface. Assume that U consists of a finite union of piecewise C^1 closed curves and isolated points.Then M is stable if and only if :

$$\delta^2_s A(Y) \geq 0 \ , \forall s \in \mathcal{A} \ .$$

Proof The sufficiency follows as in the proof of Proposition (3.3) [1].Every normal variation of M in S^3 yields a variation of Y which is in $\mathfrak{A}$.To prove the necessity,let $s \in \mathfrak{A}$ and let $\psi := -\langle s,X\rangle$.Then $x_\varepsilon := \cos(\varepsilon\psi)x + \sin(\varepsilon\psi)\nu$,where ν is the unit normal to M in S^3,is a smooth normal variation of x.Let γ_ε denote the conformal Gauss map of x_ε.As shown in the reference cited above, $\gamma' := \frac{d}{d\varepsilon}(\gamma_\varepsilon)_{\varepsilon=0}$ is determined uniquely on M* from ψ by the equation

$$\langle J((s|_{M*})^\perp),X\rangle \equiv 0 \quad \text{on } \Sigma^* .$$

On the other hand, X and Y are given by

$$X = (x,1)$$

$$Y = HX + (\nu , 0)$$

where we write an ordered pair for an element of $\mathbb{E}^5_1$ identified with $\mathbb{R}^4 \oplus \mathbb{R}^1$. Differentiation gives $\gamma' = H'X + HX' + (\nu',0)$.Using the fact that X is isotropic,we find $\langle\gamma',X\rangle = \langle(\nu',0) ,(x,1)\rangle = -\langle x',\nu\rangle_{S^3} = -\psi$.

It follows that we may write

$$s - \gamma' = V \in C^\infty(Y^*TS^4_1)$$

with

$$V|_{M*} \in C^\infty(Y^*T\Sigma^*) .$$

Clearly $\delta^2_s A(Y) = \delta^2_{\gamma'+V}A(Y)$ and so it is enough to show

$$\delta^2_\gamma A(Y) = \delta^2_{\gamma'+V}A(Y) .$$

This may be verified as follows.

Let $\beta(\cdot,\cdot)$ denote the bilinear form associated with the second variation operator,so that

$$\delta^2_{\gamma'+V}A(Y) = \beta(\gamma',\gamma') + 2\beta(\gamma',V) + \beta(V,V) .$$

For $W=\gamma'$or V we must show $\beta(W,V) = 0$.

Let $M_\varepsilon,\varepsilon > 0$, be an exhaustion of M* by relatively compact,smoothly bounded subdomains with $M_{\varepsilon'} \subset M_\varepsilon$ for $\varepsilon' > \varepsilon$.We will later impose a further condition on M_ε.Let $Y_{\sigma,\tau} = Y + \sigma V + \tau W$ + (higher order) be a smooth variation.

Let $\tilde{V} = \partial_\sigma(Y_{\sigma,\tau})$ and $\tilde{W} = \partial_\tau(Y_{\sigma,\tau})$.Then a slight modification of the second variation formula for area (see [2] pg.515) yields

$$\delta^2 A(Y_{0,0}) = \lim_{\varepsilon\downarrow 0} \left(\int_{\partial M_\varepsilon} Y^* \mathrm{div} W^T \, dA(V,\cdot) + \int_{\partial M_\varepsilon} Y^* dA([\widetilde{V},\widetilde{W}], \cdot) \right).$$

Here W^T denotes the tangential projection of W to Σ^*.We claim that the integrand in the first integral has a continuous extension to all of M.This may be seen by writing

$$\mathrm{div} W^T dA = d*\langle W, \cdot \rangle$$

and using the weak conformality of Y to extend the * operator continuously.

By careful choice of the domains M_ε , we may arrange that each arc α of U corresponds to two arcs of ∂M_ε traversed in opposite directions.Their contributions to the integral above will cancel in the limit.Since the second integral above poses no problem,we may let ε approach zero to obtain $\beta(w,V) = 0$. q.e.d.

Remark R.Bryant has shown that if M is not totally umbilic,then U is a closed set with empty interior.

3.Conformal invariants of immersed surfaces

The extrinsic conformal invariants of an immersed surface $M \subset S^3$ were derived by Blaschke[3]. We briefly summarize them here.

Let $p \in M$ be a point which is not umbilic.Let Y denote the conformal Gauss map of M and let ds_Y^2 denote the metric induced by Y.In a neighborhood of Y(p), there exists an orthonormal (w.r.t.ds_Y^2) frame for Σ,$\{e_1,e_2\}$,which point along the principal directions of M.The normal bundle of Σ is framed by its two isotropic directions,denoted by a and b. These are normalized by the conditions (i) $\langle a,b\rangle \equiv 1$ and (ii) $Y \pm a$ represent the curvature spheres of M. The frame $\{e_1,e_2,a,b\}$ is the most natural possible along a neighborhood in Σ^*.

Set $e_3 = \frac{1}{\sqrt{2}}(a+b)$, $e_4 = \frac{1}{\sqrt{2}}(a-b)$ .Let $\{\theta^j\}_{j=1...4}$ be the frame dual to $\{e_j\}_{j=1...4}$ and let $\{\theta_i^j\}_{1\le i,j\le 4}$ denote the connection one forms given by

$$d\theta^i = \sum_j \theta^j \wedge \theta_j^i \, .$$

Formula (109) pg.318 of Blaschke[3] can then be expressed :

$$\theta_1^2 =: -q\theta^1 + \overline{q}\theta^2 ,$$

$$\theta_3^4 = -2*\theta_1^2 = 2(\overline{q}\theta^1 + q\theta^2) \, .$$

These are respectively the connection forms for the tangent and normal bundles of Σ relative to the chosen frame.

Let Δ denote the laplacian for ds^2_Y.Then a scalar invariant η may be defined by the equation

$$\Delta Y + 2Y = 2\eta a.$$

This means that η may be thought of as a scalar mean curvature of Σ.

As shown in Blaschke[3],$\{q,\bar{q}\}$ are the two independent conformal invariants of M of order three.Further if $q_i:=e_i(q)$, $\bar{q}_i:=e_i(\bar{q})$;i=1,2, then $\{q_i,\bar{q}_i,\eta\}_{i=1,2}$ are the five independent fourth order conformal invariants of M.M is a Willmore surface precicely when $\eta\equiv 0$.

The curvature,K;and normal curvature $K^\perp$ of Σ are given by

$$d\theta_1^2 = -K\theta^1\wedge\theta^2 ,$$

$$d\theta_3^4 = K^\perp\theta^1\wedge\theta^2 ,$$

i.e.,

$$K = -(\bar{q}_1 + q_2 + q^2 + \bar{q}^2) ,$$

$$K^\perp = 2(q_1 - \bar{q}_2) .$$

The curvatures defined above are important conformal invariants of M.Let (u,v) be an isothermal coordinate on a neighborhood in M and let z=u+iv.Then $Q:=\langle Y_{zz}, Y_{zz}\rangle\, dz^4$ defines an invariant quartic differential on M.It was shown by Bryant that Q is holomorphic if M is a Willmore surface. The author[1] showed that the norm of Q,w.r.t. ds^2_Y ,is given by

$$|Q|^2 = (1 - K)^2 + (K^\perp)^2 .$$

4.Spectral theory and the Jacobi operator

Lemma 4.1 Let M be a compact two dimensional Riemannian manifold and let α be a smooth one form on M with $d\alpha = f *1$. Then $S[\mu] := -*(d\mu\wedge\alpha) + \frac{1}{2}f\mu$, $\mu\in C^\infty(M)$, defines a skew symmetric (w.r.t. the usual L^2 inner product for functions), first order operator.

Proof The proof is obvious and is left to the reader.

Lemma 4.2 Define the second order self adjoint operator

$$L[\mu] := \Delta\mu + \mu(-2\overline{q}_1 - 2q_2 + q^2 + \overline{q}^2 + 1 - K)$$

and the first order skew symmetric operator

$$A[\mu] := -4*(d\mu \wedge \theta_1^2) + 2\mu K \quad .$$

Then for $\xi = \lambda a + \mu b$, $J(\xi)$ is given by

$$J(\xi) = \Lambda a + Nb$$

where

$$\Lambda = L[\lambda] - A[\lambda] - \frac{\mu}{2}|Q|^2 \ , \tag{4.1}$$

$$N = L[\mu] + A[\mu] + 2\lambda \ . \tag{4.2}$$

Proof Let D denote the Levi-Civita connection in S_1^4 .The tangential and normal components of $D\xi$ will be denoted respectively by D^T and $D^\perp$. We have shown[1] that J is given by,

$$J[\xi] := \Box\xi + 2\xi + F(\xi) \ ,$$

where

$$\Box\xi = \sum_{i=1,2} (D^\perp_{e_i} D^\perp_{e_i} - D^\perp_{\nabla_{e_i} e_i})\xi$$

and F is defined by

$$\langle F(\xi), \eta\rangle = \langle D^T\xi , D^T\eta\rangle \ ; \ \xi,\eta \in C^\infty N(Y).$$

Note that

$$\Box(\lambda a + \mu b) = (\Delta\lambda)a + 2D^\perp_{\nabla\lambda}a + \lambda\Box a$$
$$+ (\Delta\mu)b + 2D^\perp_{\nabla\mu}b + \mu\Box b \ .$$

Using the formulas (108) and (109) pg.318 of Blaschke[3] one finds that F is given by

$$F(a) = (K - 1)a + 2b \ ,$$

$$F(b) = \frac{1}{2}|Q|^2 a + (K-1)b \ .$$

Further computations using the formulas of Blaschke mentioned above then yield (4.1) and (4.2). q.e.d.

Because of (4,2) and the fact that X and a are parallel,the equation $\langle J(\xi),X\rangle=0$ becomes N=0 which can be expressed $-\lambda=\frac{1}{2}(L[\mu]+A[\mu])$ .We then obtain from (1.1),

$$\delta_\xi^2 A(Y) = -\int \langle \xi, J(\xi)\rangle$$

$$= -\int \mu\Lambda$$

$$= \frac{1}{2}\int \mu((L - A)(L+A)[\mu] - |Q|^2\mu) .$$

Note that the operator Γ defined by

$$\Gamma = (L-A)(L+A) - |Q|^2$$

is fourth order,elliptic,self-adjoint and conformally invariant.The usual theorie for self adjoint elliptic operators can now be applied. If M is closed,there exists an infinite sequence of eigenvalues $\alpha_1 \leq \alpha_2 \leq \alpha_3 \leq \cdots \to \infty$ and a complete orthonormal system of corresponding eigenfuctions satisfying

$$\Gamma[\mu_j] = \alpha_j\mu_j . \tag{4.3}$$

This can be interperted as saying that there exists an infinite sequence of normal sections $\{\xi_j\}_{j=1\ldots\infty}$ satisfying

$$J[\xi_j] + \alpha_j\langle\xi_j,b\rangle a = 0 .$$

The index of M is defined by Index(M):= # $\{\alpha_j \mid \alpha_j < 0\}$.

If $\Omega\subset M$ is a smoothly bounded,relatively compact subdomain,then we consider the eigenvalue problem (4.3) together with the boundary conditions

$$\mu_j|_{\partial\Omega} = 0 ,$$

$$\partial_n \mu_j = 0 \text{ ,on } \partial\Omega .$$

Let $\Omega \subset M$ be as above. A deformation of ε type is a family of smoothly bounded subdomains Ω_t, $t \in [a,b]$, depending smoothly on t, with Ω_t contained in Ω_s for t>s and Area(Ω_t)<ε. For such a deformation we let

$$\nu(t) = \dim \left\{ \psi \in H_0^2(\Omega_t) : \Gamma[\psi] = 0 \right\} .$$

As a direct consequence of the above and the Morse-Smale index theorem [1], we have the following.

Theorem4.3 Let M be an oriented Willmore surface. Then there exists an $\varepsilon > 0$ such that if Ω_a is an umbilic free subdomain in M and Ω_t is a deformation of ε type, then

$$\mathrm{Index}(M) \geq \sum_{a \leq t \leq b} \nu(t) \quad .$$

References

1. Palmer, B., The conformal Gauss map and the stability of Willmore surfaces, *Annals of Global Analysis and Geometry*, **9**, No.3(1991), 305-317.

2. Spivak, M., *A Comprehensive Introduction to Differential Geometry*, Vol.4, (Publish or Perish, Berkely, 1975.)

3. Blaschke, W., *Vorlesung über Differentialgeometrie und geometrische Grundlagen von Einsteins Relatitivitätstheorie*, Vol.3, (Springer, Berlin, etc., 1929.)

4. Smale, S., On the Morse index theorem, *J.Math.Mech.*, **14**, (1965).

The Problem of Plateau (pp. 229–236)
ed. Th. M. Rassias

A JORDAN ARC IN $\mathbf{R}^m$ WITH POSITIVE m-DIMENSIONAL LEBESGUE MEASURE

HAROLD R. PARKS* & RICHARD M. SCHORI
Department of Mathematics
Oregon State University
Corvallis, Oregon 97331
U.S.A.

ABSTRACT

Methods of point-set topology are used to construct an example of a homeomorphism from the unit interval onto a subset of $\mathbf{R}^m$ having positive m-dimensional Lebesgue measure. This generalizes an example of W. F. Osgood.

1. Introduction

It is well known from Peano's example of a space-filling curve that the continuous image of an interval or a circle is not as nice as one might naively hope.[1] One might then hope that a homeomorphic image of an interval or a circle would be better behaved. In the 1930's Jesse Douglas, sometimes in collaboration with P. Franklin, constructed a series of examples of Jordan curves in $\mathbf{R}^3$ which bound no surface of finite area.[2,3,4,5]

In 1903 W. F. Osgood constructed an example of a Jordan arc in the plane having positive area.[6] Thus even the homeomorphic image of an interval or a circle can be very badly behaved from the point of view of measure theory and analysis.

In this note we show how to use techniques familiar to point-set topologists to construct an example of a Jordan arc in $\mathbf{R}^m$ having positive m-dimensional Lebesgue measure.

2. Cantor Sets of Positive Measure

Topologists generally agree on the following:

DEFINITION. A *Cantor set* is a compact, totally disconnected, perfect metric space.

This is reasonable because of the following well-known theorem.[7]

* partially supported by Office of Naval Research U.R.I. Grant #N00014–86–K–0687

THEOREM 1. *Any two Cantor sets are homeomorphic.*

The familiar Cantor set construction of removing middle thirds is readily modified to produce a Cantor set in $\mathbf{R}$ having positive one dimensional measure: Starting with the unit interval, instead of removing the middle third of each remaining interval at every stage of the construction, one removes a central open interval of length 2^{-2k} from each of the 2^{k-1} intervals remaining at the kth stage of the construction. By summing the lengths of the intervals removed, one sees that a Cantor set with Lebesgue measure 1/2 has been constructed.

By taking Cartesian products one can immediately obtain a Cantor set in $\mathbf{R}^m$ of positive m-dimensional Lebesgue measure. (One notes that $\{0,1\}^\omega$ with the product topology in a Cantor set, and then one applies Theorem 1.) However, for our purpose of constructing a Jordan arc of positive measure, we will find it convenient to give a direct construction of a Cantor set in $\mathbf{R}^m$ with positive m-dimensional Lebesgue measure.

Let $m \geq 2$ be an integer. We will construct a Cantor set in

$$D = [0,1] \times [0,1] \times \cdots \times [0,1] = [0,1]^m \subset \mathbf{R}^m$$

with m-dimensional Lebesgue measure 1/2 :

(1) Let $U^1 \subset D$ be the union of m relatively open slabs parallel to the coordinate $(m-1)$-planes such that U^1 has m-dimensional Lebesgue measure 1/4 and the $\mu = 2^m$ components of $D \backslash U^1$ are all congruent hypercubes. Denote these μ components of $D \backslash U^1$ by D_{s_1}, where $s_1 = 1, 2, \ldots, \mu$, such that D_1 contains $(0,0,\ldots,0)$ and D_μ contains $(1,0,0,\ldots,0)$.

(2) For each s_1, let $U_{s_1} \subset D_{s_1}$ be the union of m relatively open slabs parallel to the coordinate $(m-1)$-planes such that U_{s_1} has m-dimensional Lebesgue measure 2^{-m-3} and where the $\mu = 2^m$ components of $D_{s_1} \backslash U_{s_1}$ are all congruent hypercubes. Denote the μ components of $D_{s_1} \backslash U_{s_1}$ by $D_{s_1 s_2}$, where $s_2 = 1, 2, \ldots, \mu$, in the same relative order as the components D_{s_1} in D (that is, the linear map L_{s_1} which is a composition of a homothety and a translation and which maps D to D_{s_1} should map D_{s_2} to $D_{s_1 s_2}$).

Set

$$D^2 = \bigcup_{s_1, s_2 = 1}^{\mu} D_{s_1 s_2}, \qquad U^2 = \bigcup_{s_1 = 1}^{\mu} U_{s_1},$$

and note that $\mathcal{L}^m(U^2) = 1/8$, where $\mathcal{L}^m$ denotes the m-dimensional Lebesgue measure.

The above steps (1) and (2) of the construction are illustrated in Figure 1 for the case $m = 2$.

The inductive step of the construction is the following: If $D_{s_1 s_2 \ldots s_n}$ has been constructed, then let $U_{s_1 s_2 \ldots s_n} \subset D_{s_1 s_2 \ldots s_n}$ be the union of m relatively open slabs parallel to the coordinate $(m-1)$-planes such that $\mathcal{L}^m(U_{s_1 s_2 \ldots s_n}) = 2^{-nm} . 2^{-n-2}$ and

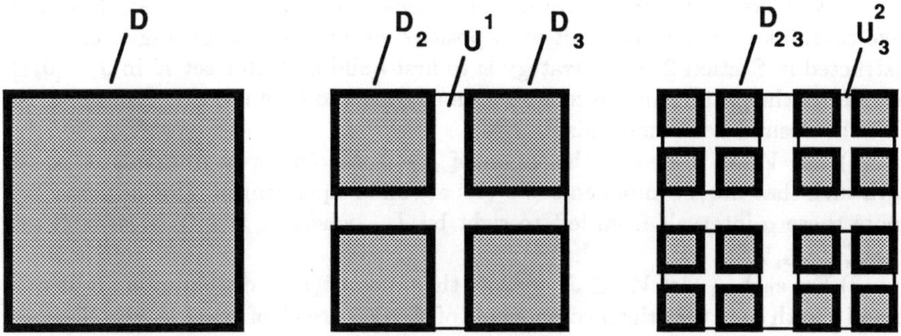

Construction of a Cantor set in the plane
Figure 1

where the $\mu = 2^m$ components of $D_{s_1 s_2 \ldots s_n} \backslash U_{s_1 s_2 \ldots s_n}$ are all congruent hypercubes. Denote the μ components of $D_{s_1 s_2 \ldots s_n} \backslash U_{s_1 s_2 \ldots s_n}$ by $D_{s_1 s_2 \ldots s_{n+1}}$ in the same relative order as above. Set

$$D^{n+1} = \bigcup_{s_1, s_2, \ldots, s_{n+1}=1}^{\mu} D_{s_1 s_2 \ldots s_{n+1}},$$

$$U^{n+1} = \bigcup_{s_1, s_2, \ldots, s_n=1}^{\mu} U_{s_1 s_2 \ldots s_n},$$

and note that $\mathcal{L}^m(U^{n+1}) = 2^{-n-2}$.

The desired Cantor set is

$$C = \bigcap_{n=1}^{\infty} D^n.$$

We note that

$$\mathcal{L}^m(C) = 1 - \sum_{n=1}^{\infty} \mathcal{L}^m(U^n) = 1/2.$$

Also note that the points of C can be thought of as infinite sequences from $\{1, 2, \ldots, \mu\}$: If $\langle s_n \rangle$ is such a sequence, then the point corresponding to $\langle s_n \rangle$

is $\bigcap_{n=1}^{\infty} D_{s_1 s_2 \ldots s_n}$, and this correspondence is a homeomorphism between C and $\{1, 2, \ldots, \mu\}^{\omega}$ with the product topology.

3. Running a Jordan Arc Through the Cantor Set.

We will now construct a Jordan arc (homeomorphic image of $[0,1]$) whose image contains C, the Cantor set with positive m-dimensional Lebesgue measure constructed in Section 2. Our strategy is to first build a Cantor set K in $J = [0,1]$ that mimics the pattern in the construction of C and to then use this Cantor set to define the desired homeomorphism.

(1) Let $V^1 \subset J$ denote the union of $\mu - 1$ disjoint open intervals of equal length such that the μ components of $J \backslash V^1$ are all of equal length. (Recall $\mu = 2^m$.) Denote these μ intervals from left to right by J_{s_1}, where $s_1 = 1, 2, \ldots, \mu$, and set $J^1 = \bigcup_{s_1=1}^{\mu} J_{s_1}$.

(2) For each s_1, let $V_{s_1} \subset J_{s_1}$ denote the union of $\mu - 1$ disjoint open intervals of equal length such that the μ components of $J_{s_1} \backslash V_{s_1}$ are all of equal length. Denote the μ intervals from left to right by $J_{s_1 s_2}$, where $s_2 = 1, 2, \ldots, \mu$. Set

$$J^2 = \bigcup_{s_1, s_2 = 1}^{\mu} J_{s_1 s_2}, \qquad V^2 = \bigcup_{s_1=1}^{\mu} V_{s_1}.$$

The above steps (1) and (2) of the construction are illustrated in Figure 2 for the case $m = 2$.

The inductive step of the construction is the following: If $J_{s_1 s_2 \ldots s_n}$ has been constructed, then let $V_{s_1 s_2 \ldots s_n} \subset J_{s_1 s_2 \ldots s_n}$ denote the union of $\mu - 1$ disjoint open intervals of equal length such that the μ components of $J_{s_1 s_2 \ldots s_n} \backslash V_{s_1 s_2 \ldots s_n}$ are all of equal length. Denote the μ intervals from left to right by $J_{s_1 s_2 \ldots s_{n+1}}$, where $s_{n+1} = 1, 2, \ldots, \mu$. Set

$$J^{n+1} = \bigcup_{s_1, s_2, \ldots, s_{n+1} = 1}^{\mu} J_{s_1 s_2 \ldots s_{n+1}},$$

$$V^{n+1} = \bigcup_{s_1, s_2, \ldots, s_n = 1}^{\mu} V_{s_1 s_2 \ldots s_n}.$$

The desired Cantor set is

$$K = \bigcap_{n=1}^{\infty} J^n.$$

Note that, as with C, identifying the infinite sequence $\langle s_n \rangle$ of elements of $\{1, 2, \ldots, \mu\}$ with the point $\bigcap_{n=1}^{\infty} J_{s_1 s_2 \ldots s_n}$, defines a homeomorphism between K and $\{1, 2, \ldots, \mu\}^{\omega}$ with the product topology.

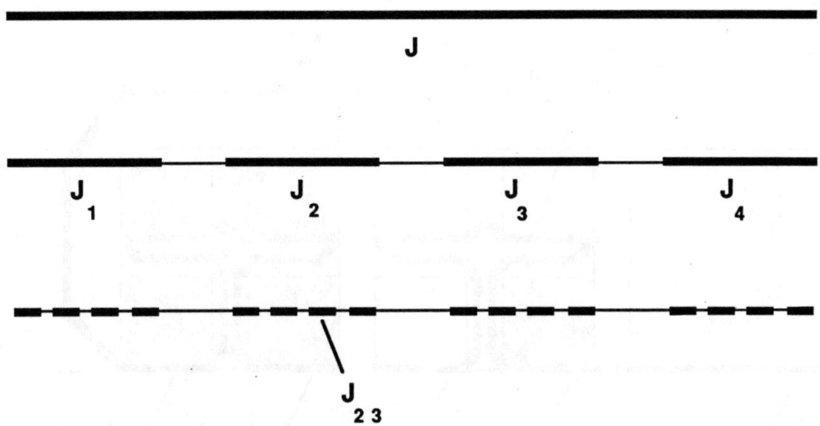

Construction of a Cantor set in the interval
mimicking the construction in the plane
Figure 2

The homeomorphism is now obtained as follows:

(0) Let $f_0 : J \to D$ map J onto the *base* of D by $f_0(t) = (t, 0, 0, \ldots, 0)$, where the *base* $b(E)$ of a hypercube E in $\mathbf{R}^m$ with sides parallel to the coordinate $(m-1)$-planes is the set of points in E with each coordinate but the first having its value equal to the minimum value of that coordinate occuring in E.

(1) Let $f_1 : J \to D$ be an embedding such that

(i) $f_1(0) = (0, 0, \ldots, 0)$,

(ii) $f_1(1) = (1, 0, 0, \ldots, 0)$,

(iii) $f_1(J_{s_1}) = b(D_{s_1}),\ \ s_1 = 1, 2, \ldots, \mu$,

(iv) $f_1(V^1) \subset U^1$.

Only condition (iv) can cause difficulty, and then only in case $m = 2$, but if D_2 and D_3 are numbered as in Figure 1, we can use the embedding f_1 which is illustrated in Figure 3.

Figure 4 illustrates the embeddings f_0 and f_1 in case $m = 3$.

(2) Let $f_2 : J \to D$ be an embedding which agrees with f_1 on V^1 and for

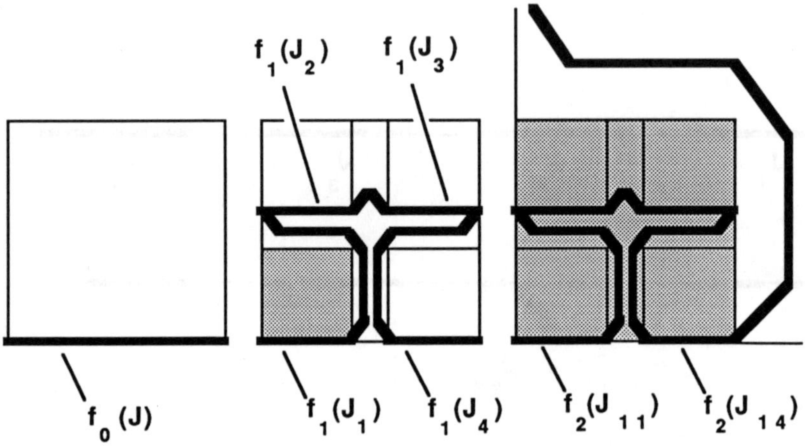

Embeddings of the interval into the plane
Figure 3

which $f_2|_{J_{s_1}}$, is defined relative to D_{s_1} just as f_1 was defined relative to D. That is, if L_{s_1} is the linear map which is a composition of a homothety and a translation and which maps D to D_{s_1} (as in Step (2) of the construction of a Cantor set with positive m-dimensional Lebesgue measure given in Section 2) and $\ell_{s_1} : \mathbf{R} \to \mathbf{R}$ is the increasing, affine function with $\ell_{s_1}(J_{s_1}) = J$, then we have

(i) $f_2|_{J_{s_1}} = L_{s_1} \circ f_1 \circ \ell_{s_1}$, for $s_1 = 1, 2, \ldots, \mu$,

(ii) $f_2|_{V^1} = f_1$.

For the case $m = 2$, part of such an embedding is illustrated in Figure 3.

The inductive step is as follows: If $f_n : J \to D$ has already been defined, then define $f_{n+1} : J \to D$ by

(i) $f_{n+1}|_{J_{s_1 s_2 \ldots s_n}} = L_{s_1 s_2 \ldots s_n} \circ f_1 \circ \ell_{s_1 s_2 \ldots s_n}$,

where $L_{s_1 s_2 \ldots s_n}$ is the composition of a homothety and translation with $L_{s_1 s_2 \ldots s_n}(D) = D_{s_1 s_2 \ldots s_n}$ and $\ell_{s_1 s_2 \ldots s_n}$ is the increasing, affine function with $\ell_{s_1 s_2 \ldots s_n}(J_{s_1 s_2 \ldots s_n}) = J$,

(ii) $f_{n+1}|_{V^n} = f_n$.

We now define $f : J \to D$. The unit interval J is the union of the Cantor set

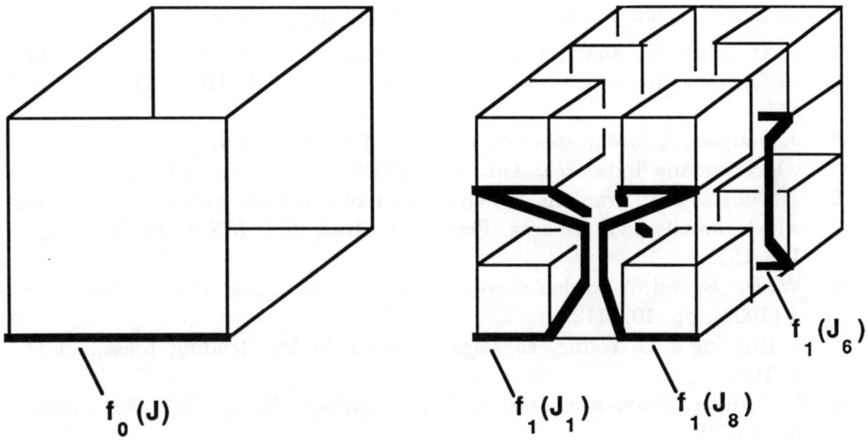

Embeddings of the interval into 3-space
Figure 4

K and the union of the complementary open intervals $\cup_{n=1}^{\infty} V^n$. Since f_{n+1} agrees with f_n on $\cup_{i=1}^{n} V^i$, then f defined on $\cup_{n=1}^{\infty} V^n$ by $f(x) = f_n(x)$ for $x \in V^n$ is well-defined. If $x \in K$ is represented by the sequence $\langle s_n \rangle$, then $f(x)$ is the point of C with the same sequence representation. That is, if $x = \cap_{n=1}^{\infty} J_{s_1 s_2 \ldots s_n}$, then $f(x) = \cap_{n=1}^{\infty} D_{s_1 s_2 \ldots s_n}$. With this description f is quite obviously both one-to-one and continuous and since J is compact, it follows that f is an embedding. One can also note that the sequence of embeddings f_n converges uniformly to f.

Finally, we note that if $f : J \to \mathbf{R}^m$ has an image with positive m-dimensional Lebesgue measure, then assuming f is continuous so $f(J)$ is measurable[8] $f \times id : J \times J^k \to \mathbf{R}^{m+k}$ has an image with positive $(m+k)$-dimensional Lebesgue measure. Thus we have

THEOREM 2. *For positive integers $1 \leq p \leq q$ there exists an embedding $f : [0,1]^p \to [0,1]^q$ such that the image of f has positive q-dimensional Lebesgue measure.*

REMARK. If one wishes to ensure that a mapping not increase the measure theoretic dimension of subsets of Euclidean space, then it is necessary to impose a

Lipschitz condition (of order one) on the map.[9]

4. References

1. G. Peano, Sur une courbe, qui remplit toute une aire plane, *Math. Ann.* **36** (1890), pp. 157–160.
2. J. Douglas, A Jordan space curve which bounds no finite simply connected area, *Proc. Nat. Acad. Sci. U.S.A.* **19**, (1933), pp. 269–271.
3. J. Douglas, An analytic closed space curve which bounds no orientable surface of finite area, *Proc. Nat. Acad. Sci. U.S.A.* **19** (1933), pp. 448–451.
4. J. Douglas, A Jordan space curve no arc of which can form part of a contour which bounds finite area, *Ann. Math.*(2) **35** (1933), pp. 100–103.
5. J. Douglas & P. Franklin, A step-polygon of a denumerable infinity of sides which bounds no finite area, *Proc. Nat. Acad. Sci. U.S.A.* **19** (1933), pp. 188–191.
6. W. F. Osgood, A Jordan curve of positive area, *Trans. Amer. Math. Soc.* **4** (1903), pp. 107–112.
7. J. Hocking & G. Young, *Topology* (Addison-Wesley, Reading, Mass., 1961), p. 100.
8. H. Federer, *Geometric Measure Theory* (Springer-Verlag, New York, 1969), pp. 69–70.
9. K. T. Smith, *Primer of Modern Analysis* (Springer-Verlag, New York, 1983), p. 372.

The Problem of Plateau (pp. 237–244)
ed. Th. M. Rassias

SOME PROBLEMS AND REMARKS ON THE EIGENVALUES OF THE LAPLACIAN AND MINIMAL SURFACES

THEMISTOCLES M. RASSIAS
Department of Mathematics, University of La Verne,
P.O.Box 51105, Kifissia, Athens 14510,
Greece

ABSTRACT

It is our object to discuss a few of the open problems on the eigenvalues of the Laplacian, on the number and characterization of minimal surfaces spanning a curve in R^3, on the Riemann mapping theorem from the point of view of minimal surfaces as well as on the analogous surfaces in spacetime, in particular manifolds with Lorentz metric (such surfaces are called maximal surfaces). Most of these problems are well known and some of these problems have resisted solution for a long time.

I. Eigenvalues of the Laplacian

The eigenvalue problem for the Laplace operator in two dimensions is classical in mathematics and physics. Computational method for estimating the eigenvalues are still of active research, particularly from the point of view of applications. It is important to study inequalities for eigenvalues of the Laplacian when the spectrum, i.e. those values of λ for which a non-trivial solution exists, is discrete. In the following we pose a few problems whose solution is useful for the computation and characterization of the Morse-Smale index for a minimal surface in R^3 (cf. [1],[2]).

1. Let D be a bounded domain in R^2. Let $\{\lambda_n\}$ be the spectrum of the Laplacian acting of functions with zero boundary data (counting with multiplicity). It is an old conjecture of Pólya that

$$\frac{\lambda_n}{n} \geq \frac{4\pi}{A(D)} ,$$

where $A(D)$ denotes the area of D.

Remark. Pólya had verified the conjecture in the case that D can tile R^2.

2. Let D be a bounded domain in R^2. Assume that $\lambda_1(D)$ and $\lambda_2(D)$ are the first and second (non zero) eigenvalues of the Laplacian for functions with zero boundary values. Prove that

$$\frac{\lambda_2(D)}{\lambda_1(D)} \leq \frac{\lambda_2(\Delta)}{\lambda_1(\Delta)},$$

where Δ is the disc in R^2, and that equality implies D is a disc.

3. Let D be a bounded convex domain in R^2. Study the qualitative behavior of the nodal line of the second eigenfunction F_2 for the Laplacian with zero boundary conditions. It is a longstanding conjecture that the nodal line of F_2 cannot enclose a compact subregion of D.

4. Let M be a compact manifold and F_n, $n = 1, 2, \ldots$ be the eigenfunctions for the Laplacian on M. Prove that the number of critical points of F_n increases with n.

References

1. Th. M. Rassias, *Eigenvalues of the Laplacian*, Mechanics, Analysis and Geometry: 200 Years after Lagrange (ed: M. Francaviglia), Elsevier Science Publishers B. V., Amsterdam, 1991, pp. 315–332.
2. S.-T. Yau (ed), *Seminar on Differential Geometry*, Princeton University Press, Princeton, 1982.

II. An Eigenvalue Problem

Consider the eigenvalue problem

$$r^2\frac{d^2F}{dr^2} + r\frac{dF}{dr} + \lambda \cdot \frac{(r^4+1)^2}{r^2} \cdot F(r) = (n^2 - 32) \cdot (r) \tag{E}$$

where $r \in [0, r_0]$, r_0 is a fixed real number and $n = 1, 2, 3, \ldots$, with boundary condition $F(r_0) = 0$ and $F \in C_p^2[0, r_0]$ where C_p^2 stands for piece-wise C^2 functions. Compute the sum (or an upper bound of the sum) of the dimensions of the eigenspaces corresponding to the negative eigenvalues of this eigenvalue problem as a function of r_0.

Remark. The above computation is useful in order to study the Morse-Smale index of the Richmond's minimal surface in R^3. A *conformal parametric representation* of a porth of Richmond's minimal surface in R^3 is a mapping

$$I : D^2_{r_0} \to R^3 \quad \text{for each} \quad r_0 > 0 \ ,$$

where

$$D^2_{r_0} = \{(u,v) \in R^2 : u^2 + v^2 < r_0^2\} \ ,$$

given by

$$I(u,v) = (x(u,v), y(u,v), z(u,v))$$

where $(u,v) \in D^2_{r_0}$ and

$$\begin{cases} x(u,v) & = \frac{1}{3}u^3 - uv^2 + \frac{u}{u^2+v^2} \\ y(u,v) & = \frac{1}{3}v^3 - u^2v - \frac{v}{u^2+v^2} \\ z(u,v) & = 2u \ . \end{cases}$$

This is a surface studied from various points of view by many mathematics, among them H. W. Richmond (the inventor of the surface in 1904), H. Schwarz and others (cf. [1–3]). In [2] the following theorem is proved:

Theorem. *The negative eigenvalues of* (E) *with boundary condition* $F(r_0) = 0$ *and* $F \in C_p^2[0, r_0]$ *happen only for* $n = 0,1,2,3,4,5$. *No negative eigenvalues occur in* (E) *if*

(i) $\lambda < -32r_0^2$ *and* $n = 0$
(ii) $\lambda < -31r_0^2$ *and* $n = 1$
(iii) $\lambda < -28r_0^2$ *and* $n = 2$
(iv) $\lambda < -23r_0^2$ *and* $n = 3$
(v) $\lambda < -16r_0^2$ *and* $n = 4$
(vi) $\lambda < -7r_0^2$ *and* $n = 5$

References

1. Th. M. Rassias, *Query No. 257* Notice of the Amer. Math. Soc. 29(4) (1982), p. 326.

2. Th. M. Rassias, *On the nulling and the Morse-Smale index of the Richmond's minimal surface in R^3*, Tamkang Journal of Mathematics 16 (1985) 49–57.
3. Th. M. Rassias, *Foundations of Global Nonlinear Analysis* Teubner-Texte zur Mathematik, Band 86, Leipzig, 1986.

III. On the Number and Characterization of Minimal Surfaces Spanning a Curve

1. It is longstanding problem to examine whether or not for any C^1 Jordon curve Γ in the Euclidean space R^3 the number of minimal surfaces bounded by Γ is odd.

Remarks. The problem was first posed by T. Radó. The problem whether for *reasonable boundary data* the problem of Plateau will always have only a finite number of solutions resist solution today. In this place belong the exciting results of F. Tomi [5] on the finiteness of the number of surfaces of absolutely minimal area spanning an analytic curve in R^3 and of R. Böhme and A. J. Tromba [1] on the generic finiteness. For further results and historical remarks in this subject the reader is referred to the encyclopedic book of J. C. C. Nitsche [2] and the recent monograph of M. Sturwe [4] (see also Th. M. Rassias [3]) as well as to a number of recent books by A. Fomenko.

References

1. R. Böhme and A. J. Tromba, *The index theorem for classical minimal surfaces*, Ann. of Math. 113 (1981), 447–4.
2. J. C. C. Nitsche, *Vorlesunge über Minimalflächen*, Grundlehren 199, Springer Verlag, Berlin-Heidelberg-New York, 1975.
3. Th. M. Rassias, *Foundations of Global Nonlinear Analysis*, Teubner-Texte zur Mathematik, Band 86, Leipzig, 1986.
4. M. Struwe, *Plateau's Problem and the Calculus of Variations* , Mathematical Notes 35, Princeton University Press, Princeton, 1988.
5. F. Tomi, *On the local uniqueness of the problem of least area*, Arch. Rat. Mech. Anal. 52 (1973), 312–318.

2. Examine if there exists a Jordon curve which bounds a continuum of solutions of Plateau's problem, all having the same value of Dirichlet's integral? ([1]).

Reference.

1. J. C. C. Nitsche, *Lectures on Minimal Surfaces*, Volume 1, Cambridge University Press, Cambridge, 1989.

3. Give an example of a Jordan curve Γ in R^3 spanning five different minimal surfaces of the type of the disc, all of which are explicitly known. Is there any characterization of this property in the sense of Morse theory on Hilbert (or Banach) manifolds as developed by R. Palais and S. Smale [1]?

Remark. I had first posed this problem in 1977 (see [2]).

References

1. R. Palais and S. Smale, *A generalized Morse theory*, Bull. Amer. Math. Soc. 70 (1964), 165–171.
2. Th. M. Rassias, *Query No. 118*, Notices of the Amer. Math. Soc. 24(2) (1977), p. 136.

4. i) Characterize the Morse type of a minimal surface by properties similar to the number of conjugate points (or the number of negative characteristic roots in the associated problem) in the case of single integral problem in the Calculus of Variations.
ii) Prove that the kth type number of a minimal surface bounded by a Jordan curve Γ is either zero for all k, or zero for all $k \neq j$ and 1 for $k = j$.

Remark. Max Shiffman had posed the above problem in [1].

Reference

1. M. Shiffman, *The Plateau problem for non-relative minima*, Annals of Math. 40 (1939), 834–854.

IV. Riemann Mapping Theorem from the Point of View of Minimal Surfaces

Riemann's problem of mapping a simply connected plane region whose boundary consists of more than a single point conformally on a circle as normal region can be reduced to the study of two problems:

1) *the interior problem* that concerns the map of the interior points and
2) *the boundary problem* that concerns the behavior of the map on the boundary.

It was Riemann who studied the first problem by using techniques of the Dirichlet principle and Schwarz and Neumann who gave proofs for the case of regions with restricted boundaries. Later, Osgood gave a satisfactory answer to the general case using methods due to Poincaré. The second problem was solved for analytic boundaries by Schwarz and in other special cases by Picard. The general case was treated by Osgood [2] and by Carathéodory [1]. In 1977, I outlined a method to apply the theory of minimal surfaces for a generalization of the Riemann mapping theorem. Consider the minimal surface equation given by

$$(1+f_y^2)f_{xx} - 2f_x f_y f_{xy} + (1+f_x^2)f_{yy} = 0 . \qquad (*)$$

The surface is assumed in the non-parametric form and Plateau's problem is regarded as a generalized Dirichlet problem, with (*) replacing Laplace's equation. According to Weierstrass [4] a parametric form of the solution for the minimal surface equation (*) is given by

$$x = \text{Re } F_1(w) \qquad (1)$$

$$y = \text{Re } F_2(w) \qquad (2)$$

$$z = \text{Re } F_3(w) \qquad (3)$$

where $F_1(w), F_2(w), F_3(w)$ are any analytic functions satisfying

$$F_1'^2(w) + F_2'^2(w) + F_3'^2(w) = 0 . \qquad (4)$$

Set

$$\psi_1(w) = \tfrac{1}{2}(F_1'(w) + iF_2'(w)) \qquad (5)$$

and

$$\psi_2(w) = \tfrac{1}{2}F_3'(w) . \qquad (6)$$

It follows that

$$x + iy = \int \psi_1(w)dw - \overline{\int \frac{\psi_2^2(w)}{\psi_1(w)}dw} . \qquad (7)$$

Denote the right-hand side of (7) by $U(\psi_1)$, then one has the following result [3].

Theorem. *Let Γ be a simple closed analytic curve in the z-plane. Then there exists a regular function $\psi_1(w)$ defined in $\Omega = \{w : |w| \geq 1\}$, such that $U(\psi_1)$ maps Ω simply onto the closed domain exterior to Γ and such that infinity is mapped into infinity, for a fixed regular function $\psi_2(w)$ defined in Ω.*

Remark. If $\psi_2(w) \equiv 0$, then the above theorem implies the Riemann mapping theorem, as a special case. It will be nice if one gives a simpler proof of the Riemann mapping theorem applying the theory of minimal surfaces.

References

1. C. Carathéodory, *Über die gegenseitige Beziehung der Ränder bei der Konformen Abbildung des Innern einer Jordanschen Kurve auf einen Kreis*, Mathematische Annalen, 73 (1913), 305–320.
2. W. S. Osgood, *On the transformation of the boundary in the case of conformal mapping*, Bull. Amer. Math. Soc. 9 (1903), 233–235.
3. Th. M. Rassias, *On some new properties of harmonic mappings*, Serdica Bulgaricae Mathematicae Publicationes 13 (1987), 133–136.
4. K. Weierstrass, *Mathematische Werke*, 3 Bände, Mayer und Müller, Berlin (1903).

V. Minimal Surfaces in Physics

In Riemannian manifolds, minimal surfaces and their generalizations to higher dimensions have been studied in physics and other technological sciences. The analogous surfaces in spacetimes, in particular manifolds with Lorentz metric, are *maximal surfaces*. Maximal surfaces have been very important for questions concerning the understanding of the n-body problem in a gravitational field as well as of the dynamics of the gravitational field. Many examples of minimal surfaces have been given in physics (see for example Bruce L. Reinhart [2]). There are closed spacetimes without any maximal surfaces. Maximal surfaces in closed spacetimes like the Friedman models (cf. C. W. Misner, K. Thorne and J. A. Wheeler [1]) occur rarely, unless the universe is static. In fact we can say that the spacetime either represents an ever expanding universe without any maximal spacelike surface, as for example the *open* Friedman models, or the spacetime possesses exactly one maximal surface as it happens in the *closed* Friedman models. In 1944, A. Lichnerowicz explained the role of maximal and constant mean curvature hypersurfaces in studying problems of general theory of relativity. In 1980, A. Fischer, J. Marsden and V. Moncrief shown why the existence of constant mean curvature hypersurfaces is needed for one to analyze the structure of the singularities in the space of solutions of Einstein's equations. It seems to be an *open problem* to examine if variational theory can be used to study maximal hypersurfaces that might be *nonsmooth*. In 1980, J. Marsden and F. J. Tipler have proved that maximal and constant mean curvature hypersurfaces are stable under perturbation of the spacetime with some generic hypotheses.

References

1. C. W. Misner, K. Thorne and J. A. Wheeler, *Gravitation*, W. H. Freeman, San Francisco, 1973.
2. Bruce L. Reinhart, *Maximal foliations of extended Schwarzschild space*, J. Math.

Physics 14(6) (1973), 719.

3. Th. M. Rassias, *Morse theory and Plateau's problem*, Selected Studies: Physics-Astrophysics, Mathematics, History of Science, A volume dedicated to the memory of Albert Einstein (eds: Th. M. Rassias and G. M. Rassias), North-Holland Publishing Company, Amsterdam, 1982, pp. 261–292.

The Problem of Plateau (pp. 245–257)
ed. Th. M. Rassias

RECENT DEVELOPMENTS ON THE STRUCTURE OF COMPACT SURFACES WITH PLANAR BOUNDARY**

RICARDO SA EARP*
Mathematics Department, Pontifícia Universidade Católica
22453 – Rio de Janeiro
Brazil

ABSTRACT

In this survey article we will summarize recent results concerning the structure of compact connected surfaces $M \subset \mathbf{R}^3$ with boundary a simple closed planar curve γ lying in the horizontal xy-plane $\mathcal{H}$. We assume M is transverse to $\mathcal{H}$ along γ and we suppose M has non-vanishing mean curvature vector. When M has certain prescribed mean curvature we establish the structure of M with respect to $\mathcal{H}$. The foremost interest is the constant mean curvature configurations. We also briefly discuss Bryant Weingarten configurations and extensions to hyperbolic three space.

0. Introduction

Throughout this paper we will bring to focus the following geometric structure:

Consider M a connected compact surface in $\mathbf{R}^3$ with boundary a simple closed planar curve γ lying in the horizontal xy-plane $\mathcal{H}$. Assume M is embedded and is transverse to $\mathcal{H}$ along γ. Moreover suppose M has non-vanishing mean curvature vector $\overrightarrow{H}$.

Our goal is to point out certain geometric structure arising from the above containing the class of constant mean curvature to prove that fundamental principles lead to global information on the configuration of M. We shall analyse the so-called $H(x, y)$-surfaces and we discuss Bryant–Weingarten surfaces

*This work is supported by S.C.T./P.R. and by CNPq, Brazil.

**Part of this paper was contained in a talk given by the author on the Workshop on Minimal Surfaces, Granada 5th to 10th September 1991.

defined in the next lines. First, we will state the basic results related to constant mean curvature surfaces, and we will set some open questions concerning our subject: Indeed, if M is a *constant mean curvature surface* then basic property as *ellipticity* leading to the *Graph Lemma* and basic Flux Formula giving the *Mean Curvature Vector Lemma* imply restrictions on the global structure of M (cf. [3]). For example:

a) If M is topologically a disk we prove that the global configuration of M is restricted to the so-called *Sacis's cap** (see Fig. 0); i.e., it satisfies (cf. [3]):
   1) $M \cap \mathcal{H}^+$ is connected, where $\mathcal{H}^+ = \{z \geq 0\}$.
   2) $M \cap \mathcal{H}$ is the union of closed simple curves with pairwise disjoint interiors.
   3) $M \cap \mathcal{H}^- - \{\gamma\}$ is either empty or is the union of graphs over the connected curves in $M \cap \mathcal{H}$, each one intersecting the convex hull of D, where $\mathcal{H}^- = \{z \leq 0\}$ and D is the bounded domain the $\mathcal{H}$ with $\partial D = \gamma$.

SACI'S CAP

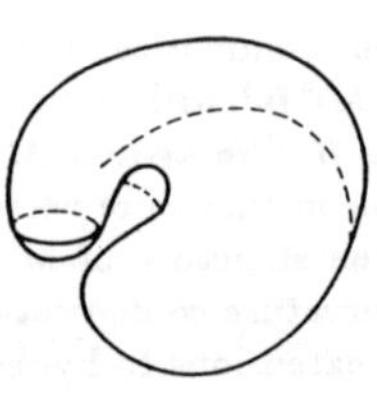

SACI

Fig. 0.

Furthermore, analogous structure theorem holds for arbitrarily genus (cf. [7]). This result is part of Theorem 1 in paragraph 1.

---

*Saci is a well-known genius in brazilian folklore.

b) If D is *convex* then M is contained in a halfspace (cf. [7]). This property yields if γ is a circle then M is a spherical cap, by applying the *Alexandrov Reflection Principle*, to prove M inherits the symmetries of γ. We will explain the Alexandrov techniques in paragraph 1 (cf. [0], [7], [14]).

Question 1: Is there a *non trivial* Saci's cap, i.e., $M \cap \text{ext}D \neq \phi$ of constant mean curvature?

Question 2: Is there a genus $g > 0$ embedded constant mean curvature surface with boundary a *convex* curve $\gamma \subset \mathcal{H}$, entirely contained in the *halfspace* $\mathcal{H}^+$?

The above question was raised by Harold Rosenberg in the Differential Geometry Seminar at Paris VII.

Question 3: If M is a constant mean curvature surface with boundary a circle γ of radius R then is it true that M is a spherical cap of mean curvature H, $H \leq \frac{1}{R}$?

Actually, when $H = \frac{1}{R}$ (the extremal situation) this question is a belief of Brezis and Coron inquired in [2]: We show that it is true in a joint work with F. Brito (cf. [3]) by applying the Flux Formula, and the fact that the umbilic points of M are isolated (see [10]). Recall that if M is embedded, and transverse to $\mathcal{H}$ along γ, then the answer is still affirmative. Another partial result is obtained by L. Barbosa in [1].

In order to generalize the preceding results, we shall treat a geometric structure arising from the class of *Prescribed Mean Curvature surfaces* and we shall discuss one arising from the class of *Weingarten surfaces*:

A) **Prescribed Mean Curvature Surfaces:** We say M is a $H(x,y)$-surface if M has *constant mean curvature along each vertical line r cutting orthogonally $\mathcal{H}$*; i.e., the mean curvature H does not depend on the third coordinate. Our main result for this geometric structure is the following (see paragraph 1):

Theorem 1. (cf. [4]) Let M be an embedded connected compact smooth surface in $\mathbf{R}^3$ with boundary a simple closed curve γ lying in the horizontal xy-plane $\mathcal{H}$. Assume M is a $H(x,y)$-surface with non-vanishing mean curvature. Suppose M along γ, is transverse to $\mathcal{H}$ and is locally contained in $\mathcal{H}^+$. Assume that the mean curvature vector $\overrightarrow{H}$ restricted to the open lower halfspace int $\mathcal{H}^-$ is non horizontal, then essentially* the following assertions hold.

1. $M \cap \mathcal{H} - \{\gamma\}$ is contained in ext D.

*i.e., assuming global transversallity.

2. $M \cap \mathcal{H}^-$ is either γ or is the union of graphs over a planar domain $\Omega \subset \mathcal{H}$, described in 2.1 (considering Assertion 3 below).
   2.1. Ω is a collection of pairwise disjoint multi-connected domains, say $\{\mathcal{R}_1, \ldots, \mathcal{R}_s\}$, in such a way that $(M \cap \mathcal{H}^+) \cup D \cup \dot{\bigcup}_{i=1}^{s} \mathcal{R}_i$, is a closed embedded surface (non smooth on $\partial(D \cup \dot{\bigcup}_{i=1}^{s} \mathcal{R}_i)$).
3. $M \cap \mathcal{H}^+$ is connected.
4. $\sum_{i=1}^{s} g_i \leq g$, where g is the genus of M and g_i is the genus of $\mathcal{R}_i$ (given in 2.1). Moreover, the null homotopic curves in M lying in $\mathcal{H}$, are exactly the boundary of topological disks in $\dot{\bigcup}_{i=1}^{s} \mathcal{R}_i$.
5. Any theoretical configuration prescribed in 1,2,3,4 may be constructed, with mean curvature H arbitrarily close to a given positive constant.

Furthermore if M has constant mean curvature then the following statement still hold (cf. [7]):*

6. No component of $\partial(M \cap \mathcal{H}^+)$ lies completely outside of the convex hull of γ, and all of them are null-homotopic in ext D.

We remark that there are two basic lemmas underlying Theorem 1, both of them arising from the constant mean curvature situation: The Graph Lemma and the Flux Formula (See paragraph 1).

Question 4: If M is a closed connected $H(x,y)$-surface, embedded in $\mathbf{R}^3$, with *non-vanishing mean curvature*, then has M a plane of symmetry?**

B) **Weingarten Surfaces:** We say M is a Bryant-Weingarten elliptic surface if M satisfies *the Weingarten relation* $H = f(H^2 - K)$, and satisfies the *ellipticity condition*

$$4(H^2 - K)(f'(H^2 - K))^2 < 1,$$

where f is a positive smooth function defined in a open interval containing $[0, \infty)$. (Global and local properties of these geometric structures are investigate by R. Bryant in a paper entitled "Complex Analysis and a class of Weingarten Surfaces").

---

*When H is constant we do not need the hypothesis that $\vec{H}$ restricted to the open lower halfspace int $\mathcal{H}^-$ is non horizontal (cf. Sec. 1.5 and [7]).

**The answer to question 4 is no. Recently, R. Kusner showed us examples in a personal communication (during the Workshop on Granada).

We claim there is also a structure theorem with respect to a Bryant-Weingarten elliptic surface M with boundary a *convex* curve γ: In fact, the geometric principles situated behind this theorem are the *Mean Curvature Vector Lemma Principle along* γ, the *Vertical Alexandrov Reflection Principle, and the Graph Lemma Principle* which are explained in paragraph 1. Besides this study is part of a paper in final step of preparation (cf. [5]). We announce below an amazing consequence, giving a characterization of the spherical cap:

Theorem 2. (cf. [5]) Let γ be a circle lying in a plane $\mathcal{H}$ and let M be a compact connected disk type surface with boundary γ. Assume M is Bryant-Weingarten elliptic, and transverse to $\mathcal{H}$ along γ. Then M is a spherical cap.

Question 5: Is there a Flux Formula for Bryant-Weingarten elliptic surfaces?

Recall that the answer is affirmative for constant mean curvature surfaces (cf. [3], [5]). Moreover, the answer is still affirmative for positive constant Gauss curvature: This is announced by H. Rosenberg, in a personal communication. Notice that if Question 5 is true, then there is a structure theorem characterizing Bryant-Weingarten elliptic Saci's cap.

1. Prescribed Mean Curvature Configurations

We believe that the core of our geometric investigations is the so-called *Alexandrov Reflection Principle* (cf. [0], [7], [14]). To such an extent, we will proceed first to explain Alexandrov's techniques:

1.1. *Alexandrov reflection principle techniques*

For the sake of simplicity and for the readers benefit we will illustrate *the Alexandrov Reflection Principle* techniques by considering one archetype situation as follows: Let M be a connected compact surface with boundary a circle $\gamma = S^1$ lying in the horizontal plane $\mathcal{H}$. Suppose M is embedded and contained in the upper halfspace $\mathcal{H}^+$. We will apply the *Alexandrov Reflection Principle* by moving vertical planes to prove that *if* M *has constant mean curvature then* M *inherits the symmetries of* γ, i.e., M is a spherical cap (by using the Delaunay's classification of constant mean curvature rotationally symmetric embedded surfaces (cf. [8])): The idea is to show that M is invariant by reflection in every vertical plane π that is a plane of symmetry of γ. Suppose, on the contrary, that M is not symmetric in π. Then there are points, p, q on M such that the line $[p, q]$ joining p to q is orthogonal to π, p and q are on opposite sides of π, and $d(p, \pi) > d(q, \pi)$. Thus the symmetry p^* of p through

π is on the other side of q, i.e., $[p^*,p] \supset [q,p]$. Now, consider the closed surface $\hat{M} = M \cup D$ (D is the disk in $\mathcal{H}$, $\partial D = S^1$). Clearly $\hat{M}$ is a boundary of a solid V in $\mathcal{H}^+$. Let $\pi(t)$ for $t \geq 0$ be a 1-parameter family of planes parallel to π, on the same side of π as p, with $\pi(0) = \pi$, t equals to the distance between $\pi(t)$ and π. For $t = T$ sufficiently large, $\pi(T)$ is disjoint from M. Let t go from T to zero and consider the first point of contact of the family $\pi(t)$ with M, suppose, occuring at a point p_1, for $t = t_1$ (i.e., $\pi(t) \cap M = \phi$, $t > t_1$ and $\pi(t_1) \cap M \neq \phi$, $p_1 \varepsilon \pi(t_1) \cap M$). Denote by $\pi^+(t)$ the halfspace of $\mathbf{R}^3 - \pi(t)$ foliated by $\pi(\bar{t})$, $\bar{t} \geq t$ (i.e., $\pi^+(t) \cap \pi = \phi$). Let $M(t)^*$ denote the symmetry of $M \cap \pi^+(t)$ about $\pi(t)$. Clearly for $\varepsilon > 0$ small enough $M^*(t_1 - \varepsilon)$ is entirely contained inside the solid V, $M^*(t_1 - \varepsilon)$ is a graph over π for M is a graph around p_1 (see Fig. 1A). As the planes $\pi(t)$ approach π starting from $\pi(T)$, consider the first (non trivial) point of contact of $M^*(t)$ with M, occuring, say, at $t = t_0$; i.e., $M^*(t)$ is inside V for $t > t_0$, $M^*(t_0)$ is a graph over π, and $M^*(t)$ "escapes" from V intercecting $\mathbf{R}^3 - V$ for $t < t_0$. Remark that this first point of contact should occur before the planes $\pi(t)$ reach π (by the assumptions in the beginning of the proof). Thus this point cannot be the image of an interior point (see Fig. 1A) of M since this would imply $M^*(t_0) \subset M$ by the interior maximum principle (cf. [8], [9]). Then M would be invariant by reflection in $\pi(t_0)$, which is impossible because $\pi(t_0)$ is not a symmetry plane of M (i.e., $S^1 \cap (\mathbf{R}^3 - \pi^+(t_0))$ is not a graph over $\pi(t_0)$). Another possibility is that M becomes orthogonal to $\pi(t)$ before an interior point of contact is reached (see Fig. 1B). Then the boundary maximum principle would imply M is invariant by reflection in $\pi(t_0)$. Both possibilities yield $M \cap \pi^+(t_0)$, $\pi(t_0) \neq \pi$, is a graph over π, giving the desired contradiction.

We call *Vertical Alexandrov Reflection Principle* the Alexandrov technique by moving vertical planes, as explained in the preceding paragraph. Similarly, we call *Horizontal Alexandrov Reflection Principle*, the analogous procedure by moving horizontal planes (i.e., planes parallel to $\mathcal{H}$).

1.2. *The graph lemma*

Consider M an embedded connected compact smooth surface in $\mathbf{R}^3$ cutting transversally a plane $\mathcal{H}$ along its boundary ∂M, with $\partial M = M \cap \mathcal{H}$ (see Fig. 2A). It is easy to see that there is a canonical way to choose a finite number of bounded multi-connected planar regions $\mathcal{R}_1, \ldots, \mathcal{R}_k$ in $\mathcal{H}$ such that $\hat{M} = M \cup \bigcup_{i=1}^{k} \mathcal{R}_i$ is a connected closed embedded (not smooth over $\partial\mathcal{R}_i$) surface in $\mathbf{R}^3$. We shall call the union of the vertical solid cylinders over $\mathcal{R}_1, \ldots, \mathcal{R}_k$ th *solid cylinder over* ∂M (see Figs. 2A and 2B. cf. [3], [4]). The Graph Lemma

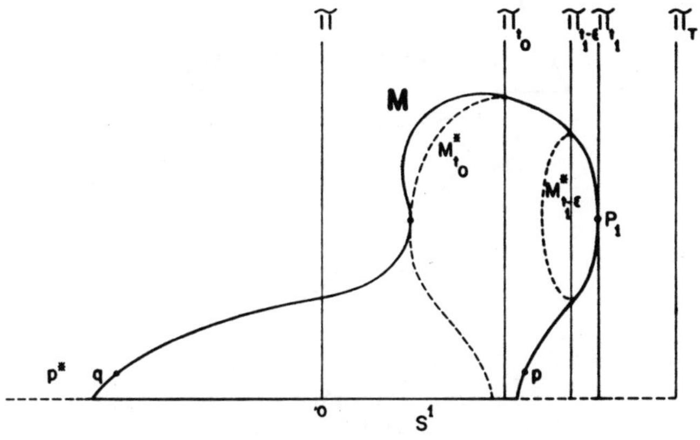

Fig. 1A.

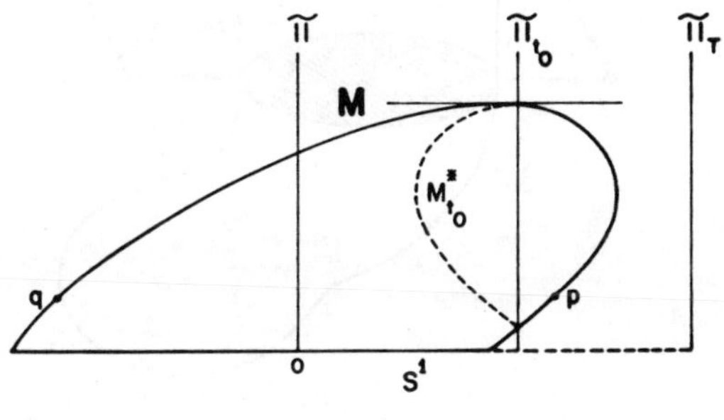

Fig. 1B.

for constant mean curvature surfaces asserts the following (cf. [3]): *If M is a connected compact embedded surface contained in the lower halfspace $\mathcal{H}^-$, with $\partial M \subset \mathcal{H}$, and if M has a continuation to a constant mean curvature surface N, $N \supset M$, with $\partial N \supset \mathcal{H}^+$ and ∂N not intersecting the solid cylinder over ∂M then M is a graph over a planar region in $\mathcal{H}$.*

We will outline the proof considering the following archetype configuration: Consider M given by Fig. 2A. Suppose M has a continuation to a constant mean curvature surface given by Fig. 3. Notice that one may attach to N a disk D obtaining a closed embedded surface $N \cup D$ enclosing a solid V not smooth over ∂N. Now one may apply the *Horizontal Alexandrov Reflection Principle* to $N \cup D$ by moving a 1-parameter family of horizontal planes $\mathcal{H}(t)$,

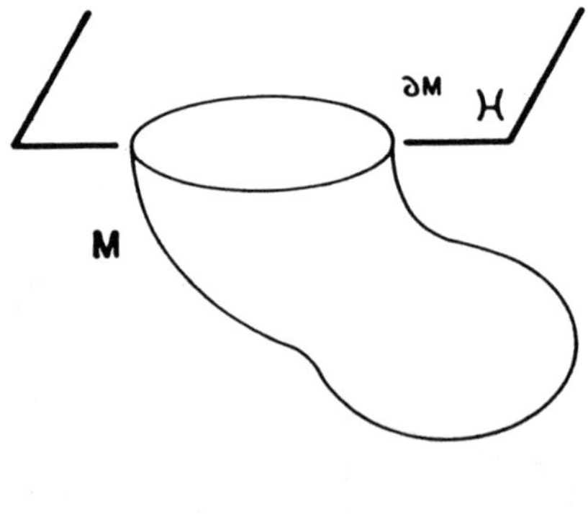

Fig. 2A.

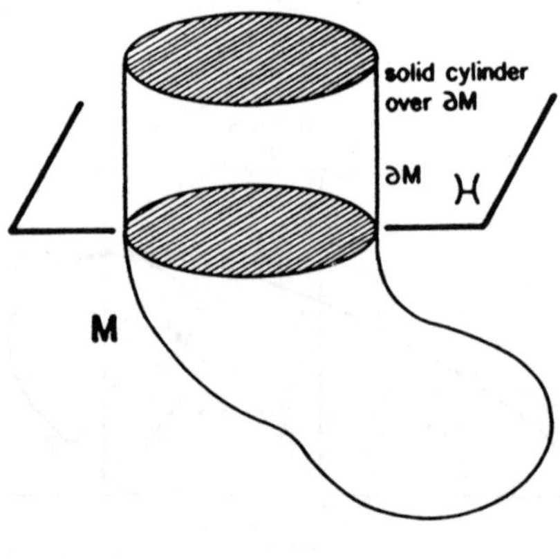

Fig. 2B.

$-\infty < t \leq 0$ towards $\mathcal{H}$, $\mathcal{H} = \mathcal{H}(0)$ (through $\mathcal{H}^-$) to ensure M is a graph. Indeed, if the first point of contact occurs for $t = t_0$, then the part of M below $\mathcal{H}(t)$ say $M(t)$ for $t \leq t_0$ is a graph over M, $M^*(t)$ (the symmetric part of $M(t)$ with respect to $\mathcal{H}(t)$) is entirely contained in the solid V for $t \leq t_0$. Furthermore, if the first point of contact occurs at a point $p \in \mathcal{H}^+ \cap N$, it should belong to the solid cylinder over ∂M, hence, away from the non-smooth part ∂N of the closed surface $N \cup D$ for $M^*(t)$ is contained in V, $t \leq t_0$ (see Fig. 3B).

1.3. *The flux formula and the mean curvature vector lemma*

Let $\overrightarrow{k} = (0,0,1)$ be the killing vector field generating vertical translations, which preserves area and volume. Denote by A the area of our surface M, and

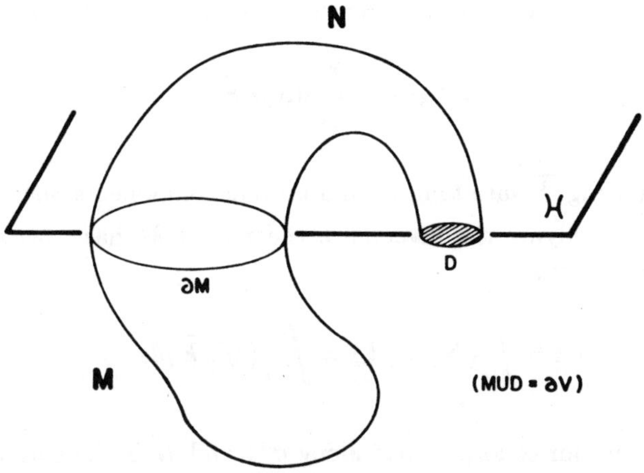

Fig. 3A.

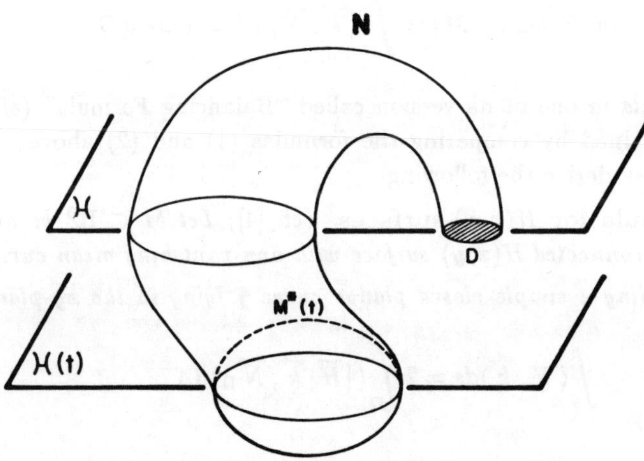

Fig. 3B.

$M(t)$ is the one-parameter family of isometric surfaces having $M(0) = M$ and velocity vector field $\overrightarrow{k}$; i.e., $M(t)$ is the one-parameter family obtained by vertical translation, $A(t)$ is the area of $M(t)$. Recall that the *first variation formula of area* as M is deformed isometrically along $\overrightarrow{k}$ yields (cf. [8]):

$$0 = A'(0) = \int_M \operatorname{div}_M \overrightarrow{k}$$

Now, by decomposing $\overrightarrow{k}$ into tangent and normal components, using the definition of $\overrightarrow{H}$, by applying the divergence theorem if M has constant mean curvature H, one derives

$$-2H \int_M \langle \overrightarrow{N}, \overrightarrow{k} \rangle dA = \int_{\partial M} \langle \overrightarrow{\eta}, \overrightarrow{k} \rangle ds \tag{1}$$

where $\overrightarrow{\eta}$ is the interior conormal field along ∂M and $\overrightarrow{N}$ is the normal field of M in the mean curvature direction; i.e., $\overrightarrow{N} = \frac{\overrightarrow{H}}{H}$. If M is embedded in $\mathbf{R}^3$, it is easy to show that the flux of $\overrightarrow{k}$ across the orientable cycle $M \cup D$ (D is the bounded domain in $\mathcal{H}$ with $\partial D = \partial M$) is zero and this implies

$$(\text{Flux of } \overrightarrow{k} \text{ across } M) = \int_M \langle \overrightarrow{k}, \overrightarrow{N} \rangle dA = \pm \text{area}(D) \tag{2}$$

The *Flux Formula* in one of its version called "Balancing Formula" (cf. [7], [11], [12]) is obtained by comparing the formulas (1) and (2) above. More generally, one may derive the following:

Flux Formula for $H(x,y)$-surfaces. (cf. [4]) *Let $M \subset \mathbf{R}^3$ be an immersed compact connected $H(x,y)$-surface with non-vanishing mean curvature vector $\overrightarrow{H}$, bounding a simple closed planar curve γ lying in the xy-plane $\mathcal{H}$. Then,*

$$\int_\gamma \langle \overrightarrow{\eta}, \overrightarrow{k} \rangle ds = 2 \int_D \langle |\overrightarrow{H}| \overrightarrow{k}, \overrightarrow{N}_D \rangle dA$$

where

a) $\overrightarrow{k} = (0,0,1)$ *is the unit constant field along the z-axis.*

b) $\overrightarrow{\eta}$ *is the unit conormal field of M along γ.*

c) $\overrightarrow{N}_D$ *is equal to* $\overrightarrow{k}$ *is* $\langle \overrightarrow{H}, \overrightarrow{v} \times \overrightarrow{\eta} \rangle > 0$ *and* $\overrightarrow{N}_D$ *is equal to* $-\overrightarrow{k}$ *if* $\langle \overrightarrow{H}, \overrightarrow{v} \times \overrightarrow{\eta} \rangle < 0$, *where v is the velocity vector field of γ when γ is oriented clockwise.*

A geometrical consequence of the Flux Formula, when M is transverse to $\mathcal{H}$ along γ, is that *the normal projection in $\mathcal{H}$* of the mean curvature $\overrightarrow{H}$ of M

along γ, points through the interior of the domain in $\mathcal{H}$ bounded by γ. We say M satisfies the *Mean Curvature Vector Lemma along* γ.

1.4. *The central decomposition*

Let $C_1, \ldots, C_P$ be the closed Jordan curves of $M \cap \text{int}D$. For each C_j, let $C_j^+(\varepsilon)$ be the planar curve on M, mean C_j, obtained by intersecting M with the plane $\{z = \varepsilon\}$. Similarly, let $C_j^-(\varepsilon)$ be the curve in $M \cap \{z = -\varepsilon\}$ that is near C_j. We form an embedded surface N by removing from M the annuli bounded by $C_j^+(\varepsilon) \cup C_j^-(\varepsilon)$ and attaching the horizontal planar domain $D_j^+ \cup D_j^-$ bounded by $C_j^+(\varepsilon) \cup C_j^-(\varepsilon)$. Also we attach D to M along γ. To ensure that N is embedded, one uses different values of ε, when several C_j are concentric. N is a properly embedded submanifold (with corners) of $\mathbf{R}^3$, $\partial N = \phi$, hence each connected component of N separates $\mathbf{R}^3$ into two connected components. Let M_γ be the component of N that contains γ. We orient M_γ by the mean curvature vector $\overrightarrow{H}$ of M. Notice that this makes sense since, abstractly, M_γ contains a connected submanifold L of M (hence $\overrightarrow{H}$ is defined along L) to which one has attached D and certain discs $D_j^\pm(\varepsilon)$. Clearly $\overrightarrow{H}$ extends across these disks to define a normal field to M_γ, for M_γ is connected. The corners of M_γ along the boundaries of the disks, do not affect this. The preceding construction is called the *central decomposition* (cf. [4], [7], [9]).

Clearly, one may construct a similar *external decomposition* by considering $M \cap \text{ext}D$ instead of $M \cap \text{int}D$ (cf. [5]).

1.5. *Theorem* 1: *Sketch of the proof*

Step 1: By using the Flux Formula one shows that the configuration $M \cap \text{int}D \neq \phi$ and $M \cap \text{ext}\, D = \phi$ is impossible. Indeed, by applying the Flux Formula to each M and M_1 (the connected component of $M \cap \mathcal{H}^+$ that contains γ), one can estimate

$$\int_\gamma \langle \overrightarrow{\eta}, \overrightarrow{k} \rangle ds$$

by two different ways, leading to a contradiction.

Step 2: A straightforward argument using the Mean Curvature Vector Lemma shows that one is in a position to apply the Graph Lemma to prove

$$M_\gamma \cap \bigcup_{j=1}^{P} C_j^-(\varepsilon) = \phi\,.$$

Now, using again the Flux Formula one may derive that $M_\gamma \cap \text{int}D = \phi$. This implies $M \cap \text{int}D = \phi$, proving Assertion 1.

Step 3: The proof of Assertion 2 follows from Assertion 1 by the Mean Curvature Vector Lemma, and by applying the Graph Lemma to $M \cap \mathcal{H}^-$.

Step 4: One derives Assertion 3 by a topological argument considering that M has non-vanishing mean curvature vector and that $M \cup D$ is embedded (Assertion 1).

Step 5: The proof of Assertion 4 follows from Assertions 1,2,3, by carrying out a straightforward two dimensional topological argument.

Step 6: One may derive Assertion 6 (when M has constant mean curvature) from Assertion 1, by applying the Vertical Alexandrov Reflection Principle. Finally, the configurations claimed in Assertion 5 may be constructed following Kapouleas' work (cf. [11]).

2. Extensions to Hyperbolic Three Space

Consider the upper halfspace model of the hyperbolic three space $\mathbf{H}^3$, i.e., $\mathbf{H}^3 = \{(x,y,z), z > 0\}$ with the hyperbolic metric $ds^2 = \frac{dx^2+dy^2+dz^2}{z^2}$. Furthermore, consider the following identifications: $\mathcal{H}$ is a geodesic plane (say $\mathcal{H} = \{y = 0\}$), $M \subset \mathbf{H}^3$ is a $H(x,z)$-surface with non-vanishing mean curvature vector $\vec{H}$, bounding a simple closed planar curve γ lying in the xz-geodesic plane $\mathcal{H}$. We claim that all the results stated in the preceding paragraphs hold for $M \subset \mathbf{H}^3$. As an example of the underlying geometric principles, one easily sees that the Vertical Alexandrov Principle here is a reflection across hyperbolic spheres with center lying in the straightline $\mathcal{H} \cap \{z = 0\}$, and with arbitrary radius (possibly ∞). Also, we identify the Horizontal Alexandrov Principle with reflection about vertical planes parallel to $\mathcal{H}$. Finally, there is an analogous Flux Formula for $H(x,z)$-surfaces in $\mathbf{H}^3$ (see [6] for further details).

References

0. A. Alexandrov, *Uniqueness theorems for surfaces in the alrge I*, AMS Translations, series 2, vol. 21 (1962) 341–354.
1. J. L. Barbosa, *Constant mean curvature surfaces with planar boundary*, preprint
2. H. Brezis and J. M. Coron, *Multiple Solutions of H-Systems and Rellich's Conjecture*, Comm. Pure and Appl. Math. **37** (1984) 149–187.
3. F. Brito and R. S. Earp, *Geometric configurations of constant mean curvature surfaces with planar boundary.* Anais Acad. Bras. Cien. (1) **63** (1991).

4. F. Brito and R. S. Earp, *On the configurations of certain prescribed mean curvature surfaces with planar boundary*, preprint.
5. F. Brito and R. S. Earp, *On the configurations of Bryant-Weingarten elliptic surfaces bounded by a planar convex curve*, in preparation.
6. F. Brito and R. S. Earp, *A note on the structure of certain surfaces bounded by a planar curve in the hyperbolic three space*, in preparation.
7. F. Brito, R. S. Earp, W. Meeks and H. Rosenberg, *Structure theorems for constant mean curvature surfaces bounded by a planar curve*, Indiana U. Math. Journ. **40** (1991) 333-393.
8. R. S. Earp and H. Rosenberg, *Some remarks on surfaces of prescribed mean curvature*, Symposium in honor of Manfredo P. do Carmo, Pitman Mono. and Survey in Pure and Appl. Math., 1991, pp. 123–148.
9. R. S. Earp and H. Rosenberg, *Some structure theorems for complete constant mean curvature surfaces with boundary a convex curve*, to appear in AMS Proc..
10. H. Hopf, *Differential geometry in the large*, Lecture Notes in Math. **1000**, Springer.
11. N. Kapouleas, *Complete constant mean curvature surfaces in Euclidean Three-Spaces*. Ann. of Math. **131** (1990) 239–230.
12. N. Korevaar, R. Kusner and B. Solomon, *The structure of complete embedded surfaces with constant mean curvature*, J. Diff. Geom. **30** (1989) 465–503.
13. N. Korevaar, R. Kusner, W. Meeks and B. Solomon, *Constant mean curvature surfaces in hyperbolic space*, to appear in Amer. J. Math..
14. W. Meeks, *The topology and geometry of embedded surfaces of constant mean curvature*, J. Diff. Geom. **27** (1988) 539–552.

The Problem of Plateau (pp. 258–284)
ed. Th. M. Rassias

THE PARAMETRIC PLATEAU PROBLEM AND RELATED TOPICS

GERHARD STRÖHMER

Department of Mathematics, The University of Iowa
Iowa City, Iowa 52242-1466, USA

ABSTRACT

The paper gives an overview of the development of several areas of research associated with minimal surfaces and surfaces of prescribed mean curvature in which substantial progress has been made in recent years. A large part is devoted to questions related to unstable surfaces, in particular different methods and their relationships are discussed, and some open problems are described.

1. Foreword

This article presents selected topics related to Plateau's problem, some of which will be discussed in detail. The choice of these topics is naturally subjective, influenced by the author's taste and experience.

We will not discuss the treatment of this problem by means of geometric measure theory, but refer the reader to the excellent introduction [M7] by F. Morgan, the book [A3] by Almgren, and the paper [A4] by Almgren and Simon.

A different approach, using Cacciopoli sets and functions of bounded variation, developed by De Giorgi and Giusti, among others, is described very well in Giusti's book [G3]. These references are by no means exhaustive, but should give the reader a rather good idea of the approaches to Plateau's problem other than the ones to be presented here.

The most important references for the classical aspects of Plateau's problem are Nitsche's encyclopedic work "Vorlesungen über Minimalflächen" [N3], and Courant's book "Dirichlet's principle" [C5]. Other articles of a largely expository nature that should be of interest to the reader are those by Hildebrandt [H14], Jost [J3], Meeks [M2], Nitsche [N5], and Rassias [R3].

Chapter 2 of this paper gives an introduction to our subject, which is largely based on well-known methods, but also tries to shed some new light on some of its aspects. In chapter 3 we describe some of the more recent developments in the area, while in chapter 4 we make a rather detailed comparison of some methods used to study unstable critical points.

2. Introduction

2.1 Some Remarks About Variational Problems

The topics we consider in this paper are more or less loosely related to equilibrium configurations of soap films. In the absence of exterior forces such a film will always try to attain a shape that is, subject to the constraints imposed on it by the experimental setup, of minimal surface area compared with neighboring surfaces. This positions our problem within the area of the calculus of variations.

In general we will describe variational problems by a set of admissible objects A, endowed with a topology and a differentiable structure, and a functional $I : A \longrightarrow \mathbb{R}$. If I attains a local minimum at $x_0 \in A$, then for any differentiable curve $x(t)$ in A going through x_0 at $t = 0$, for which $f(t) = I(x(t))$ also is differentiable at $t = 0$ we naturally have $f'(0) = 0$. (Such curves will usually be referred to as variations.) In a sense this means that the gradient of I at x_0 vanishes, i.e., x_0 is a critical point of I. In each of our problems this will be equivalent to some system of partial differential equations for x_0, which are called the Euler equations of the problem.

Thinking of the gradient as given by directional derivatives is naturally a very familiar point of view. By restricting to a set of directional vectors which are very regular, but dense in the space we are considering, we can define a gradient even if the functional is not differentiable in this space in the usual functional-analytic sense. Critical points of our functionals then often are more regular than the average point in the space, and the functional I often possesses derivatives in the functional-analytic sense there.

For the considerations of sections 2.4 and 3.2 it is, however, often necessary to choose a space in which I is differentiable.

2.2 The Problem of Plateau and the Problem of Least Area

In this section we will look, with varying attention to detail, at what, despite some modifications, can still be called the classical method of treating the problem of least area, which requires to find a surface of least area with a given boundary. Any solution of this problem must naturally fulfill its Euler equation, which in this case implies the surface has to have a vanishing mean curvature. Such surfaces are called minimal surfaces, regardless of whether they minimize area or not. Now the problem of finding a minimal surface with given boundary is usually called the Plateau problem, but by extension this term is also used more broadly to describe the problem of least area and some related problems, and this is how it is used in the title of this paper.

Our approach to Plateau's problem is parametric, that means we deal with parametrizations of our surfaces rather than the surfaces themselves. We will confine ourselves to surfaces that can be parametrized by the unit disc $B = \{(u, v) \in \mathbb{R}^2 | u^2 + v^2 < 1\}$.

For any open set $\Omega \subset \mathbb{R}^n$ we denote its boundary by $\partial\Omega$, its closure by $\bar{\Omega}$, and then define $S^1 = \partial B$.

For any set $C \subset \mathbb{R}^n$ let $C^o(C)$ be the space of continuous functions defined on C.

The boundary of the surfaces we consider will be required to cover an oriented, closed Jordan curve $\Gamma \subset \mathbb{R}^n$. By an oriented closed Jordan Γ curve we mean a non-empty subset Γ of

$$\{g : S^1 \longrightarrow \mathbb{R}^n | g \in C^o(S^1), \quad g \text{ is injective}\}$$

characterized by the conditions that for $g_1, g_2 \in \Gamma$ we have that $g_1(S^1) = g_2(S^1)$, and that $g_1 \circ g_2^{-1}$ preserves the mathematical sense of rotation of S^1. Then let $|\Gamma| = g(S^1)$ for any $g \in \Gamma$.

Now we can define the set of admissible surfaces for our mathematical model for the soap film experiment. Let

$$A(\Gamma) = \{x \in C^o(\bar{B}) \cap C^2(B) | x(S^1) \subset |\Gamma| \ , \ \text{ and } g^{-1}(x|S^1)$$

$$\text{is a weakly monotonic mapping of } \quad S^1 \quad \text{for any } g \in \Gamma\} \ .$$

A weakly monotonic mapping is any function from S^1 into itself that winds around the circle once and preserves sense of rotation, but may have intervals on which it is constant. For any open $\Omega \subset \mathbb{R}^n$ we denote by $C^{k+\alpha}(\Omega)$ the set of all functions that are $k-$times continuously differentiable on Ω, and whose derivatives are locally Hölder continuous with exponent $\alpha \in (0,1)$. Then by $C^{k+\alpha}(\bar{\Omega})$ we mean the subset of $C^{k+\alpha}(\Omega)$ whose elements have $k-$th derivatives that are Hölder-continuous up to the boundary. We do not usually specify the target space of our mappings, as it will ordinarily become clear from the context.

Finally, a Jordan curve Γ is said to belong to $C^{k+\alpha}$, for $\alpha \in [0,1]$, $k \geq 1$, if there is an element in Γ that has this regularity as a function of the angle, and whose derivative is everywhere unequal zero.

The set $A(\Gamma)$ contains, among other objects, all parametrizations of surfaces that are images of the unit sphere and whose boundary spans all of Γ. On this set we define the nonlinear functional

$$I(x) = \int_B F(x, x_u, x_v) du dv$$

with a function $F : \mathbb{R}^{3n} \longrightarrow \mathbb{R}$, $(F = F(y,p,q))$.

For the problem of least area I naturally needs to represent the surface area, so we have $F = \sqrt{|x_u|^2|x_v|^2 - (x_u, x_v)^2}$. We will also consider other functions F which allow us, roughly speaking, to take the effects of exterior forces on the soap film into account.

In order to obtain an element x where I reaches a minimal value, we will follow what is called the direct method of the calculus of variations, that is, we will select a sequence $x_k \in A(\Gamma)$ for which $I(x_k)$ converges to the infimum of I over $A(\Gamma)$, and try to find a convergent subsequence. Now the first problem we encounter here is that I, if geometrically meaningful, has to be independent of the parametrization of our surface. This means that I itself tells us very little about the parametrization, and, even for a sequence of surfaces converging to a surface in any reasonable sense, a sequence of parametrizations of these surfaces need not converge to a parametrization of the limit surface. So we need to choose our parametrizations wisely. We must also embed $A(\Gamma)$ into a larger function space, which, in this case, as F is homogeneous of order 2, will be the Sobolev space $H_2^1(B)$. (See [A1]) Now by Poincaré's lemma the H_2^1-norm is bounded for all $x \in A(\Gamma)$ by the Dirichlet integral

$$D(x) = \frac{1}{2} \int_B |x_u|^2 + |x_v|^2 \, du dv \quad .$$

So, if we can find a sequence of parametrizations x_k with bounded Dirichlet integral, we can select a subsequence from it that converges weakly in H_2^1. In many cases one can prove I to be lower semicontinuous, and so the weak limit of $\{x_k\}$ is a minimizing function.

If such parametrizations exist, it would naturally be good to choose the ones with the lowest Dirichlet integral. The existence of such a parametrization for a regular smooth surface could be proved again by selecting a weakly convergent subsequence of a sequence of parametrizations, whose Dirichlet integral would converge to the infimum of this integral over all parametrizations. The weak limit would then have to be smooth and regular, (see [M8,M9]) if the surface is. As we will see, the result of this is a so-called conformal representation of the surface, whose existence can also be obtained from Koebe's uniformization theorem [K1]. To understand this we carry out a calculation which is more general than what we need here, but its result will be useful later. So assume x is a parametrization of a given surface that minimizes an integral

$$I(x) = \int_B F(x, x_u, x_v) \ dudv$$

among all others. In order to avoid regularity considerations, which we will do throughout the paper, we assume $x \in C^2(\bar{B})$, and $F \in C^2$ also. The method we use is adapted from Courant [C5] and Nitsche [N3].

The variations to be used here are defined by a change in the independent variables of x by means of a family of transformations $T(t, u', v')$. It must fulfill the following condition.

Condition C1: $T(t, u', v')$ is a mapping from $[-\delta, \delta] \times \bar{B}$ into $\bar{B}$ which belongs to C^2 on its domain of definition so that $T(t, ., .)$ is bijective as a mapping from $\bar{B}$ to itself, and $T(0, u', v') = (u', v')$.

Then for t in a small neighborhood of 0 the mapping $T(t, ., .)$ is a C^2–diffeomorphism, and $x(T(t, u', v'))$ is another parametrization of the same surface. After a simple transformation we therefore have $h'(0) = 0$ with

$$h(t) = \int_B F(x, x_u(T_{1u'} \circ T^{-1}) + x_v(T_{2u'} \circ T^{-1}), x_u(T_{1v'} \circ T^{-1}) + x_v(T_{2v'} \circ T^{-1}))J_T^{-1} \ dudv$$

Now

$$(\frac{\partial}{\partial t}(T_{iu'} \circ T^{-1}))(0, u, v) = (\frac{\partial}{\partial t}T_{iu}) + (T_{iu'u'}\frac{\partial}{\partial t}(T_1^{-1}) + T_{iu'v'}\frac{\partial}{\partial t}(T_2^{-1})) = \Phi_{iu}$$

at $t = 0$ with $\Phi_i = (\frac{\partial}{\partial t} T_i)(0, u', v')$, and the corresponding equation is true for v. Also $\frac{\partial}{\partial t}(J_T^{-1}) = -(\frac{\partial}{\partial t} J_T) = -(\Phi_{1u} + \Phi_{2v})$ at $t = 0$. So we get

$$h'(0) = \int_B F_p(x, x_u, x_v)(x_u\Phi_{1u} + x_v\Phi_{2u}) + F_q(x, x_u, x_v)(x_u\Phi_{1v} + x_v\Phi_{2v}) \\ - F(x, x_u, x_v)(\Phi_{1u} + \Phi_{2v})\ dudv \ .$$

Leaving out the arguments of F for the sake of brevity and assuming the equations $F_p q = F_q p$ and $F_p p + F_q q = 2F$, which are true for the Dirichlet integral, we get

$$\int_B (\Phi_{1u} - \Phi_{2v})(F_p x_u - F) + 2(\Phi_{1v} + \Phi_{2u})F_p x_v \ \ dudv = 0 \ .$$

Obviously Φ_1, Φ_2 could be any smooth functions with compact support in B, which is easily seen to imply that $f(z) = i(F_p x_u - F) + 2F_p x_v$ is a holomorphic function of $z = u + iv$. Now it is advantageous to switch to complex notation. We integrate by parts to obtain

$$\int_{S^1} (Im(f(z))u + Re(f(z))v)\Phi_1 + (-Im(f(z))v + Re(f(z))u)\Phi_2 \ ds \\ = \int_{S^1} Re(zf(z)(-i)(\Phi_1 + i\Phi_2)) \ ds = 0 \ .$$

Now $T(t, z) = ze^{it\varphi(z)}$ is a possible variation for every smooth φ, and then $\Phi_1 + i\Phi_2 = zi\varphi$, so we have

$$\int_{S^1} Re(z^2 f(z))\varphi \ ds = 0$$

for all smooth functions φ, which leads to the conclusion that $Re(z^2 f(z)) = 0$ on the boundary, and therefore in the interior, so $z^2 f(z) = ic$ with $c \in \mathbb{R}$, and naturally $c = 0$, so that $f(z) = 0$, i.e., $F_p x_v = F_q x_u = 0$ and $F_p x_u - F = F_q x_v - F = 0$. So we get the following theorem:

2.2 Theorem: If F belongs to $C^2(\mathbb{R}^{3n})$, $F_p p + F_q q = 2F$, and $F_q q = F_q p$, and if also $x \in C^2(\bar{B})$, and $g'(0) = 0$ with $g(t) = I(x(T(t,.,.)))$ for all variations fulfilling the condition $C1$, then $F_p x_v = F_q x_u = 0$ and $F_p x_u - F = F_q x_v - F = 0$.

For the Dirichlet integral this means $|x_u| = |x_v|$ and $(x_u, x_v) = 0$, so the parametrization is conformal (or isothermal). (See [C5,K1,N3])

With a conformal parametrization $\sqrt{|x_u|^2|x_v|^2-(x_u,x_v)^2}$ reduces to $\frac{1}{2}(|x_u|^2+|x_v|^2)$, so the area coincides with the Dirichlet integral, which gives us far more control of our parametrizations as they are immediately in a bounded set in $H_2^1(\Omega)$ if the area is bounded, which for a minimizing sequence it naturally is.

Now it is easy to see that we can even modify our variational problem by replacing the area by the Dirichlet integral, as D is never smaller than the area and coincides with it if we have conformal parameters. This replacement of the area by the Dirichlet integral, and then by a boundary integral, was one of Douglas' great contributions to the study of this problem (see [D1]).

The restriction to parametrizations having minimal Dirichlet integral still allows, as the reader can easily check, for conformal reparametrization, which can be controlled by specifying that three points on ∂B must be mapped into three specific points on Γ.

Now let us return to the variational problem given by $A(\Gamma)$ and I. Any sufficiently regular function x minimizing I in $A(\Gamma)$ also fulfills naturally the condition $F_p x_v = F_q x_u = 0$ and $F_p x_u = F_q x_v = F$ derived above as any reparametrization of $x \in A(\Gamma)$ also belongs to $A(\Gamma)$. There are other variations, however, as the ones considered until now leave the surface unchanged. If x is a local minimum point of I in $A(\Gamma)$ with respect to almost any reasonable topology, then for any function $\Phi \in C^1(\bar{B})$ with $\Phi|\partial B = 0$ we have with $f(t) = I(x+t\Phi)$ that $f'(t) = 0$. Then

$$0 = f'(0) = \int_B F_y \cdot \Phi + F_p \cdot \Phi_u + F_q \cdot \Phi_v \;\; dudv \; .$$

This leads to the equation

$$(**) \qquad ((F_p)(x,x_u,x_v))_u + ((F_q)(x,x_u,x_v))_v = \\ = (F_y)(x,x_u,x_v) \; ,$$

which represents these variations of I.

Specializing again to the case $I(u) = D(u)$, we obtain

$$(i) \quad \Delta u = 0$$

$$(ii) \quad |x_u| = |x_v| \;\text{ and }\; (x_u,x_v) = 0$$

for any sufficiently regular functions u minimizing D in $A(\Gamma)$. One can also prove that $x|S^1$ is injective, and that the set of branch points, i.e., the points where $|x_u| = 0$, is nowhere dense in B. Outside these points x represents a regular surface. Osserman [O2] proved that there is no branch point in B, so that x

represents a regular immersion in the interior of B, if x is a minimizing function. It is easy to calculate, using the definition, that the mean curvature of any surface fulfilling (i) and (ii) in $\mathbb{R}^3$ vanishes outside the branch points. Steps towards the proof of the existence of such a function were announced by Jesse Douglas at consecutive annual meetings of the American Mathematical Society starting in 1926, and the first complete proof [D1] was published in 1931. T. Radó [R1] published his corresponding result almost at the same time. Somewhat earlier, and using completely different methods, Garnier [G2] had proved the existence of a minimal surface, which does not necessarily have minimal area in $A(\Gamma)$, for curves Γ that are not knotted.

2.3 Surfaces of Prescribed Mean Curvature

We have already carried out some calculations that were more general than needed for the classical form of the Plateau problem, and the results we obtained give us already some hints about the directions of possible generalizations.

The conditions $F_q q = F_q p$ and $F_p p + F_q q = 2F$ needed for Theorem 2.2 restrict F considerably, if it is differentiable everywhere. We then have

$$\frac{d}{dt}[F(y,tp,tq)] = [F_p(y,tp,tq)tp + F_q(y,tp,tq)tq]/t = 2F(y,tp,tq)/t \quad ,$$

which implies that F is homogeneous of degree 2 as a function of (p,q), so by differentiability at $p = q = 0$ it has to be a quadratic form. Then $F_q q = F_q p$ immediately implies

$$F(y,p,q) = \Sigma_{i,j=1}^n \, a_{ij}(y)(p_i p_j + q_i q_j) + \Sigma_{i,j=1}^n \, b_{ij}(y) p_i q_j$$

with a symmetric matrix function a_{ij} and antisymmetric b_{ij}. This class of functions was considered in [S22]. It actually includes those related to surfaces of prescribed mean curvature in Riemannian manifolds. Here we will, however, only consider problems more general than the classical Plateau problem in the case $n = 3$, and confine ourselves to

$$F = \frac{1}{2}(|p|^2 + |q|^2) + Q(y) \cdot (p \times q) \quad [Q \in C^{1+\alpha}(\mathbb{R}^3)] \quad .$$

We denote the functional with this F by E. As one can calculate the equation (**) then takes the form

$$\Delta x = \frac{1}{2} \mathrm{div} Q \;\; x_u \times x_v \quad .$$

Due to the conformality of the parameters this equation is immediately seen to imply that the mean curvature of the surface given by x is $\frac{1}{4}\text{div}Q$. Now usually one wants to prescribe the mean curvature itself as a function $H(y)$. One can see that one can find a Q with $4H = \text{div } Q$ and $|Q(y)| < 2 - \delta$ for some $\delta > 0$, if H is of compact support in $B^3 = \{y \in \mathbb{R}^3| \ \ |y| < 1\}$ and $|H| < 1$ (See, e.g., [H12]). Then $E(x) \geq \delta D(x)$, and the H_2^1−norm of x_k must be bounded for any minimizing sequence $\{x_k\}$, which then contains a subsequence that weakly converges in H_2^1. It turns out that E is lower semi-continuous in H_2^1 and therefore possesses a minimum in the closure of $A(\Gamma)$. Then one can, with considerable effort, prove that any minimizing function belongs to $C^2(B) \cap C^o(\bar{B})$, and fulfills the equations

$$(i) \qquad \Delta x = 2H(x)\, x_u \times x_v$$

$$(ii) \qquad |x_u| = |x_v|, \ \ (x_u, x_v) = 0 \ .$$

For constant H with $|H| < (\sqrt{17} - 1)/8$ the existence of such a function was proved by Heinz [H2], for $|H| < \frac{1}{2}$ by Werner [W4], then for $|H| \leq 1$ by Hildebrandt [H12, H13]. In [H12] H is allowed to be variable. Heinz [H4] proved there are curves in B^3 for which there is no solution for $|H| > 1$. The argumentation presented here is due to Hildebrandt [H12].

2.4 Unstable Critical Points

Now the solutions of the set of equations listed above or even the absolute minima of D in $A(\Gamma)$ are by no means unique. So one of the first questions that arose after Douglas' and Radó's proofs for the existence of absolute minima of D on $A(\Gamma)$ was that of further describing the set of solutions of these equations, and the first results in this direction were already achieved in 1939 by Shiffman [S12,S13,S14] and Morse-Tompkins [M10,M11,M12]. These problems will be the main subject of this paper. Particular attention will be given to the attempts to carry over the Morse theory (See [C1,M3]) to the functionals I on $A(\Gamma)$. This theory allows to establish connections between the topological structure of a (initially finite-dimensional) manifold M and the critical points of any twice differentiable real function g defined on this manifold. In brief the theory states that the homotopy type of the level sets $M_\alpha = \{m \in M | f(m) \leq \alpha\}$ can only change if α passes a critical level, that is a level where there is a critical point, a zero of the gradient of f. Assuming there is at most one critical point at any level, and it is nondegenerate, i.e., its second derivative is a nondegenerate matrix, one can describe the change of homotopy type as follows: Let k be the number of negative eigenvalues (multiple ones counted multiply) of the second derivative at the critical point, the so-called Morse-index. Then in passing this critical level a k-sphere is glued into the level set at the critical point. An elementary conclusion from this, which can be proved without the assumption of non-degeneracy of all critical points, is the so-called mountain-pass-lemma. It states that if there are two points on the manifold, for which the infimum over all connecting continuous paths of the maxima of f along the paths is larger than the value of f at any of these points, but finite, then at the level of this infimum there is a non-minimum critical point. This is a form in which the mountain-pass lemma is often stated. If we replace the connecting paths by compact topologically connected sets joining the two points, then there is such a set of minimal elevation also, and we get a non-minimum critical point on this set. In this form the theorem was formulated, e.g., by Shiffman [S12], and Courant [C4,C5]. The non-degeneracy assumption is unnecessary here as the proof is indirect and proceeds by assuming there is no critical point at that level at all.

3. Selected Results

3.1 Surfaces Bounded by Polygons

In the case that the boundary curve Γ is a polygon, the solution set of Plateau's problem can be shown to lie in the image of a "surface valued" function of only finitely many variables. So let Γ be a polygon with $N+3$ vertices $y_1, \ldots, y_{N+3}$, which we number in accordance to the orientation of the curve. Let us pick three points $z_{N+1}, z_{N+2}, z_{N+3}$ on S^1. We can exercise our usual choice by specifying that these be mapped into $y_{N+1}, y_{N+2}, y_{N+3}$, and then define a family of restricted variational problems by

$$A(z_1, \ldots, z_n) = \{x \in C^o(\bar{B}) \cap C^2(B) | x|S^1 \longrightarrow |\Gamma|$$

$$\text{is weakly monotonic and } \; x(z_i) = y_i \;, \quad i = 1, \ldots, N+3\}$$

for $z_i \in S^1$. The points z_i naturally have to be in the right order for $A(z)(z = (z_1, \ldots, z_n))$ to be non-empty. If this is the case, $A(z)$ is convex, and as $D(x)$ is strictly convex, there can be at most one minimal function in $A(z)$, and the method used for Plateau's problem actually allow us to show there is one. Let us denote it by $S(z)$ and its Dirichlet integral by $d(z)$. Then $S(z)$ depends continuously on the points z, and $d(z)$ even differentiably on these parameters, and the minimal surfaces spanned by Γ are exactly the images of the points where the derivatives along S^1 of the function $d(z)$ are zero. This has been proved by

Courant [C4,C5], Heinz [H5], Ströhmer [S19] for minimal surfaces, surfaces of prescribed mean curvature, and for minimal surfaces in Riemannian manifolds. They used their results to prove the mountain-pass lemma for these problems.

The flaw of this construction is that nobody has been able, up to now, to prove any higher regularity of the function. Therefore it is not possible to define non-degeneracy and the Morse-index for the critical points of d.

Now this motivated Shiffman to modify the restricted variational problem in such a way as to remove the assumption of monotonicity of the boundary mapping. This then gives the following set of admissible functions in the restricted variational problem

$$A_1(z_1,\ldots,z_n) = \{x \in C^o(\bar{B}) \cap C^2(B) | x((z_i, z_{i+1})) \in L_i \text{ for } i = 1,\ldots,N+3\} \quad ,$$

where L_i is the straight line connecting y_i with y_{i+1}. So the boundaries of the admissible surfaces do not even necessarily lie on Γ. The minimizing functions on these sets also are unique, but do not necessarily agree with those of the previous problem, as Lewerenz [L2] showed.

This construction is essentially due to M. Shiffman and was described in a paper by I. Marx [51], published in 1955, in which the claim is made, without much of a proof, that the ensuing functions $S_1(z)$ and $d_1(z)$ are infinitely differentiable. In 1979 Heinz [H7,H8] was able to prove that these functions are even analytic by going back to the approach to Plateau's problem, which had been used by Garnier [G1,G2] and which involves considering monodromy groups and the Riemann-Hilbert problem, based on results of Plemelj and Schlesinger [P3,S6]. This does not allow to conclude there are only finitely many solutions, however, but it naturally allows to define non-degeneracy and Morse index for the function d_1, which, as Sauvigny [S3] showed, is the same as the Morse index of the surface as an element in a function space, which shows that in this case all questions about critical points can be reduced to the finite-dimensional case.

Heinz [H7] also obtained results about the dependence of S_1 and d_1 on the polygon itself.

3.2 Applications of Global Analysis to Plateau's Problem

In this section we consider approaches to Plateau's problem which consider I as a function defined on a suitable function space or manifold to be able to apply concepts of global functional analysis. (See also 2.1) First we shall look at the Morse theory and mountain-pass lemmas for non-polygonal boundary curves.

Marston Morse was himself thinking very much about variational problems when he developed his theory of critical points (see [B9]), and it was indeed, as far as the calculus of variations of single integrals, in particular geodesics, was concerned, an immediate success.

With multiple integrals, however, significant difficulties are encountered, and it was only in 1939 that M. Shiffman [S12] could prove his mountain-pass lemma for minimal surfaces by explicitly constructing deformations of the space of admissible functions which allowed to conclude the existence of an unstable critical point, without using much abstract theory at all. Morse and Tompkins [M10] were able to prove an analogous theorem at almost the same time, using some of Morse's abstract framework.

These results only proved the mountain-pass lemma, and it was not until the 1960's that the full Morse theory was proved for any multiple integral problem (see [P1,P2,S15]) by Palais and Smale. This theorem is valid for twice differentiable functionals on Hilbert manifolds. The most important condition to be required of this functional is the so-called Palais-Smale condition which requires that in the closure of any set in this Hilbert manifold on which the functional is bounded, while the norm of its gradient is not bounded from below, there must be a critical point of the functional. It took some time until it was realized that this theorem could be useful at all for problems which lead to second order partial differential equations like the ones we are considering here. (See, in particular, [A2].) Even then it was believed that these considerations could not be applied to Plateau's problem, and so the result for polygons described in the previous chapter were used, by means of an approximation process, to prove the mountain-pass-lemma for general boundary curves for surfaces of constant mean curvature by Heinz [H5,H6], and by Ströhmer [S19] for minimal surfaces in Riemannian manifolds. More in the spirit of Shiffman's papers, but using different methods, Ströhmer [S20,S22] proceeded by constructing rather explicit deformations for surfaces of variable prescribed mean curvature, a problem which still does not seem accessible to the general theory. Then Struwe realized that for minimal surfaces and prescribed mean curvature one can, just as J. Douglas did, reduce the problem to the boundary values alone, and consider it in the space $H^{\frac{1}{2}}$ of all boundary values of H_2^1–functions, and that the functional defined by Douglas–and its analog for nonzero mean curvature–then fulfill the Palais-Smale condition. By means of a Morse theory on convex sets (see, e.g., [S32]) he developed one can then derive the mountain-pass-lemma, and if the curve does not bound any surfaces that are degenerate critical points of the functional, the Morse theory in its entirety follows

[S28]. This latter conclusion requires a theorem by Böhme [B3], who proved that in $\mathbb{R}^3$ minimal surfaces which are non-degenerate critical points cannot have any branch points. This however also implies that this property cannot be generic in three dimensions (see [B8]). A generic property is one shared by all objects in at least a dense open set. Using techniques developed in connection with critical point theory some other statements about the structure of the solution set have been proved:

By an infinite-dimensional version of Sard's lemma, Böhme and Tomi [B6] were able to prove that for boundary curves belonging to C^∞ the set of values $D(x)$ for conformal parametrizations of minimal surfaces form a set of Lebesgue measure zero on $\mathbb{R}$. Using theorems about analytic sets they were also able to derive that the number of values of $D(x)$ even has to be finite for analytic curves. Under the same assumption Böhme [B2] had proved earlier already that there are at most finitely many connected components of the set of minimal surfaces, and Tomi [T3] had been able to show that there could be at most finitely many surfaces furnishing an absolute minimum. If all these surfaces are embedded, the same is true if $\Gamma \in C^{4+\alpha}$ only (see [H1]). The method is based on proving that the second derivative in $A(\Gamma)$ at these points is a strictly positive definite quadratic form. Then there must be a neighborhood of this surface not containing another critical point of $D(x)$. A simple compactness argument then shows the finiteness, because an infinite number of solutions would have to have a cluster point in very strong topologies, as one can show by means of suitable a-priori estimates. Now the same conclusion would hold true for the set of all minimal surfaces, using an inequality of isoperimetric type in addition, if the second derivative–or variation–of any minimal surface were only nondegenerate. One of the important theorems used in the context of Morse-theory is that non-degeneracy of all critical points is a generic property for real functions of sufficiently high differentiability (see [C1]). Now one could try to approximate $D(x)$, which would be rather dissatisfying, because the approximating functional could be very odd and would rarely be geometrically meaningful. So another possibility one would do well to investigate is to approximate the boundary curve. To that end one has to change one's perspective somewhat. While up until now we have considered the surfaces bounded by a given Jordan curve we now want to look at all surfaces of the topological type of the disc bounded by any sufficiently regular Jordan curve. This was the perspective taken by Böhme and Tromba [B7,B8]. The immediate problem one encounters here is that for this kind of approximation non-degeneracy cannot–in $\mathbb{R}^3$–be a generic property, as it implies the absence of branch points in the surface, as remarked earlier. However this type of genericity with respect to the boundary curve is attractive enough to have been pursued much further, and as Böhme and Tromba proved, there are weaker–and generic–conditions which guarantee finiteness of the number of solutions for minimal surfaces, relating to the Fredholm index, which, however, require much more sophisticated analyses of the problem. The result is that, as the title of [B7] says, the number of solutions of the classical Plateau problem is generically finite.

The index theorem, which is at the heart of this argument, was generalized to surfaces of constant mean curvature by Schüffler [S7,S8], Schüffler and Tomi [S10] and Söllner [S16], by Thiel [T2] to minimal surfaces of higher topological type.

3.3 Uniqueness

Now there are some classes of Jordan curves for which the Plateau problem actually has only one solution, and one of the theorems to that effect actually was the first one describing the set of solutions more precisely:

Radó's theorem states that there is only one solution, if the boundary curve possesses a parallel projection on a plane which in itself a closed convex Jordan curve. Under this assumption, by a theorem of H. Kneser [K3], the minimal surface also has an injective projection on this plane and then can be represented as the graph of a function, which itself fulfills the so-called minimal-surface equation, an elliptic equation which by the classical maximum principle has only one solution.

This was generalized by Gulliver and Spruck [G4] to surfaces of constant mean curvature. The convexity of the projection has to be significantly sharpened here. For variable H an analogous theorem was proved by Sauvigny [S1] under the additional assumption that the surface may not be strictly unstable, and a one-sided condition on the derivative of H in a direction normal to the plane of projection, which ensures the validity of the maximum principle.

Nitsche [N2] proved that if the total curvature of the Jordan curve Γ is smaller than or equal to 4π, then there is only one minimal surface. The argument is rather interesting: Under this assumption any minimal surfaces by the Gauss-Bonnet formula must be free of branch points, and as Barboza and do Carmo [B1] proved, it also must be stable. This naturally makes the existence of several surfaces impossible, as two stable surfaces would be separated by a mountain pass, which would imply the existence of an unstable surface. Nitsche [N5] and Sauvigny [S5] proved the same result for extreme polygons (polygons that lie on the boundary of a convex set). Nitsche's proof is based on approximation, while Sauvigny directly uses his results mentioned in 3.1. Ruchert [R4] proved a generalization to surfaces of constant mean curvature, while [S21] indicates there probably is no attractive version for surfaces of variable mean curvature.

Böhme [B4] proved that for any curvature larger than this value there is no bound on the number of surfaces any more, although for a while it is still known to be finite under certain conditions [N4]. In general very little is known about conditions which would make the number of solutions finite, but there are minimal surfaces not of the topological type of the disk, which can be subject to transformation groups and thus used to generate an infinite number of solutions, as was shown by Morgan [M5]. The surfaces he constructs are bounded by more than one curve. A different kind of uniqueness theorem was also proved by Morgan [M6]: For again a generic set of boundary curves there is exactly one absolutely minimizing surface, but there may be large numbers of locally minimizing surfaces and unstable ones as well. Finally the reader is referred to the expository article by Rassias [R3], which describes results about uniqueness in more detail.

3.4 Large Surfaces of Constant Mean Curvature

For any closed Jordan curve lying on a sphere of radius R in $\mathbb{R}^3$ there are at least two surfaces of mean curvature $1/R$ bounded by this curve: The two parts into which the sphere is divided by this curve. If the curve is contained in a smaller sphere of radius $r < R$ then any such surface that stays in a sphere of radius R around the curve actually must be in the ball of radius r (see [H12]). This is also true for surfaces of mean curvature $1/R$ whose boundary does not lie on a sphere. Now Rellich conjectured that there should always be at least two solutions, customarily referred to as the small and the large solutions. As we already discussed, the small surface was first constructed by Heinz [H2], using a-priori estimates and a continuity method.

No attempt to carry this method over to the construction of a large surface has succeeded yet, but the estimate of Serrin [S11] seems to support the assumption that this might be possible, although it is quite far away from the estimates one would need for such a method.

He proved that all surfaces of mean curvature H with $|H| \leq 1$ without self-intersection, that are bounded by a Jordan curve Γ contained in the ball B_r with $r = 1/|H|$, lie in a ball of radius one.

On the other hand Wente [W3], by a most remarkable, and very complicated, counterexample, proved that there is a much greater variety of immersed surfaces of prescribed mean curvature than there are embeddings of the same curvature.

As it probably would be necessary to have an extension of Serrin's theorem to non-embedded surfaces to make the continuity method work, this somewhat dimmed the hope that this method could be used here at all.

The same is not true for the hopes of proving Rellich's conjecture, though, as already before this latter result alternative methods had been developed. The first idea, due to Wente [W1,W2], (see also Steffen [S17]) was to obtain such a surface as a minimum for $D(x)$ with a constraint on the integral

$$\tilde{V}(x) = \int_B (x, x_u \times x_v)\, dudv \quad .$$

For any given curve this allows to prove that there is a sequence of values of H, which go to zero, for which there exists a large surface of the mean curvature H. There is no way to predetermine H, though, as it appears as a Lagrange parameter in this new variational problem. Then Struwe [S29,S30,S31,S32], and Steffen [S18], realized that under certain conditions the existence of the surface can be derived from a mountain-pass lemma. With respect to the original problem the large surface is therefore unstable. Parallel, independently, and by a somewhat different method, Brézis and Coron [B10,B11] proved a similar result. In their final version these theorems state that the problem has a large solution whenever it has a small solution, and so they have proved Rellich's conjecture completely. In these arguments one has to make use of the special structure of the functional very strongly, for example it is very important that $V(x)$ is a polynomial of degree 3 in x and its derivatives. A very good presentation of the argument and of general critical point theory in its relation to Plateau's problem can be found in [S32].

4. A Comparison of Different Approaches to Unstable Critical Points

In this section we compare the methods used for obtaining critical points of non-minimum character other than those using a reduction to finite-dimensional problems for polygonal boundary curves. All these methods are based on constructions of continuous families $T_t : A \longrightarrow A (t \in [0,1])$ of deformations of the space on which the functional is defined. This is naturally most obvious in the case of the main theorems of Morse theory, as they are concerned with the homotopy type of the level sets M_α (see 2.2). These deformations must have the properties $T_0 = id_A$, $I(T_t(x)) \leq I(T_{t'}(x))$ for $t > t'$, and equality only occurs if $T_{t'}(x)$ is a critical point. Then, if there are no critical points at the levels in $[\alpha, \beta]$, we can, under certain conditions, deform M_α into a subset of M_β, and, if T_t is properly defined, we can also deform M_α into M_β with a $k-$sphere attached, if there is one critical point of index k between the levels α and β. Now these transformations are usually obtained as semiflows of the gradient or a pseudo-gradient vectorfield (see [P1,P2,S15,T4,U1]). The construction in this case works if the Palais-Smale condition is fulfilled, which, in spite of the examples mentioned in other places, is not always the case, in particular not in a space in which the functional is twice differentiable. The "hereditable" theory outlined in [C2] allows to divorce the existence of the second derivative from the Palais-Smale condition, and the perturbation methods of [U1], and the "artificial constraints" used in [S23,S24,S25,S26], allow generalizations in other directions. Such artificial constraints had been used, e.g., by Lewy [L3], Nitsche [N1], and Williams [W5]. An interesting variant of such arguments is used in [T1]. The a-priori estimates needed for the perturbation and constraint methods are incidentally almost the same, although they are used in different ways.

Coming back to a simple case without any of these auxiliary constructions, it is fairly easy to formulate an abstract framework for the method used by Shiffman [S12] and Ströhmer [S22,S24] of explicitly constructing the deformations T_t. Then we will see that the condition C2 necessary to make this method work actually implies the Palais-Smale condition, and so the method in this form is actually weaker than the approach of Palais and Smale. The argument shows a rather interesting connection between the Palais Smale Condition, and an approximate convexity property, which is part of C2.

Condition C2: *The set A of admissible objects for our variational problem is a weakly closed subset of a reflexive Banach space B_1, which in turn is compactly contained in another, not necessarily reflexive, Banach space B_2. In addition the functional I is continuously differentiable on A in B_1, and lower semi-continuous with respect to weak convergence in B_1, and any weakly convergent sequence u_k in B_1 converges strongly to its weak limit u, if $I(u_k) \longrightarrow I(u)$ also. Any set with bounded value of I also is bounded in B_1. There is a mapping $c: A\times A\times[0,1] \longrightarrow A$ which is Lipschitz continuous on bounded sets in B_1, and $\mathbb{R}$, as a function of all its variables, continuous in B_2, and $\mathbb{R}$, fulfills $c(u_1,u_2,0) = u_2, c(u_1,u_2,1) = u_1$, and*

$$I(c(u_1,u_2,t)) \leq tI(u_1) + (1-t)I(u_2) + tC \parallel u_1 - u_2 \parallel_2 \ .$$

($\parallel \ \parallel_2$ is the norm of B_2).

Condition C2 is, e.g., fulfilled for the Plateau problem for minimal surfaces, restricted to harmonic surfaces from the outset, in $B_1 = H_2^1$, with $B_2 = C^o$, with the c constructed in [S24], or in the simple example

$$I(u) = \int_\Omega |Du|^2 + G(u)\, dx$$

with $B_1 = H_2^1$, and $B_2 = L_p$ with $p < 2n/(n-2)$, $G \in C^1(\mathbb{R})$ with $|G'(t)| \leq C(|t|^{p-1}+1), G(u) \geq 0$, and $c(u_1, u_2, t) = tu_1 + (1-t)u_2$.

Now we can show

Theorem: Condition C2 implies the Palais-Smale condition.

Proof: Assume $\{u_k\}$ is a sequence such that $I(u_k)$ is bounded and the derivative $(DI)(u_k)$ of I at u_k goes to zero.

Then we have to show there exists a subsequence of u_k that converges strongly to a critical point of I.

We can select a subsequence, also called u_k, that weakly converges to a $u \in A$, and so that also $I(u_k)$ converges. Now $I(u) \leq \lim_{k\to\infty} I(u_k) = q$, and u_k converges to u in B_2. Assume $q > I(u)$. By our convexity condition

$$I[c(u, u_k, t)] \leq tI(u) + (1-t)I(u_k) + tC \parallel u_k - u \parallel_2 \ ,$$

so, as u_k converges to B_2, we get for k sufficiently large that $I(c(u, u_k, t)) - I(u_k) \leq t(I(u) - q)/2$. On the other hand we have

$$I(c(u, u_k, t)) - I(u_k) \geq (DI(u_k))(c(u, u_k, t) - u_k) - th_k(t)$$

with $h_k(t)$ going to zero as t does. Therefore

$$t(I(u) - q)/2 \geq (DI(u_k))(c(u, u_k, t) - u_k) - th_k(t)$$

and

$$\begin{aligned} t(I(u) - q)/2 &\geq - \parallel DI(u_k) \parallel_* \parallel c(u, u_k, t) - u_k \parallel - th_k(t) \\ &\geq -C \parallel DI(u_k) \parallel_* \parallel u - u_k \parallel t - th_k(t) \ . \end{aligned}$$

$\parallel \ \parallel$ is the norm of B_1 and $\parallel \ \parallel_*$ that of its dual. Dividing by t, letting t go to zero after that, and finally letting k go to infinity, we get, as $\parallel DI(u_k) \parallel_*$ goes to zero, that $(I(u) - q)/2 \geq 0$, $I(u) \geq q$, so $I(u) = q$. By our conditions this means u_k converges strongly in B_1, and so due to the continuous differentiability of I it converges to a critical point. This proves our theorem.

It seems, however, that this type of contruction also has its advantages, the most important of them being that it is fairly compatible with other deformations that squeeze sets of admissible functions on which the functional is bounded into subsets that are compact with respect to a stronger topology, as in [S22] or [S24]. In both cases results about critical points are obtained although the Palais-Smale condition is not very likely to be fulfilled. The most important step in obtaining the results of [S24,S25] is in the proof of the a-priori estimates in [S26]. From there most of the results could also be obtained by the perturbation technique of [U1], but in case of [S20] this seems not at all clear.

5. Open Problems

a) It would be interesting to determine whether it is possible to derive the Morse theory in its entirety for surfaces in $\mathbb{R}^n$ with $n > 3$, because there non-degeneracy of surfaces as critical points of the functional becomes a generic property, but non-degeneracy does not imply the absence of branch points on the boundary any more, so some modification of the methods of [S28] would be necessary. Also it might be worthwhile to find out to what extent these results can be carried over to surfaces of non-constant mean curvature in $\mathbb{R}^3$. Another question worth looking at is whether the Gromoll-Meyer theory (see [C2]) can be generalized to minimal surfaces, or surfaces of prescribed mean curvature, or whether some other treatment of unstable critical points would be possible. Even a mountain-pass lemma in a stronger topology assuring geometric proximity of the comparison surfaces would be interesting. Here the method of imposing additional constraints as in [S24] might actually be more convenient than a perturbation method, as it would allow to restrict oneself to a set of very good functions without having to consider a whole family of approximating functions. Although the example (see [H14]) of a harmonic mapping with minimal energy which is not even continuous is not encouraging, it might be worthwhile to study whether regularizing constraints along the lines of [S24] have applications to other variational problems. Even in cases where one cannot expect the critical points to be continuous, constraints might be beneficial, although they naturally have to be chosen so as not to exclude all discontinuous functions. The a-priori estimates necessary here, however, would presumably be very difficult to obtain, as for nonlinear equations estimates for the Hölder-norms usually are the first step towards further estimates (see [L1]).

Finally we give a somewhat imprecise and probably rather difficult problem. We characterized the conformal parameters in 2.2 as parametrization of minimal

Dirichlet integral. Now it might be interesting to study parametrizations defined on arbitrary domains Ω which minimize, e.g., the integral

$$\int_{\Omega} |x_u|^p + |x_v|^p \, dx$$

for p not equal 2. These could be called p-conformal coordinates and might have interesting properties, they would, e.g., have bounded H_p^1-norm in any case in which there is any parametrization of the surface that belongs to H_p^1 at all.

References

[A1] Adams, R. A., Sobolev spaces, Academic Press, NY, 1975.

[A2] Ambrosetti, A. and P. H. Rabinowitz, Dual variational methods in critical point theory and applications, *J. Funct. Anal.* **14**, 349–381 (1973).

[A3] Almgren, F., An invitation to varifold geometry, Benjamin, New York-Amsterdam, 1966.

[A4] Almgren, F. and L. Simon, Existence of embedded solutions of Plateau's problem, *Ann. Sc. Norm. Sup. Pisa Cl. Sci.* **(4) 6**, 447–495 (1979).

[B1] Barboza, J. L. and M. do Carmo, Stable minimal surfaces, *Bull. Amer. Math. Soc.* **80**, 581–583 (1974).

[B2] Böhme, R., Die Zusammenhangskomponenten der Lösungen analytischer Plateauprobleme, *Math. Z.* **133**, 31–40 (1973).

[B3] Böhme, R., Die Jacobifelder zu Minimalflächen im $\mathbb{R}^3$, *Manusc. Math.* **16**, 51–73 (1975).

[B4] Böhme, R., A Plateau problem with many solutions for boundary curves in a given knot class (Berlin 1979), pp. 36–41, Lecture Notes in Mathematics 838, Springer, Berlin (1981).

[B5] Böhme, R., New results on the classical problem of Plateau on the existence of many solutions, *Sém. Bourbaki*, 34e année, 1981/82.

[B6] Böhme, R. and F. Tomi, Zur Struktur der Lösungsmenge des Plateauproblems, *Math. Z.* **133**, 1–29 (1973).

[B7] Böhme, R. and A. Tromba, The number of solutions to the classical Plateau problem is generically finite, *Bull. AMS* **83**, 1043–1046 (1977).

[B8] Böhme, R. and A. Tromba, The index theorem for classical minimal surfaces, *Ann. Math.* **113**, 447–499 (1981).

[B9] Bott, R., Marston Morse and his mathematical works, *Bull. AMS* **3**, 907–950 (1980).

[B10] Brézis, H. and J.-M. Coron, Sur la conjecture de Rellich pour les surfaces à courbure moyenne prescrite, *C. R. Acad. Sci. Paris*, Ser. I, **295**, 615–618 (1982).

[B11] Brézis, H. and J.-M. Coron, Multiple solutions of H-systems and Rellich's conjecture, *Comm. Pure Appl. Math.* **37**, 149–187 (1984).

[C1] Cairns, S. and M. Morse, Critical point theory in global analysis and differential topology, New York, etc., Academic Press, 1969.

[C2] Chang, K. C., Infinite dimensional Morse theory and its applications, Séminaire de Mathématiques Superieures, 97, Presses de l'université de Montreal, 1985.

[C3] Chang, K. C. and J. Eells, Unstable minimal surface coboundaries, *Acta Mathematica Sinica* **2**, 233–247 (1986).

[C4] Courant, R., Critical Points and unstable minimal surfaces, *Proc. Nat. Acad. Sci.* **27**, 51–57 (1941).

[C5] Courant, R., Dirichlet's principle, conformal mapping, and minimal surfaces, Springer, Berlin, etc., 1977.

[D1] Douglas, J., Solution of the problem of Plateau, *Trans. Amer. Math. Soc.* **33**, 263–321 (1931).

[F1] Federer, H., Geometric measure theory, Springer, Berlin, etc., 1969.

[F2] Fomenko, A. T., The Plateau problem, Gordon and Breach, NY [to appear].

[G1] Garnier, R., Solution du problème de Riemann pour les systèmes différentiels linéaires du second ordre, *Ann. Sci. Ec. Norm. Sup.* **43** (3), 177–307 (1926).

[G2] Garnier, R., Le problème de Plateau, *Ann. Sci. Ec. Norm. Sup.* **45** (3), 53–144 (1928).

[G3] Giusti, E., Minimal surfaces and functions of bounded variation, Birkhäuser, Basel, etc., 1984.

[G4] Gulliver, R. D. and J. Spruck, Surfaces of constant mean curvature which have a simple projection, *Math. Z.* **129**, 95–107 (1972).

[H1] Hardt, R. and L. Simon, Boundary regularity and embedded solutions for the oriented Plateau problem, *Ann. of Math.* **110**, 439–486 (1979).

[H2] Heinz, E., Ober die Existenz einer Fläche konstanter mittlerer Krümmung bei
vorgegebener Berandung, *Math. Ann.* **127**, 258–287 (1954).

[H3] Heinz, E., On certain non-linear elliptic equations and univalent mappings, *J. d'Analyse Mathématique* **5**, 197–272 (1956/57).

[H4] Heinz, E., On the nonexistence of a surface of constant mean curvature with finite area and prescribed rectifiable boundary, *Arch. Rat. Mech. Anal.* **35**, 249–252 (1969).

[H5] Heinz, E., On surfaces of constant mean curvature with polygonal boundaries, *Arch. Rat. Mech. Anal.* **36**, 335–347 (1970).

[H6] Heinz, E., Unstable surfaces of constant mean curvature, *Arch. Rat. Mech. Anal.* **38**, 257–267 (1970).

[H7] Heinz, E., Über die analytische Abhängigkeit der Lösungen eines linearen elliptischen Randwertproblems von Parametern, *Nachr. Akad. Wiss. Göttingen II, Math. Phys. Kl.*, 1979(1).

[H8] Heinz, E., Über eine Verallgemeinerung des Plateauschen Problems, *Manuscripta Mathematica* **28**, 81–88 (1979).

[H9] Heinz, E., Ein mit der Theorie der Minimalflächen zusammenhängendes Variationsproblem, *Nachr. Akad. Wiss. Göttingen II, Math. Phys. Klasse*, 1980(3).

[H10] Heinz, E., Minimalflächen mit polygonalem Rand, *Math. Z.* **183**, 547–564 (1983).

[H11] Heinz, E., An estimate for the total number of branch points of quasi-minimal surfaces, *Analysis* **5**, 383–390 (1985).

[H12] Hildebrandt, S., Randwertprobleme für Flächen vorgeschriebener mittlerer Krümmung und Anwendungen auf die Kapillaritätstheorie, *Math. Z.* **112**, 205–213 (1969).

[H13] Hildebrandt, S., On the Plateau problem for surfaces of constant mean curvature, *Comm. Pure Appl. Math.* **23**, 97–114 (1970).

[H14] Hildebrandt, S., Partielle Differentialgleichungen und Differentialgeometrie, *Jber. d. Dt. Math.-Verein.* **85**, 129–145 (1983).

[H15] Hildebrandt, S., H. Kaul, and K. O. Widmann, An existence theory for harmonic mappings of Riemannian manifolds, *Acta Math.* **138**, 1–16 (1977).

[J1] Jäger, W., Ein Maximumprinzip für ein System nichtlinearer Differentialgleichungen, *Nachr. Akad. Wiss. Göttingen II, Math. Phys. Klasse*, 1976(11).

[J2] Jäger, W. and H. Kaul, Uniqueness and stability of harmonic maps and their Jacobi fields, *Manuscripta Mathematica* **28**, 269–291 (1979).

[J3] Jost, J., Das Existenzproblem für Minimalflächen, *Jber. d. Dt. Math.-Verein.* **91**, 1–32 (1988).

[K1] Koebe, P., Abhandlungen zur Theorie der konformen Abbildungen III, *J. f. d. reine u. angew. Math.* **147**, 67–104 (1927).

[K2] Klötzler, R., Mehrdimensionale Variationsrechnung, Basel, etc., Birkhäuser, 1970.

[K3] Kneser, H., Lösung der Aufgabe 41, *Jahresber. Deutsche Math. Verein.* **35**, 123–124 (1926).

[L1] Ladyzhenskaya, O. A. and N. N. Ural'tseva, Linear and quasilinear elliptic equations, New York, etc., Academic Press, 1968.

[L2] Lewerenz, F., Eine Bemerkung zu den Marx-Shiffmanschen Minmalvektoren bei Polygonen, *Arch. Rat. Mech. Anal.* **75**, 199–202 (1981).

[L3] Lewy, H., Ober die Methode der Differenzengleichungen zur Lösung von Variations-und Randwertproblemen, *Math. Annalen* **98**, 107–124 (1928).

[M1] Marx, I., On the classification of unstable minimal surfaces with polygonal boundaries, *Comm. Pure Appl. Math.* **8**, 235–244 (1955).

[M2] Meeks, W. H. III, A survey of the geometric results in the classical theory of minimal surfaces, *Bol. Soc. Brasil. Mat.* **1**, 29–86 (1981).

[M3] Milnor, J., Morse theory, Princeton, N. J., Princeton University Press, 1963.

[M4] Morgan, F., Smooth curve in $\mathbb{R}^4$ bounding a continuum of area-minimizing surfaces, *Duke Math. J.* **43**, 867–870 (1976).

[M5] Morgan, F., A smooth curve in $\mathbb{R}^3$ bounding a continuum of minimal submanifolds, *Arch. Rat. Mech. Anal.* **75**, 193–197 (1980/81).

[M6] Morgan, F., Generic Uniqueness for hypersurfaces minimizing the integral of a elliptic integrand with constant coefficients, *Indiana Univ. Math. J.* **30**, 29–45 (1981).

[M7] Morgan, F., Geometric Measure Theory: A beginners guide, Academic Press, Boston, 1988.

[M8] Morrey, Jr., C. B., The problem of Plateau in a Riemannian manifold, *Ann. of Math.* **49** (2), 807-851 (1948).

[M9] Morrey, Jr., C. B., Multiple integrals in the calculus of variations, Springer, Berlin, etc., 1966.

[M10] Morse, M. and C. B. Tompkins, The existence of minimal surfaces of general critical types, *Ann. of Math.* **40**, 443–472 (1939); **42**, 331 (1941).

[M11] Morse, M. and C. B. Tompkins, Minimal surfaces of non-minimum type by a new mode of approximation, *Ann. of Math.* **42**, 62–72 (1941).

[M12] Morse, M. and C. B. Tompkins, Unstable minimal surfaces of higher topological structure, *Duke Math. J.* **8**, 350–375 (1941).

[N1] Nitsche, J. C. C., Variational problems with inequalities as boundary conditions, or, How to fashion a cheap hat for Giacometti's brother, *Arch. Rat. Mech. Anal.* **35**, 83–113 (1969).

[N2] Nitsche, J. C. C., A new uniqueness theorem for minimal surfaces, *Arch. Rat. Mech. Anal.* **52**, 319–329 (1973).

[N3] Nitsche, J. C. C., Vorlesungen über Minimalflächen, Springer, Berlin, etc., 1975.

[N4] Nitsche, J. C. C., Contours bounding at most finitely many solutions of Plateau's problem, Complex analysis and its applications, pp. 438–446, 670 Nauka Moscow, 1978.

[N5] Nitsche, J. C. C., Uniqueness and non-uniqueness for Plateau's problem–one of the last major questions, Obata, Morio, Minimal submanifolds and geodesics, Tokyo, 1978, pp. 143–161.

[O1] Osserman, R., A survey of minimal surfaces, Van Nostrand Reinhold, New York, 1969.

[O2] Osserman, R., A proof of the regularity everywhere of the classical solution to Plateau's problem, *Annals of Math.* **91**, 550–569 (1970).

[P1] Palais, R. S., Morse theory on Hilbert manifolds, *Topology* **2**, 299–340 (1963).

[P2] Palais, R. S. and S. Smale, A generalized Morse theory, *Bull. Amer. Math. Soc.* **70**, 165–172 (1964).

[P3] Plemelj, J., Riemannsche Funktionenscharen mit gegebener Monodromiegruppe, *Monatsh. für Math. u. Physik* **19**, 211–245 (1908).

[R1] Radó, T., The problem of the least area and the problem of Plateau, *Math. Z.* **32**, 763–796 (1930).

[R2] Rassias, T. M., Foundations of Global Nonlinear Analysis, Teubner, Leipzig, 1986.

[R3] Rassias, T. M., Survey of uniqueness and classification theorems for minimal surfaces in Nonlinear Analysis, ed., T. M. Rassias, World Scientific Publishers, Singapore, 1987.

[R4] Ruchert, H., Uniqueness theorem for surfaces of constant mean curvature, *Arch. Math.* **33**, 91–104 (1979).

[S1] Sauvigny, F., Flächen vorgeschriebener mittlerer Krümmung mit eineindeutiger Projektion auf eine Ebene, *Math. Z.* **180**, 41–67 (1982).

[S2] Sauvigny, F., Die zweite Variation von Minimalflächen im $\mathbb{R}^p$ mit polygonalem Rand, *Math. Z.* **189**, 167–184 (1985).

[S3] Sauvigny, F., On the Morse index of minimal surfaces in $\mathbb{R}^p$ with polygonal boundaries, *Manuscripta Math.* **53**, 167–197 (1985).

[S4] Sauvigny, F., On the total number of branch-points of quasi-minimal surfaces with polygonal surfaces, [to appear].

[S5] Sauvigny, F., Ein Eindeutigkeitssatz für Minimalflächen im $\mathbb{R}^p$ mit polygonalem Rand, *J. f. d. reine u. angew. Math.* **358**, 92–96 (1985).

[S6] Schlesinger, L., Vorlesungen über lineare Differentialgleichungen, Leipzig, 1908.

[S7] Schüffler, K., Isoliertheit und Stabilität von Flächen konstanter mittlerer Krümmung, *Manusc. Math.* **40**, 1–15 (1982).

[S8] Schüffler, K., Jacobifeler zu Flächen konstanter mittlerer Krümmung, *Archiv der Math.* **41**, 176–182 (1983).

[S9] Schüffler, K., Eine globalanalytische Behandlung des Douglas'schen Problems, *Manusc. Math.* **48**, 189–226 (1984).

[S10] Schüffler, K. and F. Tomi, Ein Indexsatz für Flächen konstanter mittlerer Krümmung, *Math. Z.* **182**, 245–257 (1983).

[S11] Serrin, J., On surfaces of constant mean curvature which span a given space curve, *Math. Z.* **112**, 77–88 (1969).

[S12] Shiffman, M., The Plateau problem for nonrelative minima, *Ann. of Math.* **40**, 834–854 (1939).

[S13] Shiffman, M., Unstable minimal surfaces with several boundaries, *Ann. of Math.* **43**, 197–222 (1942).

[S14] Shiffman, M., Unstable minimal surfaces with any rectifiable boundary, *Proc. Nat. Acad. Sci. USA* **28**, 103–108 (1942).

[S15] Smale, S., Morse theory and a nonlinear generalization of the Dirichlet problem, *Ann. Math.* **80**, 382–396 (1964).

[S16] Söllner, M., Plateau's problem for surfaces of constant mean curvature from a global point of view, *Manusc. Math.* **43**, 191–217 (1983).

[S17] Steffen, K., Flächen konstanter mittlerer Krümmung mit vorgeschriebenem Volumen oder Flächeninhalt, *Arch. Rat. Mech. Anal.* **49**, 99–128 (1972).

[S18] Steffen, K., On the nonuniqueness of surfaces with prescribed constant mean curvature spanning a given contour, *Arch. Rat. Mech. Anal.* **94**, 101–122 (1986).

[S19] Ströhmer, G., Instabile Minimalflächen in Riemannschen Mannigfaltigkeiten nichtpositiver Schnittkrümmung, *J. f. d. rein u. angew. Math.* **315**, 16–39 (1980).

[S20] Ströhmer, G., Instabile Flächen vorgeschriebener mittlerer Krümmung, *Math. Z.* **174**, 119–133 (1980).

[S21] Ströhmer, G., Einige Bemerkungen zur Frage der eindeutigen Lösbarkeit des Dirichletproblems für ein gewisses System nichtlinearer Differentialgleichungen, *Math. Z.* **175**, 161–163 (1980).

[S22] Ströhmer, G., Instabile Lösungen der Eulerschen Gleichungen gewisser Variationsprobleme, *Arch. Rat. Mech. Anal.* **79**, 219–239 (1982).

[S23] Ströhmer, G., Instabile Minimalflächen mit halbfreiem Rand, *Analysis* **2**, 315–335
(1982).

[S24] Ströhmer, G., Unstable solutions of the Euler equations of certain variational problems, *Math. Z.* **186**, 179–199 (1984).

[S25] Ströhmer, G., About the Morse theory for certain variational problems, *Math. Annalen* **2270**, 275–284 (1985).

[S26] Ströhmer, G., Some remarks about variational problems with constraints, in Rassias & Rassias, eds., Differential geometry-calculus of variations, Lecture notes in pure and applied mathematics, Vol. 100, New York: Marcel Dekker, 1985.

[S27] Struwe, M., Multiple solutions of differential equations without the Palais-Smale condition, *Math. Annalen* **261**, 399–412 (1982).

[S28] Struwe, M., On a critical point theory for minimal surfaces spanning a wire in $\mathbb{R}^n$, *J. Reine u. Angew. Math.* **349**, 1–23 (1984).

[S29] Struwe, M., Large H-surfaces via the mountain-pass lemma, *Math. Ann.* **270**, 441–459 (1985).

[S30] Struwe, M., Functional analytic aspects of the Plateau problem, *Delft Progress Report* **10**, 271–281 (1985).

[S31] Struwe, M., Nonuniqueness in the Plateau problem for surfaces of constant mean curvature, *Arch. Rat. Mech. Anal.* **93**, 135–157 (1986).

[S32] Struwe, M., Plateau's problem and the calculus of variations, Princeton University Press, Princeton, N. J., 1988; Math. Notes 35.

[T1] Tausch, E., A class of variational problems with linear growth, *Math. Z.* **164**, 159–178 (1978).

[T2] Thiel, U., The index theorem for k-fold connected minimal surfaces, *Math. Ann.* **270**, 489–501 (1985).

[T3] Tomi, F., On the local uniqueness of the problem of the least area, *Arch. Rat. Mech. Anal.* **52**, 312–318 (1973).

[T4] Tromba, A., A general approach to Morse theory, *J. Differential Geometry* **12**, 47–85 (1977).

[T5] Tromba, A., Degree theory on oriented infinite dimensional varieties and the Morse number of minimal surfaces spanning a curve in $\mathbb{R}^n$, Part I: $n \geq 4$, *Trans. AMS* **290**, 385–413 (1985); Part II: $n = 3$, *Manusc. Math.* **48**, 139–161 (1984).

[U1] Uhlenbeck, K., Morse theory by perturbation methods with applications to harmonic mappings, *Trans. Am. Math. Soc.* **267**(2), 569–583 (1981).

[W1] Wente, H. C., A general existence theorem for surfaces of constant mean curvature, *Math. Z.* **120**, 277–288 (1971).

[W2] Wente, H. C., Large solutions to the volume-constrained Plateau problem, *Arch. Rat. Mech. Anal.* **75**, 59–77 (1980).

[W3] Wente, H. C., Counterexample to a conjecture of H. Hopf, *Pacific J. Math.* **121**, 193–243 (1986).

[W4] Werner, H., Das Problem von Douglas für Flächen konstanter mittlerer Krümmung, *Math. Ann.* **133**, 303–319 (1957).

[W5] Williams, G. H., Nonlinear nonhomogeneous elliptic variational problems with a non-constant gradient constraint, *J. Math. Pures Appl.* **60**(9), 213–226 (1981).

The Problem of Plateau (pp. 285–326)
ed. Th. M. Rassias

THE ROLE OF MINIMAL AND RIGID SURFACES IN THEORETICAL PHYSICS

VICTOR TAPIA
Postfach 132
5024 Salzburg
Austria

1. Introduction

We look at the role of minimal and rigid surfaces in theoretical physics. We start by considering some mathematical preliminaries concerning extrinsic Riemannian geometry. We define an isometric embedding and consider the main theorems on local and global isometric embeddings of Riemannian manifolds into higher-dimensional flat spaces. As a result of these theorems any Riemannian space $V_n(t,s)$ can be considered, in the local case, as a submanifold of a Euclidean space $E_N(T,S)$ with $N = n(n+1)/2$, $T \geq t$, $S \geq s$. Then we consider a set of equations describing the embedding, the Gauss–Codazzi–Ricci (GCR) equations; they are necessary conditions for the embedding. We introduce the concept of class of the embedding: the minimal number of extra dimensions required to satisfy the GCR equations. Then, we consider some results for class one isometric embeddings.

Next, we consider an analytic Riemannian space $V_n(t,s)$ with analytic metric as already locally and isometrically embedded in a pseudo-Euclidean flat space $E_N(T,S)$ with global Euclidean topology, $N = n(n+1)/2$, $T \geq t$, $S \geq s$. Then, making use of of the Euclidean global character of $E_N(T,S)$ we introduce a parametrisation of the embedding consisting in writing $E_N(T,S)$ as the direct product of two manifolds, $E_N = M_n \otimes B_{N-n}$, with M_n being a maximally symmetric space. With this parametrisation we are able to fix the gauge for any covariant field theory of the embedded surface allowing, furthermore, to write them as theories of normal deformations of a background geometry. We consider extrinsic Riemannian geometry of minimal and rigid surfaces. In the case of a minimal surface our parametrisation of the embedding is equivalent to a covariant generalisation of the harmonic gauge. In the case of a rigid surface it is a novel gauge which looks similar to the harmonic gauge except that the Christoffel symbol is contracted with the Einstein tensor rather than with the metric.

We also discuss the theory of General Relativity for the gravitational interaction and their problems. We then consider string theory in which minimal surfaces are of central importance.

In previous works we have developed a model for the description of the gravitational interaction based on the use of extrinsic gravity. In this model the space-time is embedded into a higher-dimensional space and the embedding functions are considered as the dynamical variables. The model shows a number of advantages over the conventional theory of gravity, i.e., General Relativity. Just to mention a bit more, there exist consistent conservation laws. The energy-momentum tensor is proportional to the Einstein tensor. The ADM Lagrangian for extrinsic gravity has intriguing properties, etc. Here we consider the canonical structure of extrinsic gravity. The calculation relies on the use of second-order field theory. In spite of this fact, and after conveniently define some notions of second-order field theory, the shift and lapse constraints are easily calculated and they satisfy the universal algebra for parametrisation invariant theories. There exist also a number of constraints coming from the particular form of the Lagrangian. After fixing the gauge, with a particular parametrisation of the embedding, we give account of the lapse and shift constraints. For the rest of the constraints we calculate the consistency conditions giving rise to new constraints.

2. Embedded Surfaces

The concept of an abstract Riemannian manifold arises as the result of the evolution of mathematical attitudes. In the earlier period mathematicians thought more concretely of curved surfaces embedded in a flat Euclidean space [1]. The concept of an embedded surface was generalised later on to that of an abstract Riemannian manifold defined intrinsically through the metric [2]. Almost immediately after the intrinsic point of view came into favor a natural question arose: the isometric embedding problem. It is clear that any embedded surface can be considered as a particular case of an intrinsically defined manifold. However, it was not so clear if an intrinsically defined Riemannian manifold can be considered as a submanifold of some higher-dimensional Euclidean space. Today we know that the answer is yes, any intrinsically defined Riemannnian manifold can be isometrically embedded, locally and globally, in an Euclidean space of appropriate dimension and signature.

The embedding problem was first considered by L. Schläfli [3] just after Riemann presented his famous thesis. Schläfli discussed the local form of the embedding problem and he conjectured that a Riemannian manifold with positive defined and analytic metric can be locally and isometrically embedded as a submanifold of a Euclidean space E_N with $N = n(n+1)/2$. The first concrete result was presented by M. Janet [4] in 1926 who described a method of proof based on a power series development, therefore it was limited to local results. Furthermore, he required the metric to be analytic. This proof however, as Janet himself noticed, it was incomplete. He solved only the local problem for two-dimensional manifolds with analytic metric. In 1927 E. Cartan [5] extended the Janet's proof to n-dimensional, $n \geq 2$, manifolds treating it as an application of his theory of Pfaffian forms. The dimensionality requirement was $N = n(n+1)/2$ as conjectured by Schläfli. In 1931 C. Burstin [6] completed the Janet's proof and also extended it to the case in which the embedding space is a given Riemannian manifold V_N with positive defined and analytic metric. In 1956 K. Leichtweiss [7] gave a new proof based more substantially than Burstin's on the Gauss–Codazzi–Ricci equations of Riemannian geometry. In 1961 A. Friedman [8] extended the theorem to Riemannian manifolds with indefinite metric. The first global isometric embedding theorems of a V_n into a E_N were established by J. Nash [9]. The results depend crucially on the compactness of V_n. The first global results for indefinite metrics were obtained by C. J. Clarke [10] and by R. E. Greene [11].

In conclusion, any intrinsically defined Riemannian manifold has a local and a global isometric embedding in some Euclidean space. Then one can consider both approaches to Riemannian geometry, the intrinsic and the extrinsic, as completely equivalent. In the extrinsic approach to Riemannian geometry the objects characterising the geometry of the embedded surface are the functions whose restriction defines the surface. In the mathematical literature these exist many results on local and global isometric embeddings, but only a few of them are useful for applications in theoretical physics, particularly to string theories and to gravitation; our selection of mathematical topics has been guided by what we think is promising for an eventual application in physics. Since physics is normally a local affair, the study has been mainly restricted to local results. Today we know that global properties are as important as the local ones, therefore, some global results are also included.

2.1. *Isometric embeddings*

The fundamentals of extrinsic Riemannian geometry necessary for what follows can be found in the book by L. P. Eisenhart [12]. Let $V_n(t, s)$ be an

n-dimensional Riemannian manifold with analytic and non-degenerate metric $g_{\mu\nu}(x)$ with t positive and s negative eigenvalues, $t+s=n$, and let $E_N(T,S)$ be an N-dimensional flat space with global Euclidean topology with analytic and non-degenerate metric $G_{AB}(X)$ with T positive and S negative eigenvalues, $T+S=N$.

A differentiable map $f: V_n(t,s) \to V_N(T,S)$, $N \geq n, T \geq t, S \geq s$, is a C^k (C^∞, analytic), $k \geq 1$, immersion if [13]

i. f is of differentiability class $C^k(C^\infty$, analytic).
ii. rank $(df) = n$ at all points $p \in V_n$; df is the differential of f.

An immersion is not necessarily injective, therefore, $f(V_n)$ is not necessarily a manifold. An injective immersion is an embedding. The set $f(V_n)$ with the differential structure induced by the embedding is a manifold. If $f(V_n)$ has a submanifold structure equivalent to the manifold structure induced by the embedding, then f is regular. Thus, if f is a regular embedding, $f(V_n)$ is a submanifold of V_N. Summarising, an embedding is an analytic injective map $f: V_n \to E_N$, with $N \geq n$ and rank $(df) = n$.

The embedding f is isometric at a point $p \in V_n$ if

$$ds^2(V_n) = ds^2(E_N)\big|_{V_n} . \tag{2.1}$$

The isometric embedding is local or global depending on whether Eq. (2.1) holds locally or globally.

Let $x^\mu, \mu = 0, \ldots, n-1$, and $X^A, A = 0, \ldots, N-1$, be local coordinates on V_n and E_N, respectively. The map f is then given by $X^A = X^A(x^\mu)$ where rank $(X^A_\mu) = n, X^A_\mu = \partial_\mu X^A$. Equation (2.1) in local coordinates is

$$g_{\mu\nu} = G_{AB} X^A_\mu X^B_\nu , \tag{2.2}$$

with $g = \det g_{\mu\nu} \neq 0$, such that the metric of V_n is that isometrically induced by the embedding. The information about the intrinsic geometry of V_n and that about its situation as a submanifold of V_N, the extrinsic geometry, are both contained in $X^A = X^A(x^\mu)$.

The previous is called an explicit embedding: the functions $G_{AB}(X)$ and $X^A(x)$ are given and the metric $g_{\mu\nu}$ is obtained through Eq. (2.2), such that the metric of V_n is that isometrically induced by the embedding.

In an implicit embedding, by the contrary, one must find functions $G_{AB}(X)$ and $X^A(x)$ such that Eq. (2.2) is satisfied for a given $g_{\mu\nu}$. Its existence is guaranteed by the local [4,5,6,8] and global [9,10,11] isometric embedding theorems.

For the physical applications, which we are going to consider below, both local and global results will be of importance. Let us start by considering the local results.

2.2. *Local embeddings*

The first result concerns positive definite metrics and is due to M. Janet [4], E. Cartan [5] and C. Burstin [6].

Theorem. Any Riemannian manifold V_n can be analytically and isometrically embedded in E_N with $N = n(n+1)/2$.

The proof consists in a power series expansion. The generalisation to indefinite metrics is due to A. Fiedman [8].

Theorem (Local Isometric Embedding). Let $g_{\mu\nu}(x^\lambda)$ be analytic functions in a neighbourhood of $x^\lambda = 0$ and let $G_{AB}(X^C)$ be analytic functions in a neighbourhood of $X^A = 0$. If $N = n(n+1)/2$, then there exist analytic functions $X^A = X^A(x^\mu)$ in a neighbourhood of $x^\mu = 0$ satisfying the conditions

$$X^A(0) = 0\,, \qquad \mathrm{rank}(X^A_\mu(0)) = n\,, \tag{2.3a}$$

$$g_{\mu\nu}(x) = G_{AB}(X(x))X^A_\mu(x)X^B_\nu(x)\,. \tag{2.3b}$$

The proof by Friedman, as that by Burstin, is based on the general sketch by Janet, but in the rest is a new proof even in the case of positive definite metrics.

Of course the theorem establishes only the existence of functions G_{AB} and X^A satisfying Eq. (2.2), but, due to the definition of an isometric embedding it amounts to an existence theorem for the embedding.

No assumptions are made on the functions G_{AB} apart from the fact that they must be $\mathrm{Rie}(G) = 0$, i.e., they are the components of the metric of some flat pseudo-Euclidean space E_N with Euclidean global topology, but, of course, one can consider Riemannian manifolds which are, among others:

i. of constant curvature,
ii. conformally flat,
iii. Ricci flat, etc.

From now on the embedding of a Riemannian manifold V_n will be understood as the local isometric embedding of V_n into a flat space E_N with Euclidean global topology, $N = n(n+1)/2$, $T \geq t$, $S \geq s$. The same restriction will be used for the global results. In the local case we can establish this result as follows.

Theorem. Any analytic Riemannian manifold $V_n(t,s)$ can be analytically and isometrically embedded in $E_N(T,S)$ with $N = n(n+1)/2$, $T \geq t, S \geq s$.

Thus, the theorem guarantees the existence of a local analytic and isometric embedding of $V_n(t,s)$ into $E_N(T,S)$ with $N = n(n+1)/2$, $T \geq t, S \geq s$. For a space-time, $V_4(1,3)$, we obtain $E_{10}(T,S), T \geq 1, S \geq 3$.

2.3. *The signature of the embedding space*

According to the local isometric embedding theorem the signature of the embedding space $E_N(T,S)$ can be chosen quite arbitrarily, except for the restriction $T \geq t, S \geq s$, but no additional restrictions are imposed over the signature of the other additional dimensions. We have tried to figure out an argument to determine the signature of the embedding space which we are going to present here.

The signature of the embedding space will be in general given by

$$T = T(t,s)\,, \tag{2.4a}$$
$$S = S(t,s)\,. \tag{2.4b}$$

The problem is to determine the functions T and S. The first criterion they must satisfy is that of reciprocity. For a space with signature (s,t) the functions T and S must be such that

$$T(s,t) = S(t,s)\,, \tag{2.5a}$$
$$S(s,t) = T(t,s)\,. \tag{2.5b}$$

Actually the two previous conditions are only one. Combining this condition with

$$T + S = N(n = t + s)\,, \tag{2.6}$$

we obtain

$$T(t,s) + T(s,t) = N(t+s)\,, \tag{2.7a}$$
$$S(t,s) + S(s,t) = N(t+s)\,. \tag{2.7b}$$

We assume that T and S are at most quadratic in t and s. Let us start by writing

$$T(t,s) = \alpha_1 t + \alpha_2 s + \beta_1 t^2 + \beta_2 ts + \beta_3 s^2\,. \tag{2.8}$$

We know that for positive definite metrics, $t = n, s = 0$, it must be $T = n(n+1)/2$; therefore, $\alpha_1 = 1/2, \beta_1 = 1/2$. On the other hand, for a negative definite metric, $t = 0, s = n$, it must be $T = 0$; therefore, $\alpha_2 = 0, \beta_3 = 0$. Then, the expression for T reduces to

$$T(t,s) = \frac{1}{2}t + \frac{1}{2}t^2 + \beta_2 ts \,. \tag{2.9}$$

Introducing this result in Eq. (2.6) we conclude that, $\beta_2 = 1/2$, such that

$$T(t,s) = \frac{1}{2}t[1 + (t+s)] = \frac{1}{2}t(1+n)\,, \tag{2.10a}$$

and therefore S is given by a similar expression

$$S(t,s) = \frac{1}{2}s[1 + (t+s)] = \frac{1}{2}s(1+n)\,. \tag{2.10b}$$

Let us observe however that the previous formulae do not guarantee that T and S are integer numbers. This fact however finds an easy explanation. Let us consider the simple case in which $n = 2$. We will see, in Sec. 3, that in pseudo-Cartesian coordinates the metric can be written as

$$g_{\mu\nu} = \begin{vmatrix} \lambda_1 + \eta(\phi_1)^2 & \eta\phi_1\phi_2 \\ \eta\phi_1\phi_2 & \lambda_2 + \eta(\phi_2)^2 \end{vmatrix}\,, \tag{2.11}$$

where $\lambda_1, \lambda_2 = \pm 1$ and $\eta = \pm 1$ is the signature of the extra dimension. For $\phi =$ const. the metric has eigenvalues λ_1 and λ_2. We must check if for $\phi \neq$ const. this signature is preserved. The eigenvalues are given by

$$\begin{aligned}\lambda =& \frac{1}{2}[(\lambda_1 + \lambda_2) + \eta(\phi_1^2 + \phi_2^2) \\ &\pm \left((\lambda_1 - \lambda_2)^2 + 2\eta(\lambda_1 - \lambda_2)(\phi_1^2 - \phi_2^2) + (\phi_1^2 + \phi_2^2)^2\right)^{1/2}]\,.\end{aligned} \tag{2.12}$$

In the case $\lambda_1 = \lambda_2 = 1$ the previous equation reduces to

$$\lambda = 1 + \frac{1}{2}(\eta \pm 1)(\phi_1^2 + \phi_2^2)\,. \tag{2.13a}$$

Since both eigenvalues must be positive, $\eta = 1$. In the case $\lambda_1 = \lambda_2 = -1$, we obtain

$$\lambda = -1 + \frac{1}{2}(\eta \pm 1)(\phi_1^2 + \phi_2^2)\,, \tag{2.13b}$$

such that $\eta = -1$. In the case $\lambda_1 = -\lambda_2 = 1$, we have

$$\lambda = \frac{1}{2}[\eta(\phi_1^2 + \phi_2^2) \pm \left(4 + 4\eta(\phi_1^2 - \phi_2^2) + (\phi_1^2 + \phi_2^2)^2\right)^{1/2}]\,. \tag{2.13c}$$

The requirement of having one positive and one negative eigenvalue does not fix the sign of η. The same argument is also valid for the case $\lambda_1 = -\lambda_2 = -1$.

For a space-time the possible signatures are (2,8) and (3,7). We must still check if for known explicit embeddings this conjecture is satisfied. A list of explicit embeddings of several space-times was given by J. Rosen [14] and later improved by C. D. Collinson [15]. For space-times of class less than or equal to two, one is sure that no problem could arise for this conjecture. Problems can appear only for space-times of class greater than or equal to three. The only known example violating this conjecture is the Petrov space T_1, group G_4, metric 3 [14,16], which, being of class 2, has embedding signature (4,2).

2.4. *Global embeddings*

The first concrete result on global embeddings is due to J. Nash [9].

Theorem. Any compact (non-compact) $C^k, k \geq 3$, Riemannian manifold V_n has a C^k global isometric embedding into E_N with

$$N_c = n(3n + 11)/2 \,, \tag{2.14a}$$

in the compact case, and

$$N_{nc} = n(n + 1)(3n + 11)/2 \,, \tag{2.14b}$$

in the non-compact case.

In the proof of this theorem the positivity of the metric plays a crucial role and, in fact, the proof breaks down if the metric is indefinite. The extension to indefinite metrics was given by C. J. Clarke [10] which in the case of non-compact manifolds is also an improvement on Nash's result.

Theorem. Any C^∞ Riemannian manifold $V_n(t, s)$ with $C^k, k \geq 3$, the Riemannian metric can be globally and C^k isometrically embedded in a E_N (T, S) with $N = T + S$, $S = s + 1$ and

$$T = n(3n + 11)/2 \tag{2.15a}$$

for a compact $V_n(t, s)$, and

$$T = (2n^3 + 15n^2 + 37n + 6)/6 \tag{2.15b}$$

for a non-compact $V_n(t, s)$.

The next result, due to R. E. Greene [11], uses a stronger differentiability assumption.

Theorem. Any C^∞ Riemannian manifold $V_n(t,s)$ with C^∞ Riemannian metric can be globally and C^∞ isometrically embedded in $E_N(T,S)$ where $N = T + S$ and

$$T = S = n(n+5)/2\,, \tag{2.16a}$$

for a compact $V_n(t,s)$, and

$$T = S = 2(2n+1)(n+3)\,, \tag{2.16b}$$

for a non-compact $V_n(t,s)$.

In the non-compact case the improvement over Clarke's result starts only for $n \geq 20$. One unpleasant aspect of Clarke's theorem is the explicit dependence of N on the signature of $V_n(t,s)$; the local isometric embedding theorem does not do it. On the other hand, the Green's theorem introduces an artificially augmented number of time- and space-like dimensions in order to guarantee the possibility of hosting the time- and space-like dimensions of $V_n(t,s)$.

All the previous numbers are just best bounds. Based on this fact we have looked for an optimisation of the previous results. First of all, for a zero-dimensional object, a point, we need zero dimensions for the embedding, i.e., we must have

$$N = an + bn^2 + \cdots\,. \tag{2.17}$$

The curve best adjusting the previous bounds from below is given by

$$N = n(4n+5)\,. \tag{2.18}$$

Since no rigorous proof of this exists we state this result as follows.

Conjecture. Any C^∞ Riemannian manifold $V_n(t,s)$ with C^∞ Riemannian metric can be globally and C^∞ isometrically embedded in $E_N(T,S)$ where $N = T + S$ and $N = n(4n+5)$.

Let us observe that for $n = 2$ we obtain $N = 26$. We will meet this number again when studying string theories.

2.5. *The Gauss–Codazzi–Ricci equations*

In order to obtain a deeper understanding of the geometrical properties of embeddings now we describe a set of equations being necessary conditions for an isometric embedding $f : V_n \to E_N$. We follow L. P. Eisenhart [12]. The space V_n is considered as already embedded in a space E_N with $N \geq n$, with N not necessarily given by $n(n+1)/2$, and therefore is given by the equations $X^A = X^A(x^\mu)$. Since the embedding is isometric the metric of V_n is given by Eq. (2.1). From the metricity condition $\nabla_\lambda g_{\mu\nu} \equiv 0$ one concludes

$$G_{AB} X^A_\lambda Y^B_{\mu\nu} \equiv 0 \,, \tag{2.19}$$

where $Y^A_{\mu\nu}$ are the Gauss tensors defined by

$$Y^A_{\mu\nu} = X^A_{\mu;\nu} + \bar{\Gamma}^A_{CD} X^C_\mu X^D_\nu \,, \tag{2.20}$$

with $\bar{\Gamma}^A_{BC}$ the Christoffel symbols of E_N restricted to V_n. The $Y^A_{\mu\nu}$ are N symmetric tensors in V_n and are the generalisation to higher dimensions of the Gauss tensor of the second fundamental form of a V_n embedded in a V_{n+1} used in the theory of hypersurfaces.

The Christoffel symbol is given by

$$\Gamma_{\lambda,\mu\nu} = X^A_\lambda G_{AB}(X^B_{\mu\nu} + \bar{\Gamma}^B_{CD} X^C_\mu X^D_\nu) \,. \tag{2.21}$$

The Gauss tensors are then given by

$$Y^A_{\mu\nu} = H^{AC} G_{CB}(X^B_{\mu\nu} + \bar{\Gamma}^B_{CD} X^C_\mu X^D_\nu) \,, \tag{2.22}$$

where

$$H^{AB} = G^{AB} - X^A_\mu g^{\mu\nu} X^B_\nu \,. \tag{2.23}$$

Let

$$H_{AB} H^{BC} = \delta^C_A \,, \tag{2.24}$$

then

$$\Gamma_{\lambda,\mu\nu} = X^A_\lambda H_{AB} Y^B_{\mu\nu} \,. \tag{2.25}$$

For the Riemann tensor we finally obtain

$$R_{\lambda\mu\nu\rho} = H_{AB}(Y^A_{\lambda\nu} Y^B_{\mu\rho} - Y^A_{\lambda\rho} Y^B_{\mu\nu}) \,. \tag{2.26}$$

2.6. *Parametrising the GCR equations*

The X^A_μ are N vectors in V_n and, at the same time, n vectors in E_N, tangent to V_n. Therefore, the N vectors $Y^A_{\mu\nu}$ in E_N are orthogonal to V_n and then not all of them can be linearly independent. This allows us to introduce a convenient parametrisation of the GCR equations. Let B_{N-n} be the normal complement of $V_n \subset E_N$. Let N^{aA} be vectors in E_N normal to V_n, i.e., a local basis of B_{N-n} restricted to V_n.

Theorem. If the metric tensor of V_n is non-degenerate, then the N^{aA} can be chosen as non-null vectors.

This theorem was originally established by L. P. Eisenhart [12] only for $N = n + 1$, but, by recurrence, can be generalised to an arbitrary $N > n$.

The n vectors X^A_μ and the $N - n$ vectors N^{aA} are a basis for E_N. Since the vectors N^{aA} are non-null they can be orthogonalised. Then, in addition to Eq. (2.2), we have the relations

$$G_{AB}N^{aA}N^{bB} = \eta^{ab} = \mathrm{diag}(\pm, \dots, \pm), \tag{2.27a}$$

$$G_{AB}N^{aA}X^B_\mu = 0. \tag{2.27b}$$

Let us now consider how the basic vectors N^{aA} and X^A_μ change along V_n, i.e., their covariant derivatives with respect to the coordinates x^μ of V_n. They will, respectively, be vectors and tensors in V_n, but again, vectors in E_N, and therefore linear combinations of the basic vectors X^A_μ and N^A_a. Identity (2.19) means that the N vectors $Y^A_{\mu\nu}$ in E_N are orthogonal to V_n. Therefore, one can write them as a linear combination of the normal vectors N^{aA}

$$Y^A_{\mu\nu} = \Omega_{a\mu\nu}N^{aA}. \tag{2.28}$$

We can write $\Omega^a_{\mu\nu} = \eta^{ab}\Omega_{b\mu\nu}$ as

$$\Omega^a_{\mu\nu} = N^a_A Y^A_{\mu\nu}. \tag{2.29}$$

$\Omega^a_{\mu\nu}$ are $N-n$ symmetric tensors in V_n. They are called reduced Gauss tensors. Similarly one can conclude that

$$N^{aA}_{,\mu} = -\Omega^{a\nu}_\mu X^A_\nu + t^{ab}_\mu N^A_b. \tag{2.30}$$

We can write $t_{ab\mu} = \eta_{ac}\eta_{bd}t^{cd}_\mu$ as

$$t_{ab\mu} = G_{AB}N^A_a(N^B_{b,\mu} + \Gamma^B_{CD}N^C_b X^D_\mu). \tag{2.31}$$

$t_{ab\mu}$ are the components of $(N - n)(N - n - 1)/2$ vectors, antisymmetric in ab, in V_n. They correspond to the normal part of the Christoffel symbol in V_n. They are called torsion vectors because $t_{(ab)\mu} = 0$.

The tensors $\Omega^a_{\mu\nu}$ and the vectors $t_{ab\mu}$ cannot be arbitrarily prescribed. They must satisfy certain integrability conditions which are necessary conditions for the embedding. They are obtained by making use of the Ricci identity for X^a_μ. The results are the Gauss

$$R_{\lambda\mu\nu\rho} = \eta_{ab}(\Omega^a_{\lambda\nu}\Omega^b_{\mu\rho} - \Omega^a_{\lambda\rho}\Omega^b_{\mu\nu})$$
$$= H_{AB}(X^A_{\lambda\nu}X^B_{\mu\rho} - X^A_{\mu\nu}X^B_{\lambda\rho}), \tag{2.32a}$$

the Codazzi

$$\Omega^a_{\mu\nu;\lambda} - \Omega^a_{\mu\lambda;\nu} = (t^{ab}_\lambda \Omega_{b\mu\nu} - t^{ab}_\nu \Omega_{b\mu\lambda}), \tag{2.32b}$$

and the Ricci

$$t^{ab}_{\mu;\nu} - t^{ab}_{\nu;\mu}$$
$$= \eta_{cd}(t^{ac}_\mu t^{bd}_\nu - t^{ac}_\nu t^{bd}_\mu) + (\Omega^a_{\mu\lambda}\Omega^{b\lambda}_\nu - \Omega^a_{\nu\lambda}\Omega^{b\lambda}_\mu), \tag{2.32c}$$

equations, where

$$X_{A\lambda} = \eta_{AB}X^B_\lambda . \tag{2.33}$$

For simplicity we have put $G_{AB} = \eta_{AB}$.

A set of real $\Omega^a_{\mu\nu}$ and $t_{ab\mu}$ solving the GCR equations for a given metric $g_{\mu\nu}$ is called an implicit embedding of V_n into E_N. In an explicit embedding the functions $X^A(x^\mu)$ are known while $\Omega^A_{\mu\nu}$ and $t_{ab\mu}$ are calculated using Eqs. (2.29) and (2.31). If one defines $V_n(t,s)$ as a submanifold of $V_N(T,S)$ with $N = n(n+1)/2$ then the existence of a solution of the GCR equations is guarantee. In this explicit embedding we must use only the Gauss equations in order to obtain the Riemman tensor. The other two sets of equations, the Codazzi and Ricci ones, since one assumes that V_n is already embedded in E_N, do not bring new information.

The scalar curvature is given by

$$R = G^{\mu\nu\lambda\rho}M_{AB}X^A_{\mu\nu}X^B_{\lambda\rho}, \tag{2.34}$$

where

$$G^{\mu\nu\lambda\rho} = g^{\mu\nu}g^{\lambda\rho} - \frac{1}{2}(g^{\mu\lambda}g^{\nu\rho} + g^{\mu\rho}g^{\nu\lambda}), \tag{2.35}$$

is the DeWitt metric.

2.7. *The embedding class*

In many cases the number of extra dimensions, $N - n$, necessary for the embedding is well below the minimal value required by the local or global isometric embedding theorem. This is the reason for introducing the concept of the embedding class of a $V_n(t,s)$ which is defined as the minimal number of extra dimensions necessary for the embedding or the minimal number of extra dimensions necessary to satisfy the GCR equations. In the local case the embedding class runs from zero to $n(n-1)/2$. If the embedding class of a V_n is e, then it must admit e tensors $\Omega^a_{\mu\nu}$ and $e(e-1)/2$ tensors $t_{ab\mu}$ satisfying the GCR equations. For a space-time, $V_4(1,3)$, the embedding class runs from zero to six.

We finish this section with two trivial remarks. If Rie $= 0$, then the embedding class is zero. If the local embedding class is greater than $n(n-1)/2$, then the embedding is locally not minimal.

2.8. *Other results*

For $e > 1$, the tensors $\Omega^a_{\mu\nu}$ and the vectors $t_{ab\mu}$ are not defined uniquely by the embedding due to the possibility of making pseudo-rotations, at every point of V_n, of the vectors N^{aA} orthogonal to V_n. These degrees of freedom can be used to simplify the GCR equations in some special cases. Furthermore, the GCR equations are not all functionally independent. In fact, they are interrelated by the Bianchi identities for the Riemann tensor, involving derivatives of the Gauss tensors. As a consequence of these identities part of the Codazzi equations are trivially satisfied [17,18]. In local coordinates these equations were given by R. Blum [19]. In some exceptional cases it is only necessary to satisfy the Gauss equations to guarantee the embedding of a given V_n. A particular application of these identities to the case in which one of the Gauss tensors has rank greater than or equal to three is found in references [17,18]. These results include the Thomas' theorem [20] for class one embeddings.

One of the fundamental problems of embeddings is to determine the embedding class of a given V_n. For many relativistic metrics, $n = 4$, upper limits ≤ 6 are known on the embedding class [14,15,21]. In the general case a practical algorithm to solve the GCR equations, either in the sense of determining all the solutions of Einstein field equations for a given class or in the sense of determining the class of a given metric is not known. The tradition is to look for necessary algebraic conditions replacing the GCR equations, cf. the book by J. A. Schouten [22]. The purpose of these methods is to determine the class by means of algebraic manipulations with the metric, the curvature tensor and, if necessary, its covariant derivatives. However, no systematic treatment of the GCR equations have been carried out, except in the cases of metrics of embedding class one or two. Just in order to illustrate the kind of results one has e in the next section we will consider class one embeddings.

2.9. *Class one embeddings*

A space $V_n(t,s)$ is of embedding class one if and only if there exists a symmetric tensor $\Omega_{\mu\nu}$ satisfying the Gauss

$$R_{\lambda\mu\nu\rho} = \eta(\Omega_{\lambda\nu}\Omega_{\mu\rho} - \Omega_{\lambda\rho}\Omega_{\mu\nu}), \tag{2.36}$$

and the Codazzi

$$\Omega_{\mu\nu;\lambda} - \Omega_{\mu\lambda;\nu} = 0, \tag{2.37}$$

equations, where $\eta = \pm 1$. For the Ricci tensor one has

$$R_{\mu\nu} = \eta(\Omega\Omega_{\mu\nu} - \Omega_\mu^\lambda\Omega_{\lambda\nu}), \tag{2.38}$$

where $\Omega = \Omega_\mu^\mu = \text{tr}(\Omega_{\mu\nu})$.

For positive definite metrics T. Y. Thomas [20] and N. A. Rosenson [23,24,25] gave an algebraic criterion to determine whether or not the embedding class is one. While the work by Thomas is non-covariant, Rosenson derived necessary and sufficient conditions in the form of tensor equations. This criterion involves long calculations, namely, the evaluation of a big number of determinants. This criterion is extended with minor changes to indefinite metrics. A modern and generalised version of Thomas' result was given by B. O'Neill [26]. Thus, there always exists a definite, but long, criterion to check if the embedding class is one. However, there exist some examples where one can apply a necessary and sufficient condition to check if the embedding class is one. The first result on this line is due to E. Kasner [27] and involves a necessary condition.

Theorem. There are no vacuum solutions of embedding class one.

All the components of the Ricci tensor cannot be zero at the same time, and if Ric = 0, then the embedding class is different from one. For general relativity, $n = 4$, there are no non-flat vacuum space-times having an embedding in five dimensions. The embedding of a non-vacuum metric in five dimensions is minimal; the embedding of a non-flat vacuum metric in six dimensions is minimal. The previous theorem was also derived, independently and contemporarily, by J. A. Schouten and D. J. Struik [28]. However, the proofs given by Kasner and by Schouten and Struik depend on the diagonalisation of a symmetric tensor through a local rotation, an invalid procedure for indefinite metrics. The correct proof of this theorem for normal hyperbolic manifolds was given by P. Szekeres [29]. Furthermore, the Szekeres' proof fails if the signature of the metric has more than one plus sign, i.e., for non-hyperbolic manifolds.

The next result is due to L. P. Eisenhart [12].

Theorem. Any $V_n(t, s)$ of constant scalar curvature has embedding class one.

The following result, concerning the relation between the Gauss and the Codazzi equations, is due to T. Y. Thomas [20].

Theorem. If $V_n(t, s)$ is embedded in E_{n+1} and if rank $(\Omega_{\mu\nu}) \geq 4$, then the Codazzi equations follow from the Gauss equations.

The proof makes use of the Gauss equations and of the Bianchi identity [18]. In General Relativity, $n = 4$, the theorem is applicable only if the equality holds, rank $(\Omega_{\mu\nu}) = 4$, i.e., det $(\Omega_{\mu\nu}) \neq 0$. Therefore, for General Relativity one can establish the following result.

Corollary. If there exist a non-singular symmetric tensor $\Omega_{\mu\nu}$ satisfying the Gauss equations, then the space-time is of embedding class one.

Due to the algebraic simplicity of the Gauss equations for class one embeddings, all possible Gauss tensors corresponding to a given Ricci tensor can be determined starting with an adequate tetrad representation of $R_{\mu\nu}$ and $\Omega_{\mu\nu}$. This method is due to P. Szekeres [29] and gives also a criterion to know if a space is of embedding class one, and also provides the correct proof of the Thomas theorem for indefinite metrics. It goes more or less as follows. If a space is of embedding class one then Eq. (2.38) implies $[R, \Omega] = 0$. Then, $R_{\mu\nu}$ and $\Omega_{\mu\nu}$ admit the same set of eigenvectors. The set of simultaneous eigenvectors of $R_{\mu\nu}$ and $\Omega_{\mu\nu}$ provides a tetrad for the space-time. The generalisation to the case in which one or more of the eigenvectors are null is straightforward. The previous criterion can also be used to determine the class one solutions of Einstein field equations when the energy-momentum tensor is given. For perfect fluid and Maxwell like energy-momentum tensors confer the references [30,31]. These results include the Eisenhart theorem when the energy-momentum tensor is put equal to zero. More details on class one embeddings can be found in the references [30,31,32,33,34,35,36].

2.10. *Minimal surfaces*

Some of the fundamental definitions concerning minimal surfaces can be found in the references [37,38,39]. Minimal n-dimensional surfaces V_n are submanifolds of a higher dimensional flat space E_N, $N \geq n$, with global Euclidean topology; they are described in terms of an action proportional to its n-dimensional volume. This is given by

$$A = \int \sqrt{|g|}\, d^n x\,. \tag{2.39}$$

V_n is a minimal surface if A is an extremum for variations with respect to the coordinates X^A. If A is thought of as an action and $\sqrt{|g|}$ as a Lagrangian density the corresponding field equations are

$$Y^A = g^{\mu\nu} Y^A_{\mu\nu} = 0\,. \tag{2.40}$$

Usually one deals with global Cartesian coordinates in E_N, such that the second term in Eq. (2.2) does not appear and the field equations reduce to $\Box X^A = 0$; we call therefore Eq. (2.40) the covariant wave equation.

The field equations can be rewritten as

$$Y^A = g^{\mu\nu} Y^A_{\mu\nu} = g^{\mu\nu} \Omega^a_{\mu\nu} N^A_a = 0\,, \tag{2.41}$$

similar to the geodesic equation. Since the N_{aA} are linearly independent vectors what must be zero is

$$\Omega^a = g^{\mu\nu} \Omega^a_{\mu\nu} = 0\,. \tag{2.42}$$

Therefore not all the field equations are independent and a fixing of the gauge is necessary.

2.11. *Rigid surfaces*

A rigid surface is described by the action

$$S = \int R \sqrt{|g|}\, d^n x\,. \tag{2.43}$$

Such an action leads to a natural metrical elasticity of space, i.e., to the appearance of generalised forces which act against the curving of space; the concept of rigid surface seems to have been first introduced by A. D. Sakharov [40]. They find applications in the description of the fine structure of strings [41,42].

The field equations are

$$G^{\mu\nu} Y^A_{\mu\nu} = G^{\mu\nu} \Omega^a_{\mu\nu} N^A_a = 0\,. \tag{2.44}$$

Once again what must be zero is

$$G^{\mu\nu}\Omega^{a}_{\mu\nu}=0\,. \tag{2.45}$$

Not all the field equations are independent and a fixing of the gauge is necessary.

3. A Parametrisation of the Embedding

Since the actions describing minimal and rigid surfaces are geometric, they are invariant under space-time diffeomorphisms on V_n, i.e. field redifinitions, and worldsheet diffeomorphisms corresponding to local (gauge) symmetry transformations. The later is often called reparametrisation invariance with respect to the n coordinates x^μ; n first-class constraints appear due to it. Confer the books on constrained systems [43,44,45,46] for further details. In order to deal properly with this system a choice of the gauge is forced. If one views this formulation of embedded surfaces as a field theory on the worldsheet, then space-time diffeomorphisms correspond to field redefinitions, and worldsheet diffeomorphisms correspond to local (gauge) symmetry transformations.

The coordinates X^A play the role of the fields while the coordinates x^μ are the coordinates in the base space. To fix the gauge means to give to the fields an explicit dependence on the coordinates x^μ. We introduce a parametrisation of the embedding consisting in choosing n of the coordinates on E_N as those on V_n, from a field theoretical point of view the rest of the coordinates play the role of dynamical fields. Early results on this line were partially obtained by J. Krause [47,48,49]. In this gauge the geometrical properties of the model can always be referred to a background geometry. Of course, the explicit form of the gauge will depend on the model under consideration. When the previous parametrisation is applied to minimal and rigid surfaces one obtain particularly meaningful gauge fixings. The dynamical fields are the diffeomorphism measuring the deformation of V_n with respect to a background geometry M_n. Therefore, any geometrical action, or covariant field theory, can be considered as describing a field theory for normal deformations of V_n with respect to a background geometry M_n.

Let us write the coordinates on E_N as $X^A=(y^\mu,\phi^a)$, $a=1,\dots,N-n$. V_n is defined by giving X^A as functions of n parameters x^μ, the coordinates on V_n. Let us choose $y^\mu=x^\mu$, then for X^A_μ we obtain

$$X^A_\mu=(\delta^{\nu\,\leftarrow\,\text{this is not a tensor index}}_\mu\,,\,\phi^a_\mu)\,, \tag{3.1}$$

with $\phi^a_\mu = \partial_\mu \phi^a$. The metric tensor of E_N can be written as

$$G_{AB} = \begin{vmatrix} \overset{\circ}{g}_{\mu\nu} & H_{a\mu} \\ H_{b\nu} & H_{ab} \end{vmatrix} , \tag{3.2}$$

where $\overset{\circ}{g}_{\mu\nu}$ is the metric on the manifold M_n obtained by putting $\phi = \text{const}$. Furthermore V_n and M_n are diffeomorphic and the diffeomorphism between them is given by the functions ϕ. The metric induced on V_n is given by, cf. Eq. (2.2),

$$g_{H\mu\nu} = \overset{\circ}{g}_{\mu\nu} + h_{\mu\nu} , \tag{3.3}$$

$$h_{\mu\nu} = H_{ab}\phi^a_\mu \phi^b_\nu + H_{\mu a}\phi^a_\nu + H_{\nu a}\phi^a_\mu . \tag{3.4}$$

The host space E_N can always be written, at least locally, as the direct product of two manifolds, i.e.

$$E_N(T,S) = M_n(t,s) \otimes B_{N-n}(T-t, S-s) . \tag{3.5}$$

For a given V_n we choose M_n, e.g., as the most symmetric of the spaces diffeomorphic to it. In this case the metric induced on V_n takes the simple form

$$g_{H\mu\nu} = \overset{\circ}{g}_{\mu\nu} + \eta_{ab}\phi^a_\mu \phi^b_\nu . \tag{3.6}$$

Furthermore

$$Y^\lambda_{\mu\nu} = -\Gamma^\lambda_{\mu\nu} + \overset{\circ}{\Gamma}{}^\lambda_{\mu\nu} , \tag{3.7a}$$

$$Y^a_{\mu\nu} = \phi^a_{\mu;\nu} . \tag{3.7b}$$

It must be observed that $Y^\lambda_{\mu\nu}$ being the difference of two Christoffel symbols is a tensor.

We choose furthermore M_n as a flat Minkowski space. Some words of caution are in order here. In fact one cannot choose M_n as a flat Minkowski space for a space-time not diffeomorphic to it. However, all the argumentation can be generalised to other geometries, e.g., maximally symmetric space-times.

For the Riemann tensor we obtain

$$R_{\lambda\mu\nu\rho} = M_{ab}(\phi^a_{\lambda\nu}\phi^b_{\mu\rho} - \phi^a_{\mu\nu}\phi^b_{\lambda\rho}) , \tag{3.8}$$

$$M_{ab} = \eta_{ab} - \phi_{a\lambda} g^{\lambda\rho} \phi_{b\rho} , \tag{3.9}$$

$$\phi_{a\lambda} = \eta_{ab}\phi^b_\lambda , \tag{3.10}$$

where for simplicity we have put $\overset{\circ}{g}_{\mu\nu} = \eta_{\mu\nu}$.

From Eq. (3.3) we see that a simple geometrical meaning can be given to the previous parametrisation, in fact the deviation of V_n from the geometry given by M_n is due only to the second term which in turn depends only on the functions ϕ. The functions ϕ measure the departure from the background geometry M_n of the space V_n. Therefore our model can be considered as describing a field theory for normal deformations of V_n with respect to the background geometry M_n.

3.1. *Minimal surfaces*

Due to the parametrisation invariance of the minimal surface action a gauge fixing is necessary. We fix the gauge by using the parametrisation of the embedding introduced previously, i.e., by choosing n of the coordinates on E_N as those on V_n such that the rest of the coordinates play the role of dynamical fields. When using global Cartesian coordinates on E_N the previous gauge reduces to the harmonic one such that ours is a covariant generalisation of it.

In the previous gauge we have

$$\Omega^\mu = 0\,, \tag{3.11}$$

as the gauge fixing condition, and

$$\Omega^a = 0\,, \tag{3.12}$$

as true dynamical field equations. For a V_n diffeomorphic to a Minkowski space Eq. (3.11) in Cartesian coordinates reduce to the harmonic gauge $\Box x^\mu = 0$. It must be observed that Eq. (3.11) is a covariant gauge, i.e., it takes the same form in all coordinate systems contrary to what happens for the usual harmonic gauge which is not covariant due to the fact that the second term in Eq. (2.20) is lacking. Ours is therefore a covariant generalisation of the harmonic gauge. The harmonic gauge was first introduced by V. Fock [50]. Confer reference [51] for further details.

The field equations (3.12) follow directly from the gauged action

$$S_H = \int \sqrt{|g_H|}\, d^n x\,, \tag{3.13}$$

with $g_H = \det g_{H\mu\nu}$.

3.2. *Rigid surfaces*

We fix the gauge for rigid surfaces by using the parametrisation of the embedding introduced previously, i.e., by choosing n of the coordinates on

E_N as those on V_n such that the rest of the coordinates play the role of the dynamical fields. The physical meaning of this gauge is that the dynamical fields measure the departure from flatness of space-time. In the previous gauge the field equations reduce to

$$G^{\mu\nu}Y^{\lambda}_{\mu\nu} = G^{\mu\nu}(-\Gamma^{\lambda}_{\mu\nu} + \overset{\circ}{\Gamma}{}^{\lambda}_{\mu\nu}) = 0\,, \tag{3.14a}$$

$$G^{\mu\nu}Y^{a}_{\mu\nu} = G^{\mu\nu}\phi^{a}_{\mu;\nu} = 0\,. \tag{3.14b}$$

Equations (3.14a) are the (covariant) gauge fixing condition, while Eqs. (3.14b) are the true dynamical equations. They follow from the action

$$S_H = \int R_H \sqrt{-g_H}\, d^n x\,, \tag{3.15}$$

i.e., from the gauged action. The corresponding Lagrangian is

$$\mathcal{L}_H = \sqrt{-g}\, G^{\mu\nu\lambda\rho} M_{ab} \phi^{a}_{\mu\nu} \phi^{b}_{\lambda\rho}\,. \tag{3.16}$$

In reference [52] it was not emphasized enough that the procedures of taking the variation of the action in order to obtain the field equations, denoted by δ in the graph, and the fixing of the gauge, denoted by γ, commute, i.e., we can first fix the gauge and later on look for the field equations. This is illustrated by the following graph

$$\begin{array}{ccc}
\mathcal{L}_H & \xrightarrow{\ \delta\ } & \delta\mathcal{L}_H = 0 \\
\gamma\downarrow & & \downarrow\gamma \\
\mathcal{L}_G & \xrightarrow[\delta]{} & \delta(\mathcal{L}_G) = (\delta\mathcal{L}_H)_G = 0
\end{array}$$

The commutation of this diagram is valid also for minimal surfaces.

3.3. *Class one embeddings*

We include these results here in order to illustrate the extremely simple form that some formulae take with the previous parametrisation.

The metric tensor for a class one embedding is given by

$$g_{\mu\nu} = \overset{\circ}{g}_{\mu\nu} + \eta\phi_\mu\phi_\nu\,, \tag{3.18}$$

with $\eta = \pm 1$. For the contravariant metric tensor one obtains

$$g^{\mu\nu} = \overset{\circ}{g}{}^{\mu\nu} - \eta\gamma^{-1}\phi^\mu\phi^\nu\,, \tag{3.19}$$

where

$$\gamma = 1 + \eta\phi^\mu\phi_\mu = (1 - \eta\phi_\mu\nabla^\mu\phi)^{-1}\,. \tag{3.20}$$

$$\phi^\mu = \overset{\circ}{g}{}^{\mu\nu}\phi_\nu = \overset{\circ}{\nabla}{}^\mu\phi = \gamma g^{\mu\nu}\phi_\nu = \gamma\nabla^\mu\phi\,. \tag{3.21}$$

Thus

$$\phi_\mu\phi^\mu = \gamma\phi_\mu\nabla^\mu\phi\,. \tag{3.22}$$

For the Christoffel symbols one has

$$\Gamma^\lambda_{\mu\nu} = \overset{\circ}{\Gamma}{}^\lambda_{\mu\nu} + \eta\gamma^{-1}\phi^\lambda\overset{\circ}{\nabla}_\mu\phi_\nu\,. \tag{3.23}$$

Then, it follows

$$\overset{\circ}{\nabla}_\mu\phi_\nu = \gamma\nabla_\mu\phi_\nu\,, \tag{3.24}$$

such that

$$\Gamma^\lambda_{\mu\nu} = \overset{\circ}{\Gamma}{}^\lambda_{\mu\nu} + \eta\gamma\nabla^\lambda\phi\nabla_\mu\phi_\nu\,. \tag{3.25}$$

The Gauss tensor is given by

$$\Omega_{\mu\nu} = \eta\gamma^{-1/2}\overset{\circ}{\nabla}_\mu\phi_\nu = \eta\gamma^{1/2}\nabla_\mu\phi_\nu\,. \tag{3.26}$$

The Riemann tensor is

$$\begin{aligned} R_{\lambda\mu\nu\rho} &= \eta\gamma^{-1}(\overset{\circ}{\nabla}_\lambda\phi_\nu\overset{\circ}{\nabla}_\mu\phi_\rho - \overset{\circ}{\nabla}_\lambda\phi_\rho\overset{\circ}{\nabla}_\mu\phi_\nu) \\ &= \eta\gamma(\nabla_\lambda\phi_\nu\nabla_\mu\phi_\rho - \nabla_\lambda\phi_\rho\nabla_\mu\phi_\nu)\,. \end{aligned} \tag{3.27}$$

4. Gravitation

The macroscopic geometry of space-time is correctly given by Riemannian geometry. Due to this fact we will keep Riemannian geometry. The geometrical object from which all the information concerning the geometry of space-time can be derived is the metric. Therefore, most of our argumentation will be referred to the metric. The gravitational interaction is related with the geometry of space-time. Gravitation is the field theory of the geometry of space-time. The gravitational field is, by definition, responsible for the departure from flatness of space-time. The way in which the gravitational interaction acts on the space-time structure is, however, not completely understood, and, in fact, the formulation of a consistent field theory for the gravitational interaction remains as one of the main challenges of twentieth century theoretical physics.

From a predictive point of view the best theory we have in order to describe the macroscopic effects of the gravitational field is General Relativity.

General Relativity considers the metric as the fundamental field from which the gravitational interaction can be derived. This choice of the metric as the fundamental field to describe the gravitational interaction is historically prejudiced by the intrinsic approach to Riemannian geometry, which was fashionable during the years of the elaboration of General Relativity. In this approach the metric contains all the geometric information of space-time.

What differentiate General Relativity from other field theories is the fact that inertia and dynamics are inextricably joined. We believe that this is the fact which does not allow for a consistent formulation of a field theory for the gravitational field and consequently for its quantisation.

4.1. *General relativity*

General Relativity takes as its starting point the Hilbert action

$$S = \int R\sqrt{-g}\, d^n x \,, \tag{4.1}$$

which, regarded as a functional of the metric $g_{\mu\nu}$, under variation gives vacuum Einstein field equations

$$G^{\mu\nu} = 0 \,, \tag{4.2}$$

with $G^{\mu\nu}$ the Einstein tensor.

4.2. *Conservation laws*

As was mentioned above, General Relativity is the best theory we have in order to describe the macroscopic effects of the gravitational field. The same is not true from a field theoretical point of view. For example, it is not possible to define conservation laws in a consistent way. One finds, for example, that it is impossible to define unambiguously conservation laws, unless of having recourse to some *ad hoc* definitions, e.g., the ADM mass, the Bondi mass, etc. All existing definitions are not exempted from criticism. For a report on the status of the art concerning conservation laws in general relativity confer the very lucid introduction to the paper by A. A. Logunov and Yu. M. Loskutov [53].

4.3. *The role of the metric*

In the theory of special relativity the role of the metric is to guarantee the covariance of the physical laws, i.e., to provide the inertial frames to which the dynamics is referred. In General Relativity, as in any other theory of the gravitational field (in general when gravity is involved in the game), the metric represents not only the space-time structure, the inertial frames, but also provides the fields from which the gravitational interaction is derived. The role of the metric is therefore double: a inertial one, which guarantees the covariance of the physical laws, and a dynamical one giving account of the dynamics of the gravitational field.

From a field theoretical point of view the previous situation translates as follows: the metric $g_{\mu\nu}$ characterising V_n, has been raised to the role of a dynamical field. This fact, which is considered as one of the main conceptual achievements of General Relativity, turns against the theory. This forced democracy between the inertial and the gravitational role of the metric avoids the application of all the mathematical apparatus of field theory.

One concludes that the metric is therefore not a dynamical object at all. At this point it is not only advisable but conceptually strictly necessary to have an invariant decomposition of the metric in an inertial part, guaranteeing the covariance of the physical laws, plus a genuinly gravitational one giving account of the gravitational interaction. This can be done of course only if we know how the gravitational interaction acts on the geometry of space-time.

4.4. *Background geometries*

Here we show that General Relativity naturally contains the concept of a background geometry. We start writing Einstein field equations with a cosmological term

$$R_{\mu\nu} - (1/2)(R - \Lambda/2)g_{\mu\nu} = T_{\mu\nu} . \tag{4.3}$$

The Riemann tensor is written as

$$\mathrm{Rie} = \mathrm{Weyl} - (R/6)(g.g - g.g) + (1/2)(\mathrm{Ric}.g + \mathrm{Ric}.g - \mathrm{Ric}.g - \mathrm{Ric}.g) . \tag{4.4}$$

The background geometry is obtained by switching off all those which have something to do with gravitation. Since the Weyl tensor is not determined by the interaction of gravitation with matter, it represents the pure gravitational contribution. The first step is then to put Weyl = 0. The background configuration of space-time is obtained in the absence of matter, therefore, for the Ricci tensor one obtains

$$\mathrm{Ric} = (\Lambda/4)g , \tag{4.5}$$

or equivalently, for the Riemann tensor

$$\mathrm{Rie} = (\Lambda/12)(g.g - g.g) . \tag{4.6}$$

Therefore, a maximally symmetric space-time defines the background geometry. In this way the maximally symmetric space-time plays the role of the ground state of the geometry.

Minkowski space cannot be the background (state) geometry for space-times not diffeomorphic to it. In fact, this gives a change of the signature or, equivalently, of the topology. The solution is, of course, to enlarge the family of spaces admissible for the ground state of the geometry.

4.5. *Quantum gravity*

All attempts for the quantisation of General Relativity have so far met with considerable difficulties. The problems are of various types, from the conceptual to the technical ones. General Relativity is a classical theory, and so, it cannot be complete: fine for large-distance, macro phenomena, but incapable of giving the right answers for such microphysics problems as seen especially in cosmology and the very early universe. All attempts at formulating a Quantum Gravity have failed due to the highly non-linear structure of General Relativity.

So, in spite of the tremendous amount of work done on the subject there is still no generally accepted theory of Quantum Gravity. Forced by this undesirable situation one becomes inclined to the view that one has missed some important points in our conception of the gravitational field.

The impossibility, up to now, of obtaining a consistent quantisation of general relativity can be traced back to the fact of considering the metric as the fundamental field to be quantised. In this case one is, at the same time, quantising dynamical and kinematical quantities. Quantisation of kinematical quantities is a nonsense. That is due to an improper identification of the base space and the space of the fields. This separation will be provided by our embedding approach to gravitation.

5. String Theory

Since the advent of strings [54,55] minimal surfaces are playing a fundamental role in string theory since they are described by a minimal surface action.

5.1. *The Nambu–Goto action*

In analogy with the relativistic action for point particle, Y. Nambu [56] and T. Goto [57] independently proposed that the action for a free relativistic string be the area of its worldsheet. Classical solutions are then minimal surfaces in space-time.

The form of the action most commonly used was popularised by A. M. Polyakov [58,59] but was originally written down by L. Brink, P. Di Vecchia and P. Howe [60] and by S. Deser and B. Zumino [61]. It gives the same classical equations of motion as the Nambu–Goto action. It must be however observed that two classically equivalent actions need not be necessarily equivalent at the quantum mechanical level. This is due to the fact that two classically equivalent actions can have an extremum along the same classical trajectory but they, in general, will differ outside it. Quantum mechanics considers also trajectories near to the classical ones, therefore both quantum mechanics are in principle different.

The Nambu–Goto action for string theory is given by

$$S_{NG} = A = \int \sqrt{|g|}\, d^n x \,. \tag{5.1}$$

We saw that not all the field equations (4.41) for a minimal surface are linearly independent. In order to deal with this underdetermination we must fix the gauge.

5.2. *Gauge fixing*

Due to the parametrisation invariance of the Nambu–Goto action a gauge fixing is necessary. The simplest choice of the gauge is the conformal one [58,62]. Most recently another gauge has been advocated for the quantisation of string theories [63,64,65,66,67]: the harmonic one. According to our previous results the harmonic gauge is just a particular case of our parametrisation of the embedding.

5.3. *Consistent dimensions for the string*

One of the predictions of string theories with the greatest impact is the fact that the theory will be consistent only if the ambient space E_N has dimension $N = 26$. This gives rise to some problems related to the global isometric embedding theorem. In fact, the partition function for the string is constructed by summing over all the possible two-dimensional surfaces. According to the global embedding theorem by R. E. Greene [11] in order to globally describe any two-dimensional surface one needs 28 dimensions, in contrast with the 26 predicted by string theory. Therefore, one is not summing up over all possible two-dimensional surfaces unless the global isometric embedding theorem gives a lower dimensionality requirement. This was the motivation to look for an improvement of the global embedding theorem in Sec. 2.

6. Extrinsic Gravity

The use of embeddings in physics, more particularly in the physics of gravitation, is as old as General Relativity. In fact, just after Einstein presented his final version of General Relativity [68], E. Kasner [27,69] and J. A. Schouten and D. J. Struik [28] used embeddings in their studies of General Relativity. Besides considering the embeddings as a mathematical tool there exists the possibility of giving them a physical status in the physics of gravitation.

6.1. *The embedding approach to general relativity*

For the action (2.43) considered as a functional of the embedding functions the field equations assume the explicit form

$$\partial_\mu(\sqrt{-g}\,G^{\mu\nu}\phi^a_\nu) = \sqrt{-g}\,G^{\mu\nu}Y^A_{\mu\nu} = 0\,. \tag{6.1}$$

Of course, they are a valid generalisation of the Einstein field equations.

The first attempt to give a physical interpretation to these equations in General Relativity is due to T. Regge and C. Teitelboim [70]. As observed by them, Eqs. (5.1) do not imply $G^{\mu\nu} = 0$ due to the identities (2.19) which show that in the generic case only six among the ten equations (5.1) are linearly indepedent. Regge and Teitelboim argue that this difficulty is not unsurmountable and could be circumvented by imposing, in an *ad hoc* fashion, the additional constraint $G_{\perp\alpha} = 0$, where the symbol $\perp$ refers to the unit normal to V_3 lying on V_4. Their argumentation is however wrong. From Eqs. (5.1) only six of them are really linearly independent. Einstein field equations are ten linearly independent equations; it is only after a 1+3 splitting that only six equations are left, the other four becoming constraints.

As a way out of this difficulty S. Deser *et al.* [71] proposed to find another action functional $I[X]$ with field equations equivalent to Einstein ones. However, it is very difficult to establish the equivalence without introducing some arbitrariness in the game.

6.2. *The gravitational field*

As mentioned previously, the gravitational interaction can be identified as responsible for the departure from flatness of space-time. We must now find a way of putting this definition in mathematical terms. We identify the fields responsible for the gravitational interaction as the functions measuring the departure from flatness of space-time. With the formalism developed previously the concept of departure from flatness can be given a mathematical meaning. For this is enough to consider a plane tangent to the given surface. The deviation from flatness is then measured by the length of the normal from the surface to the plane. From our parametrisation of the embedding one realises that these functions are the embedding functions, the diffeomorphism between the flat Minkowski space $M_4(1,3)$ and the curve space-time $V_4(1,3)$. Therefore, one can think of gravitation as a field theory for the embedding functions. Therefore, we must consider the space-time $V_4(1,3)$ as embedded in a ten-dimensional flat space $E_{10}(T,S), T \geq 1, S \geq 3$, with global Euclidean topology.

The functions ϕ are considered as the fundamental fields responsible for the dynamics of the gravitational interaction. Now the components of the metric are no longer the basic variables but rather derived objects constructed from the more basic fields ϕ. According to this the metric of the space-time is not a fundamental but more an effective field produced by the gravitational interaction.

The next step is to look for a Lagrangian such that under variation with respect to the fields ϕ gives field equations equivalent to the Einstein ones at least for weak fields where we know they work well enough. We have found that a rigid surface like Lagrangian satisfies these criteria. The corresponding field theory is developed in the next section. The first quality control for the resulting field equations would be for a static spherically symmetric field. This is due to the fact that two of the three classical tests of general relativity, which we know to be correct at first order, are based on the Schwarzschild solution.

6.3. *Extrinsic gravity*

The idea of regarding our four-dimensional space-time as a worldsheet or membrane in some higher-dimensional space-time is not a new one [72,73,74]. We develop a model for the gravitational interaction in which the gravitational field is identified with the functions describing the embedding of space-time in a ten-dimensional flat space rather than with the metric. The choice of ten dimensions is dictated by the local isometric embedding theorem. It must be emphasized that no physical meaning is given to the extra dimensions, the embedding is used just as a mathematical tool in order to write the metric in a particular way. The Lagrangian for the model is the Hilbert one but is considered this time as a functional of the embedding functions rather than of the metric. The field equations are the Einstein equations contracted with the ten Gauss tensors of the embedding. For a static spherically symmetric field the correct asymptotic behavior, the Schwarzschild solution, is obtained. Only six among the ten field equations are linearly independent, therefore a fixing of the gauge is necessary. We fix the gauge with a particular parametrisation of the embedding in which the dynamical fields measure the departure from flatness of space-time. The ungauged energy-momentum tensor is identically zero; after fixing the gauge it acquires non-null values which are physically relevant for the formulation of consistent conservation laws. In fact, it turns out to be proportional to the Einstein tensor. For more details confer the references [52,75,76].

Our model is based on the action given by Eq. (2.43) but considered this time as a functional of the embedding functions. The corresponding field equations are

$$G^{\mu\nu} Y^A_{\mu\nu} = 0 \,. \tag{6.2}$$

Of course, they are a valid generalisation of the Einstein field equations. Contrarily to what was done previously we will consider Eqs. (6.2) as the true field equations for the gravitational field. We must accept the fact that equations (6.2) are the correct field equations following from Eq. (2.43) when considered as a functional of the embedding functions.

The fact that Eqs. (6.2) are not equivalent to the Einstein equations is not strange. This fact follows from the particular way the chain rule adopts for a field transformation in which the old fields (the metric) depend on the new fields (the embedding functions), i.e., only through its derivatives.

We are dealing with a second-order Lagrangian, therefore generalised momenta must be defined in the following way [77]

$$\Pi^\mu_A(\mathcal{L}) = \frac{\partial \mathcal{L}}{\partial X^A_\mu} - d_\nu \left(\frac{\partial \mathcal{L}}{\partial X^A_{\mu\nu}} \right) , \tag{6.3a}$$

$$N^{\mu\nu}_A(\mathcal{L}) = \frac{\partial \mathcal{L}}{\partial X^A_{\mu\nu}} \,. \tag{6.3b}$$

The energy-momentum tensor generalises to

$$\mathcal{H}^\mu_\nu(\mathcal{L}) = X^A_\nu \Pi^\mu_A(\mathcal{L}) + X^A_{\nu\lambda} N^{\lambda\mu}_A(\mathcal{L}) - \delta^\mu_\nu \mathcal{L} \,. \tag{6.4}$$

For our Lagrangian we obtain

$$\Pi^\mu_A(\mathcal{L}) = -2\sqrt{-g}\, G^{\mu\lambda} X_{A\lambda} \,, \tag{6.5a}$$

$$N^{\mu\nu}_A(\mathcal{L}) = 2\sqrt{-g}\, M_{AB} G^{\mu\nu\gamma\delta} X^B_{\gamma\delta} \,. \tag{6.5b}$$

For the energy-momentum tensor we obtain

$$\mathcal{H}^\mu_\nu(\mathcal{L}) \equiv 0 \,. \tag{6.6}$$

Condition (6.6) leads unavoidably to the existence of constraints, which also follow from the fact that only $n(n-1)/2$ of the field equations are linearly independent. In order to deal properly with this system we must fix the gauge; this will be done after looking for the static spherically symmetric solution.

Equation (6.6) is the most important property of our model. In fact only when the energy-momentum tensor is identically zero can one obtain physically relevant conservation laws. After the fixing of the gauge the energy-momentum tensor acquires a non-null value. Only then one can construct conservation laws based on the energy-momentum tensor. The crucial point is that the starting energy-momentum tensor must be identically zero. This point was emphasized in the fifties as a must for any physically sensible theory [78,79,80] but later on it fell to disuse.

We have seen that from the second-order Lagrangian $\mathcal{L}$ we obtain the field equations (6.2) which are not fourth-order equations, as expected, but merely second-order equations. This is in agreement with the result obtained by H. Rund [81] who extensively studied variational problems on subspaces of a Riemannian manifold. The occurrence of a second-order field equation from a second-order Lagrangian indicates that the Lagrangian entails some kind of degeneracy [81].

6.4. *Static spherically symmetric solution*

In order to establish a connection with gravitation, and with the predictions of General Relativity, we must check if the predictions of our model are in reasonable agreement with the physical world. S. Deser *et al.* [71] claimed that for the static spherically symmetric field no manifest conflict with observation is found such as a solution giving the wrong precession of the perihelion but they did not give any explicit calculation of this. Next we check that the static spherically symmetric solution to Eqs. (6.5) asymptotically coincides with the Schwarzschild solution.

Static spherically symmetric fields are of embedding class two, confer the references [28,69], i.e., we need only two extra dimensions for the embedding. The complete explicit embedding of static spherically symmetric space-times was given by Y. K. Gupta and P. Goel [17]. The convenient line element to start with is

$$ds^2 = A^2(r)d\tau^2 - B^2(r)dr^2 - r^2 d\Omega^2 , \qquad (6.7)$$

$$d\Omega^2 = d\theta^2 + \sin^2\theta d\varphi^2 . \qquad (6.8)$$

We introduce a new time coordinate defined as $t = A\tau$, such that the line element takes the form

$$ds^2 = dt^2 - 2L'tdtdr + (L'^2t^2 - B^2)dr^2 - r^2d\Omega^2\,, \tag{6.9}$$

with $L = \ln A$. The embedding is then given by

$$ds^2 = dt^2 - (dx^2 + dy^2 + dz^2) + d\phi_+^2 - d\phi_-^2\,, \tag{6.10}$$

with

$$\phi_\pm = \psi_1 \pm \psi_2\,, \tag{6.11}$$

$$\psi_1 = \frac{\lambda}{2}\left[\frac{t^2}{A} + \int \frac{B^2 - 1}{A'}dr\right], \tag{6.12a}$$

$$\psi_2 = -\frac{1}{2\lambda}A\,. \tag{6.12b}$$

The field equations reduce to

$$(-1 + B^2 - 2rL' - 2rB^2L')(L'' + L'^2) + 2(BB' - B^2L' + rBB'L')L' = 0\,, \tag{6.13a}$$

$$(1 - B^2)^2(1 + 2rL')(L'' + L'^2) + 8rBB'L'^2 + 2(1 - B^2)(2BB' - 2B^2L' + rBB'L') = 0\,. \tag{6.13b}$$

The asymptotically flat ansatz

$$A^2 \approx 1 + \frac{\alpha}{r} + \cdots\,, \tag{6.14a}$$

$$B^2 \approx 1 + \frac{\beta}{r} + \cdots\,, \tag{6.14b}$$

gives as the only non-trivial solution

$$\alpha + \beta = 0\,. \tag{6.15}$$

In order to recover the Newtonian limit we must have $\alpha = 2m$, i.e., at first-order our solution coincides with the Schwarzschild one and therefore gives the same physical predictions.

6.5. *The normal gauge*

We fix the gauge by using the parametrisation of the embedding introduced previously. The physical meaning of this gauge is that the dynamical fields measure the departure from flatness of space-time. In this gauge furthermore the geometrical properties of the model are always referred to a background geometry. For our model we choose furthermore M_n as a flat Minkowski space. Some words of caution are in order here. In fact one cannot choose M_n as a flat Minkowski space for a space-time not diffeomorphic to it. However, all the argumentation can be generalised to other geometries, e.g., maximally symmetric space-times.

The metric reduces to

$$g_{\mu\nu} = \overset{\circ}{g}_{\mu\nu} + h_{\mu\nu} \,. \tag{6.16}$$

The metric is therefore decomposed in the sum of a flat metric, the inertial part, plus a rest, a dynamical part, measuring the deviation from it, representing the gravitational contribution. Furthermore, with the previous interpretation of the gravitational field an invariant decomposition of the metric in inertial plus gravitational parts is naturally obtained.

In the previous gauge the Lagrangian reduces to

$$\mathcal{L}_H = \sqrt{-g}\, G^{\mu\nu\lambda\rho} M_{ab} \phi^a_{\mu\nu} \phi^b_{\lambda\rho} \,. \tag{6.17}$$

The corresponding generalised momenta are

$$\Pi^\mu_a(\mathcal{L}_H) = -2\sqrt{-g}\, G^{\mu\lambda} \phi_{a\lambda} \,, \tag{6.18a}$$

$$N^{\mu\nu}_a(\mathcal{L}_H) = 2\sqrt{-g}\, M_{ab} G^{\mu\nu\gamma\delta} \phi^b_{\gamma\delta} \,. \tag{6.18b}$$

We obtain the following energy-momentum tensor

$$\mathcal{H}^\mu_\nu(\mathcal{L}_H) = 2\sqrt{-g}\, \overset{\circ}{g}_{\nu\sigma} G^{\sigma\mu} \,. \tag{6.19}$$

Our model is now a field theory over a Minkowski space. Therefore the natural metric with which to raise and lower indices is the background one, $\overset{\circ}{g}_{\mu\nu}$. For the energy-momentum tensor we obtain

$$\mathcal{H}^{\mu\nu}(\mathcal{L}_H) = \overset{\circ}{g}^{\mu\lambda} \mathcal{H}^\mu_\lambda(\mathcal{L}_H) = 2\sqrt{-g}\, G^{\mu\nu} \,, \tag{6.20}$$

which now is effectively a symmetric tensor density. This is not all the energetics content of the model. In fact, from a canonical point of view, i.e., after a 1+3 splitting, the Lagrangian given by Eq. (6.17) is a constrained one, confer the reference [82]. These canonical constraints are due only to the particular structure of the Lagrangian. Therefore a subsequent fixing of the gauge will be necessary. After this fixing of the gauge the energy-momentum tensor will acquire further contributions. This will be reported somewhere else.

The background flat geometry appears naturally as the inertial part of the metric. This background geometry is absolutely necessary for the description of the gravitational field. Therefore, a flat space-time is the base space over which the field theory describing the dynamics of the gravitational interaction must be formulated. A flat space-time provides the inertial frames to which the dynamics of the gravitational field is referred. With the flat background geometry playing such a fundamental role the dynamics of the gravitational field can be described in terms of a field theory in a flat space-time.

In the field theory we developed for the gravitational interaction the gravitational field is identified with the functions measuring the departure from flatness of space-time. These are the embedding functions ϕ, the diffeomorphism between the curved and the flat, Minkowski, space-times. The metric is no longer the basic field but rather a derived object constructed from the more basic fields ϕ, i.e., an effective field produced by the gravitational field. In this way the background Minkowski space-time appears naturally and is the base space over which the field theory for the gravitational field must be formulated.

Various current problems of General Relativity are not present in this model. First of all the gravitational field acquires an objective reality which enables one to return to the concept of Minkowski space as the fundamental space, since it is only in this space that there are true conservation laws for the energy, the linear and the angular momenta. The need for this property in a sensible theory of gravitation was already emphasized by A. A. Logunov and Yu. M. Loskutov [53]. The energy-momentum tensor is now really a symmetric tensor density, it is proportional to the Einstein tensor.

The quantisation of the gravitational field is in principle possible since it is reduced to the quantisation of a field theory in a Minkowski space. What remains for us to do is not very difficult: that is, the quantisation of a four-dimensional surface embedded in a higher-dimensional space. Furthermore, since the advent of strings in theoretical physics we already have acquired some experience on how to quantise two-dimensional surfaces embedded in higher-dimensional spaces. We must only generalise these quantisation methods for the case of four-dimensional space-times embedded in ten-dimensional flat spaces. However, the quantisation dream will be possible only if we are able to give a canonical structure to the previous theory.

The Hilbert Lagrangian depends on the second-order derivatives of the embedding functions in a non-trivial way; they cannot be removed by subtracting a divergence as, e.g., when going from the Hilbert to the gamma-gamma Lagrangian when the metric is considered as the fundamental field [83]. Even when the canonical formalism for a higher-order Lagrangian can be satisfactorily developed in complete parallel with the first-order case [77,82], the existence of a first-order equivalent Lagrangian would make life much easier. In the canonical formalism this requirement over the dependence on first-order derivatives is restricted only to the time derivatives while space derivatives can appear up to an arbitrary order. We already know of a Lagrangian holding this property: the ADM Lagrangian, where time derivatives appear at first-order while the space-like ones do it at the second one. This property holds however with respect to the metric as the fundamental field. What is extremely relevant of the ADM Lagrangian is that, in spite of the fact that the metric depends on the embedding functions through their first-order derivatives, and therefore one would wait for mixed time and space third-order derivatives, this property still holds with respect to the embedding functions considered as the fundamental fields. In fact the ADM Lagrangian when written in terms of the embedding functions contains only first-order time derivatives and second-order space-like ones.

The ADM Lagrangian presents some interesting properties. The energy-momentum tensor is identically zero. In such a case one would wait that, after fixing the gauge, the formulation of consistent conservation laws would be possible. However, this is not true since the starting Lagrangian is not a scalar density. In fact, after fixing the gauge, the energy-momentum tensor is no more proportional to the Einstein tensor.

In spite of all these good properties of the ADM Lagrangian, we have not been able to provide this model with a canonical structure due to the appearance of some awkard calculations. Firstly, the shift constraints are easily calculated while the lapse constraint relies on an involved matrix calculation, therefore we lack of an explicit expression for the lapse constraint. Secondly, after fixing the gauge the Legendre transformation is invertible, however the velocities cannot be easily solved for due to the appearance of a similar calculation.

Due to the previous facts we have decided to come back to the full, second-order, Hilbert Lagrangian. Before determining the canonical structure of this model it is useful to consider ADM variables.

7. ADM Variables

Even when here we are not going to make an ADM analysis, the ADM variables show to be useful for what follows.

7.1. *ADM variables*

In the ADM formalism [84] the metric $g_{\mu\nu}$ is parametrised in terms of new variables N, N_i, γ_{ij}, given by

$$g_{00} = N^2 - \gamma^{ij} N_i N_j \,, \tag{7.1a}$$
$$g_{0i} = -N_i \,, \tag{7.1b}$$
$$g_{ij} = -\gamma_{ij} \,, \tag{7.1c}$$

with the relations for the inverse tensors

$$g^{00} = 1/N^2 \,, \tag{7.2a}$$
$$g^{0i} = -\gamma^{ij} N_j / N^2 \,, \tag{7.2b}$$
$$g^{ij} = -\gamma^{ij} + \gamma^{ik}\gamma^{jl} N_k N_l / N^2 \,. \tag{7.2c}$$

where γ^{ij} is the reciprocal to γ_{ij}. Our notation differs from the ADM one in the following: four-dimensional quantities are unmarked while three-dimensional quantities are related with the γ metric.

One further useful relation is that the determinant of the metric is given by

$$g = \det(g_{\mu\nu}) = -N^2 \gamma < 0 \,, \tag{7.3}$$

where

$$\gamma = \det(\gamma_{ij}) > 0 \,. \tag{7.4}$$

7.2. *The ADM variables in term of the embedding functions*

In terms of the embedding functions X^A the ADM variables are given by

$$\gamma_{ij} = -X^A_i X_{Aj} \,, \tag{7.5a}$$
$$N_i = -\dot{X}^A X_{Aj} \,, \tag{7.5b}$$
$$N^2 = \tilde{N}_{AB} \dot{X}^A \dot{X}^B \,, \tag{7.5c}$$

where

$$N_{AB} = \eta_{AB} + \gamma^{ij} X_{Ai} X_{Bj} \,. \tag{7.6}$$

The matrix N_{AB} posses two interesting properties

$$N_{AB} X^B_i \equiv 0 \,, \tag{7.7a}$$

and therefore

$$N_{AB} N^B_C = N_{AC} \,. \tag{7.7b}$$

It can be furthermore shown that X^A_i are the only null eigenvectors of N_{AB}.

8. 1+3 Splitting of the Hilbert Lagrangian

From a field theoretical point of view the action (2.43) shows quite an interesting structure since it is a second-order constrained system.

Let us start by observing that the canonical momenta are given by

$$\pi_A(\mathcal{L}_H) = \dot{\partial}_A \mathcal{L}_H - d_i(\dot{\partial}^i_A \mathcal{L}_H) - (\ddot{\partial}_A \mathcal{L}_H)^{\cdot} = -2\sqrt{-g}\, G^{0\mu} X_{A\mu}\,, \tag{8.1a}$$

$$N_A(\mathcal{L}_H) = \ddot{\partial}_A \mathcal{L}_H = 2\sqrt{-g}\, N_{AB} G^{00ij} X^B_{ij} = 2N^{-1}\sqrt{-\gamma}\, N_{AB}\gamma^{ij} X^B_{ij}\,. \tag{8.1b}$$

The last expression for $N_A(\mathcal{L}_H)$ is obtained thanks to the use of ADM variables.

Since no second-order time derivatives are involved in (8.1b) we have the primary constraints

$$\xi_A(\mathcal{L}_H) = N_A(\mathcal{L}_H) - 2N^{-1}\sqrt{-\gamma}\, N_{AB}\gamma^{ij} X^B_{ij} \approx 0\,. \tag{8.2}$$

The Poisson bracket of these constraints is given by

$$\{\xi_A, \xi_B\} \approx 0\,, \tag{8.3}$$

such that all the primary constraints are first-class. We must therefore consider the primary Hamiltonian

$$H_1 = H_C + \int_\Sigma \lambda^A \xi_A d^3x\,. \tag{8.4}$$

It must be observed however that $H_C \equiv 0$.

Let us remind the reader that in this case the Poisson bracket is defined by

$$\{F, G\} = \int_{\Sigma(t)} \Big[\frac{\delta^x \mathcal{F}(z)}{\delta\phi^A(z)}\frac{\delta^x \mathcal{G}(z)}{\delta\pi_A(z)} + \frac{\delta^x \mathcal{F}(z)}{\delta\dot{\phi}^A(z)}\frac{\delta^x \mathcal{G}(z)}{\delta N_A(z)} - (\mathcal{F} \leftrightarrow \mathcal{G})\Big] d\Sigma(z)\,. \tag{8.5}$$

However, the Poisson bracket of two quantities in two different points of the space-like sections, as is the case for the constraints, is defined by

$$\begin{aligned}
&\{\mathcal{F}_1(x), \mathcal{F}_2(y)\} \\
&= \int_\Sigma \Bigg[\frac{\delta^{(3)}}{\delta\phi^A(z)}\Big(\delta^{(3)}(z-x)\mathcal{F}_1(z)\Big)\frac{\delta^{(3)}}{\delta\pi_A(z)}\Big(\delta^{(3)}(z-y)\mathcal{F}_2(z)\Big) \\
&\quad + \frac{\delta^{(3)}}{\delta\dot{\phi}^A(z)}\Big(\delta^{(3)}(z-x)\mathcal{F}_1(z)\Big)\frac{\delta^{(3)}}{\delta N_A(z)}\Big(\delta^{(3)}(z-y)\mathcal{F}_2(z)\Big) \\
&\quad - \Big(\delta^{(3)}(z-x)\mathcal{F}_1(z) \leftrightarrow \delta^{(3)}(z-y)\mathcal{F}_2(z)\Big)\Bigg] d^3z\,.
\end{aligned} \tag{8.6}$$

The consistency conditions are

$$\dot{\xi}_A = \{\xi_A, H_1\} \approx 0\,, \tag{8.7}$$

but they are weakly satisfied as a consequence of Eq. (8.3).

The next step is therefore to fix the gauge. A convenient set of gauge fixing conditions is

$$\chi_A(\mathcal{L}_H) = \pi_A(\mathcal{L}_H) + 2\sqrt{-g}\,G^{0\mu}X_{A\mu} \approx 0\,. \tag{8.8}$$

In fact one can now verify that

$$\text{rank}\begin{vmatrix} \{\xi,\xi\} & \{\xi,\chi\} \\ \{\chi,\xi\} & \{\chi,\chi\} \end{vmatrix} = n(n-1)\,. \tag{8.9}$$

Therefore we still need $2n$ gauge fixing conditions.

Before going to fix the gauge let us observe however the existence of lapse and shift constraints. As a consequence of this the missing $2n$ gauge fixing conditions are only coordinate conditions.

Let us now observe that

$$\eta^{AB}N_{AC}N_{BD} = N_{CD}\,. \tag{8.10}$$

Therefore

$$\eta^{AB}N_A N_B = 4N^{-2}\gamma M_{AB}\gamma^{ij}X^A_{ij}\gamma^{kl}X^B_{kl}\,. \tag{8.11}$$

The lapse and shift constraints are given by

$$\mathcal{H}_\perp(\mathcal{L}_H) = \eta^{AB}\pi_A\pi_B - \gamma \approx 0\,, \tag{8.12a}$$

$$\mathcal{H}_i(\mathcal{L}_H) = X^A_i\pi_A + \dot{X}^A_i N_A \approx 0\,. \tag{8.12b}$$

One can now verify that the algebra of constraints is given by

$$\begin{aligned}\{\mathcal{H}_\perp(x), \mathcal{H}_\perp(y)\} &= 4\gamma(y)\mathcal{H}_i(y)\gamma^{ij}(y)\frac{d}{dy^j}\delta^{(3)}(x-y) \\ &\quad - 4\gamma(x)\mathcal{H}_i(x)\gamma^{ij}(x)\frac{d}{dx^j}\delta^{(3)}(y-x)\,,\end{aligned} \tag{8.13a}$$

$$\{\mathcal{H}_\perp(x), \mathcal{H}_i(y)\} = \mathcal{H}_\perp(x)\frac{d}{dx^j}\delta^{(3)}(y-x) - \mathcal{H}_\perp(y)\frac{d}{dy^j}\delta^{(3)}(x-y)\,, \tag{8.13b}$$

$$\{\mathcal{H}_i(x), \mathcal{H}_j(y)\} = \mathcal{H}_j(x)\frac{d}{dx^i}\delta^{(3)}(y-x) - \mathcal{H}_i(y)\frac{d}{dy^j}\delta^{(3)}(x-y)\,, \tag{8.13c}$$

9. 1+3 Splitting of the Hilbert Lagrangian in the Normal Gauge

In this case the momenta reduce to

$$\pi_a(\mathcal{L}_{H,G}) = -2\sqrt{-g}\, G^{0\mu}\phi_{a\mu}\,, \tag{9.1a}$$
$$N_a(\mathcal{L}_{H,G}) = 2\sqrt{-g}\, N_{ab}G^{00ij}\,\phi^b_{ij}\,. \tag{9.1b}$$

The primary constraints are

$$\xi_a(\mathcal{L}_{H,G}) = N_a(\mathcal{L}_{H,G}) - 2\sqrt{-g}\, N_{ab}G^{00ij}\,\phi^b_{ij} \approx 0\,. \tag{9.2}$$

Now the primary Hamiltonian is

$$H_1 = H_C + \int_\Sigma \lambda^a \xi_a d^3x\,. \tag{9.3}$$

It must now be observed that $H_C = G_{00}$.

Now the consistency conditions

$$\dot{\xi}_a = \{\xi_a, H_1\} \approx 0\,, \tag{9.4}$$

leads directly to the secondary constraints

$$\chi_a(\mathcal{L}_{H,G}) = \pi_a(\mathcal{L}_{H,G}) + 2\sqrt{-g}\, G^{0\mu}\phi_{a\mu} \approx 0\,. \tag{9.5}$$

One can now verify that

$$\det \begin{vmatrix} \{\xi,\xi\} & \{\xi,\chi\} \\ \{\chi,\xi\} & \{\chi,\chi\} \end{vmatrix} \neq 0\,. \tag{9.6}$$

Therefore the extended set of constraints $\{\xi,\chi\}$ is of second-class. One can define Dirac brackets and a reduced phase space.

More detailed results will appear in our forthcoming work [76].

Conclusions

We have looked at the role of minimal and rigid surfaces in theoretical physics. We obtained a natural parametrisation of embedded surfaces which completely fixes the gauge for minimal and rigid surfaces which turns out to be a covariant generalisation of the harmonic one.

We have furthermore seen that the use of extrinsic Riemannian geometry in gravitation allows the formulation of a consistent field theory for the gravitational interaction showing several advantages over the conventional theory.

References

1. K. F. Gauss, *Disquisitiones generales circa superficies curvas*, 1827; English translation: *General Investigations of Curved Surfaces of 1827 and 1825*, by C. J. Morehead and A. M. Hiltebeitel, Raven Press, New York, 1965.
2. B. Riemann, *Uber die Hypotesen welche der Geometrie zu Grunde liegen*, Gott. Abh. **13** (1868) 1.
3. L. Schläfli, *Nota alla memoria del Sig. Beltrami, "Sugli spazii di curvatura constante,"* Ann. di Mat. Pura et Appl. (2nd ser.) **5** (1873) 170.
4. M. Janet, *Sur la possibilite du plonger un espace Riemannien donne dans un espace Euclidien*, Ann. Soc. Polon. Math. **5** (1926) 38.
5. E. Cartan, *Sur la possibilite du plonger un espace riemannien donne dans espace euclidien*, Ann. Soc. Polon. Math. **6** (1927) 1.
6. C. Burstin, *Ein Beitrag zum Problem der Einbettung der Riemannschen Raume in Euklididischen Raumen*, Rec. Math. Moscou (Mat. Sbornik) **38** (1931).
7. K. Leichtweiss, *Das problem von Cauchy in der mehrdimensionalen Differentialgeometrie I*, Math. Ann. **130** (1956) 442.
8. A. Friedman, *Local isometric embedding of Riemannian manifolds with indefinite metric*, J. Math. Mech. **10** (1961) 625.
9. J. Nash, *The imbedding problem for Riemannian manifolds*, Ann. Math. **63** (1956) 20.
10. C. J. Clarke, *On the isometric global embedding of pseudo-Riemannian manifold*, Proc. Roy. Soc. London **A314** (1970) 417.
11. R. E. Greene, *Isometric embedding of Riemannian and pseudo-Riemannian manifolds*, Memoirs Am. Math. Soc. **97** (1970).
12. L. P. Eisenhart, *Riemannian Geometry*, Princeton Univ. Press, Princeton, New Jersey, 1926.
13. Y. Choquet-Bruhat, C. DeWitt-Morette with M. Dillard-Bleick, *Analysis, Manifolds and Physics*, North-Holland, Amsterdam, 1982.
14. J. Rosen, *Embedding of various relativistic Riemannian spaces in pseudo-Euclidean spaces*, Rev. Mod. Phys. **37** (1965) 204.
15. C. D. Collinson, *Embedding of the plane-fronted waves and other space-times*, J. Math. Phys. **9** (1968) 403.

16. A. Z. Petrov, *Recent Developments in General Relativity*, Pergamon Press, London, 1962.
17. Y. K. Gupta, and P. Goel, *Class II analogue of T. Y. Thomas's theorem and different types of embeddings of static spherically symmetric space-times*, GRG **6** (1975) 499.
18. H. Goenner, *On the interdependence of the Gauss-Codazzi-Ricci equations of local isometric embedding*, GRG **8** (1977) 139.
19. R. Blum, Can. J. Math. **7** (1955) 445.
20. T. Y. Thomas, *Riemannian spaces of class one and their characterisation*, Acta Math. **67** (1936) 169.
21. A. Friedman, *Isometric embedding of Riemannian manifolds into Euclidean spaces*, Rev. Mod. Phys. **37** (1965) 201.
22. J. A. Schouten, *Ricci-Calculus*, Springer-Verlag, Berlin, 1954.
23. N. A. Rosenson, *O Riemannowich prostranstwach klassa I*, part I, Izvest. Akad. Nauk SSSR **4** (1940) 181.
24. N. A. Rosenson, *O Riemannowich prostranstwach klassa I*, part II, Izvest. Akad. Nauk SSSR **5** (1941) 325.
25. N. A. Rosenson, *O Riemannowich prostranstwach klassa I*, part III, Izvest. Akad. Nauk SSSR **7** (1943) 253.
26. B. O'Neill, *An algebraic criterion for immersion* Pacific J. Math. **9** (1959) 1239.
27. E. Kasner, *The impossibility of Einstein fields immersed in flat space of five dimensions*, Am. J. Math. **43** (1921) 126.
28. J. A. Schouten and D. J. Struik, *On some properties of general manifolds relating to Einstein's theory of gravitation*, Am. J. Math. **34** (1921) 223.
29. P. Szekeres, *Embedding properties of general relativistic manifolds*, Il Nuovo Cimento **43** (1966) 1062.
30. H. Stephani, *Uber Losungen der Einsteinschen Feldgleichungen die sich in einen funfdimensionalen flache Raum einbetten lassen*, Commun. Math. Phys. **4** (1967) 137.
31. H. Stephani, *Konform flache gravitations felder*, Commun. Math. Phys. **5** (1967) 337.
32. M. Kohler and K. L. Chao, *Zentralsymmetische statische Schwerefelder mit Raumen der Klasse I*, Z. Naturforsch. **20a** (1965) 1537.
34. D. Kramer, G. Neugebauer and H. Stephani, *Konstruktion und Charakterisierung von Gravitationsfeldern*, Forschr. Physik **20** (1972) 1.
35. A. Barnes, *Space-times of embedding class one in general relativity*, GRG **5** (1966) 147.

36. D. Kramer, H. Stephani, E. Herlt and M. MacCallum, *Exact solutions of Einstein's field equations*, Cambridge Univ. Press, Cambridge, England, 1980.
37. Th. M. Rassias, *Sur la multiplicite du premier bord conjugue d'une hypersurface minimale de R^n, $n > 3$*, C. R. Acad. Sciences Paris **284** (1977) 497.
38. Th. M. Rassias, *Morse theory and Plateau's problem*, Selected Studies: Physics, Astrophysics, Mathematics, History of Science, North-Holland, Amsterdam, 1982.
39. Th. M. Rassias, *Foundations of Global Non-Linear Analysis*, Teubner, Leipzig, 1986.
40. A. D. Sakharov, Dokl. Akad. Nauk SSSR **177** (1967) 70; English translation: Sov. Phys. Dokl. **12** (1968) 1040.
41. A. M. Polyakov, Nucl. Phys. **B268** (1986) 406.
42. F. Alonso and D. Espriu, Nucl. Phys. **B283** (1987) 393.
43. P. A. M. Dirac, *Lectures on Quantum Mechanics*, Yeshiva University, New York, 1964.
44. A. Hanson, T. Regge and C. Teitelboim, *Constrained Hamiltonian Systems*, Academia Nazionale dei Lincei, Rome, 1976.
45. E. C. G. Sudarshan and N. Mukunda, *Classical Dynamics: A Modern Perspective*, Wiley, New York, 1974).
46. K. Sundermeyer, *Constrained Dynamics*, Springer Verlag, Berlin, 1982.
47. J. Krause, *Embedding approach to a restricted covariance group in general relativity*, Il Nuovo Cimento **B18** (1973) 302.
48. J. Krause, *The de Donder coordinate condition and minimal class -1 spacetime*, J. Math. Phys. **16** (1975) 1090.
49. J. Krause, *Structure of the Einstein tensor for class-1 embedded spacetime*, Il Nuovo Cimento **B32** (1976) 381.
50. V. Fock, *The Theory of Space, Time and Gravitation*, Pergamon Press, New York, 1959.
51. V. Tapia, *The geometrical meaning of the harmonic gauge for minimal surfaces*, Il Nuovo Cimento **B103** (1989) 441.
52. V. Tapia, *Gravitation a la string, Class. Quantum Grav.* **6** (1989) L49.
53. A. A. Logunov and Yu. M. Loskutov, *Fiz. Elem. Chastits At. Yadra* **18** (1987) 429; English translation: Sov. J. Part. Nucl. **18** (1987) 179.
54. J. H. Schwarz, Phys. Reports **89** (1982) 223.
55. M. B. Green, J. H. Schwarz and E. Witten, *Superstrings*, Cambridge Univ. Press, Cambridge, 1987.
56. Y. Nambu, *Lectures at the Copenhagen Symposium*, 1970.

57. T. Goto, Prog. Theor. Phys. **46** (1971) 1560.
58. A. M. Polyakov, Phys. Lett. **B103** (1981) 207.
59. A. M. Polyakov, Phys. Lett. **B103** (1981) 211.
60. L. Brink, P. Di Vecchia and P. Howe, Phys. Lett. **B65** (1976) 471.
61. S. Deser and B. Zumino, Phys. Lett. **B65** (1976) 369.
62. K. Fujikawa, Phys. Rev. **D25** (1982) 2584.
63. L. Baulieu, W. Siegel and B. Zwiebach, Nucl. Phys. **B287** (1987) 93.
64. W. Siegel and B. Zwiebach, Nucl. Phys. **B288** (1987) 332.
65. D. W. Düsedau, Phys. Lett. **B188** (1987) 51.
66. L. Baulieu and A. Bilal, Phys. Lett. **B192** (1987) 339.
67. A. Rebhan and U. Kraemmer, Phys. Lett. **B196** (1987) 477.
68. A. Einstein, *Zur Allgemeinen Relativitatstheorie*, Preuss. Akad. Wiss. Berlin, Sitzber. (1915) 778.
69. E. Kasner, *Finite representation of the solar gravitational field in flat space of six dimensions*, Amer. J. Math. **43** (1921) 130.
70. T. Regge and C. Teitelboim, *General relativity a la string: a progress report*, Proceedings of the First Marcel Grossman Meeting, ed. R. Ruffini, North-Holland, Amsterdam, 1977.
71. S. Deser, F. A. E. Pirani and D. C. Robinson, *New embedding model of general relativity*, Phys. Rev. **D14** (1976) 3301.
72. M. D. Maia, *On Kaluza-Klein relativity*, GRG **18** (1986) 695, and references therein.
73. M. Pavšič, *Einstein's gravity from a first order Lagrangian in an embedding space*, Phys. Lett. **A116** (1986) 1, and references therein.
74. G. W. Gibbons and D. L. Wiltshire, *Spacetime as a membrane in higher dimensions*, Nucl. Phys. **B287** (1987) 717.
75. V. A. Franke and V. Tapia, *The ADM Lagrangian in extrinsic gravity*, Il Nuovo Cimento B (1991), to appear.
76. V. Tapia, *the Canonical Structure of Extrinsic Gravity*, Il Nuovo Cimento B (1991), to appear.
77. M. Ferraris and M. Francaviglia, J. Math. Phys. **26** (1985) 1243.
78. P. A. M. Dirac, Phys. Rev. **73** (1948) 1092.
79. P. A. M. Dirac, Canad. J. Math. **3** (1951) 1.
80. P. G. Bergmann and J. H. M. Brunings, Rev. Mod. Phys. **21** (1949) 480.
81. H. Rund, *Invariant Theory of Variational Problems of Subspaces of a Riemannian Manifold*, Vandenhoock Rupret, Göttingen, 1971.
82. V. Tapia, *Second-order field theory and non-standard Lagrangians*, Il Nuovo Cimento **B101** (1988) 183.

83. L. Landau and E. M. Lifshitz, *Teoriya Polya*, Nauka, Moskva, 1973; English translation: The Classical Theory of Fields, Pergamon, Oxford, 1979.
84. R. Arnowitt, S. Deser and C. W. Misner, *The dynamics of general relativity*, Gravitation: An Introduction to Current Research, ed. L. Witten, Wiley, New York, 1962.

The Problem of Plateau (pp. 327–335)
ed. Th. M. Rassias

ON THE NUMBER OF RIGID MINIMAL IMMERSIONS BETWEEN SPHERES

GABOR TOTH

Department of Mathematics, Rutgers University
Camden, New Jersey, 08102 U.S.A.

Abstract

We show, by a Baire Category argument applied to the parameter space of all minimal immersions between spheres, that linearly rigid minimal immersions abound for sufficiently high degree.

1 Introduction and Preliminaries

It is well known that a (full) homothetic immersion $f : S^m \to S^n$ between Euclidean spheres is minimal iff the components of f are (linearly independent) spherical harmonics on S^m of a fixed order k, or equivalently,

$$\Delta f = \lambda_k f,$$

where $\lambda_k = k(k+m-1)$ is the kth eigenvalue of the Laplace-Beltrami operator Δ on S^m. In this case, k is said to be the (algebraic) degree of f. The metric induced by f on S^m then has (constant) curvature m/λ_k.

The universal example is given by the *standard minimal immersion*

$$f_{\lambda_k} : S^m \to S^{n(\lambda_k)}$$

whose components comprise an orthonormal basis $\{f^i_{\lambda_k}\}_{i=0}^{n(\lambda_k)}$ in the space $\mathcal{H}^k_{S^m}$ of spherical harmonics on S^m of order k endowed with the normalized L^2-scalar product

$$\langle \mu, \mu' \rangle = \frac{n(\lambda_k)+1}{\mathrm{vol}\,(S^m)} \int_{S^m} \mu\mu' \cdot v_{S^m}, \ \mu, \mu' \in \mathcal{H}^k_{S^m},$$

where

$$n(\lambda_k)+1=\dim \mathcal{H}^k_{S^m}=(m+2k-1)\frac{(m+k-2)!}{k!(m-1)!}.$$

Here f_{λ_k} is universal in the sense that, for any full minimal immersion $f: S^m \to S^n$ of degree k, we have

$$f=A\cdot f_{\lambda_k}$$

for some (uniquely determined) $(n+1)\times(n(\lambda_k)+1)$-matrix of maximal rank. The Do Carmo-Wallach parametrization associates to f the symmetric matrix

$$\langle f\rangle_{\lambda_k}=A^{\top}A-I\in S^2(\mathbf{R}^{n(\lambda_k)+1}),$$

where I is the identity. Two full minimal immersions $f': S^m \to S^n$ and $f'': S^m \to S^n$ of degree k are said to be *range* (resp. *domain*) *equivalent* if there exists $U\in O(n+1)$ (resp. $a\in SO(m+1)$) such that $f''=U\cdot f'$ (resp. $f''=f'\circ a$). If $f''=U\cdot f'\circ a$ for some $U\in O(n+1)$ and $a\in SO(m+1)$ then f' and f'' are said to be *equivalent*. Clealy, $\langle f\rangle_{\lambda_k}$ depends only on the range equivalence class of f. As in [2,6], we obtain that the space of range equivalence classes of full minimal immersions $f: S^m\to S^n$, of degree k, (hence $n\le n(\lambda_k)$), is parametrized by the compact convex body

$$\mathcal{L}_{\lambda_k}=\{C\in\mathcal{F}_{\lambda_k}|C+I\ge 0\},\tag{1}$$

where the linear subspace $\mathcal{F}_{\lambda_k}\subset S^2(\mathbf{R}^{n(\lambda_k)+1})$ is the orthogonal complement of the set of all projections

$$\text{proj}\,[(f_{\lambda_k})_*(X)]=(f_{\lambda_k})_*(X)\cdot(f_{\lambda_k})_*(X)\in S^2(\mathbf{R}^{n(\lambda_k)+1}),\ X\in T(S^m).\tag{2}$$

(Here, $(f_{\lambda_k})_*(X)$ is shifted to the origin of $\mathbf{R}^{n(\lambda_k)+1}$ and $\cdot$ stands for the symmetric tensor product.) By birth, $f_{\lambda_k}: S^m \to S^{n(\lambda_k)}$ is equivariant with respect to the homomorphism $\rho_{\lambda_k}: SO(m+1)\to SO(n(\lambda_k)+1)$ that is just the orthogonal $SO(m+1)$-module structure on $\mathcal{H}^k_{S^m}\cong\mathbf{R}^{n(\lambda_k)+1}$, where the isomorphism is given by the orthonormal basis chosen for f_{λ_k}. Taking the symmetric square of the representation given by ρ_{λ_k}, we obtain that $\mathcal{F}_{\lambda_k}$ is an $SO(m+1)$-submodule of $S^2(\mathbf{R}^{n(\lambda_k)+1})$ with $\mathcal{L}_{\lambda_k}\subset\mathcal{F}_{\lambda_k}$ an $SO(m+1)$-invariant subspace. In fact,

$$a\cdot\langle f\rangle_{\lambda_k}=\langle f\circ a^{-1}\rangle_{\lambda_k},\ a\in SO(m+1).\tag{3}$$

In particular, the $SO(m+1)$-orbit of $\langle f\rangle_{\lambda_k}$ corresponds to those full minimal immersions that are (domain) equivalent to f. A full (minimal) immersion

$f : S^m \to S^n$ of degree k is said to be *linearly rigid* [2,6] if whenever A is an $(n+1)\times(n+1)$-matrix that satisfies $A(f(S^m)) \subset S^n$ and $A \cdot f : S^m \to S^n$ is a homothetic (minimal) immersion (of degree k) then $A \in O(n+1)$. Linear rigidity, when applied to f_{λ_k}, is clearly equivalent to $\mathcal{L}_{\lambda_k} = \mathcal{F}_{\lambda_k} = \{0\}$. For $m = 2$, linear rigidity of f_{λ_k} was proved by Calabi in [1]. Do Carmo and Wallach [2,6] showed that f_{λ_k} is linearly rigid for $k \leq 3$. Finally, they also proved that, for $m \geq 3$ and $k \geq 4$, f_{λ_k} is not linearly rigid. More precisely, for $m \geq 3$, we have

$$\mathcal{F}_{\lambda_k} \otimes_{\mathbf{R}} \mathbf{C} \supset S^2(\mathcal{H}^k_{S^m})/\{\sum_{j=0}^{k} \mathcal{H}^{2j}_{S^m} \oplus \sum_{j=1}^{k-1} V^{(2j,2,0,\dots,0)}_{SO(m+1)}\} = \sum_{(a,b)\in\Delta,\, a,b \text{ even}} V^{(a,b,0,\dots,0)}_{SO(m+1)}$$

as $SO(m+1)$-modules, where $\Delta \subset \mathbf{R}^2$ is the closed triangular domain with vertices $(4,4)$, (k,k) and $(2k-4,4)$. In these formulas we use standard terminology in representation theory, namely, $V^{\rho}_{SO(m+1)}$ is the complex irreducible $SO(m+1)$-module with highest weight ρ (whose components are with respect to the standard maximal torus) and, for the moment, the spherical harmonics are complex valued. In particular, by the Weyl dimension formula, we obtain

$$\begin{aligned} \dim \mathcal{L}_{\lambda_k} &= \dim \mathcal{F}_{\lambda_k} \geq \frac{1}{2}(n(\lambda_k)+1)(n(\lambda_k)+2) - \sum_{j=0}^{k}(n(\lambda_{2j})+1) \\ &- \frac{1}{2}(m+1)(m-2)\sum_{j=1}^{k-1}\frac{2j-1}{2j+1}\frac{2j+m}{2j+m-2}(n(\lambda_{2j})+1). \qquad (4) \end{aligned}$$

Remark As shown by Muto [4] the lower estimate is sharp for $m = 3$ and $k = 4$, i.e., in this case, $\dim \mathcal{L}_{\lambda_2} = 18$.

The purpose of this paper is to show that linearly rigid minimal immersions abound.

Theorem 1 *Let $m \geq 3$ and $k \geq 4$ and assume that*

$$\begin{aligned} (n(\lambda_k)+1)(n(\lambda_k)+2) &> (n(\lambda_k)-m)(m(m+1)+2) + 2\sum_{j=0}^{k}(n(\lambda_{2j})+1) \\ &+ (m+1)(m-2)\sum_{j=1}^{k-1}\frac{2j-1}{2j+1}\frac{2j+m}{2j+m-2}(n(\lambda_{2j})+1) \qquad (5) \end{aligned}$$

Then there exist $\aleph_1$ mutually inequivalent full linearly rigid minimal immersions $f : S^m \to S^n$ of degree k. Moreover, for each $m \geq 3$, there exists $k(m) \geq 4$ such that (5) holds for $k \geq k(m)$.

Remarks 1. The last assertion of Theorem 1 is clear. If fact, for fixed $m \geq 3$, all the terms in (5) are polynomials in k with positive coefficients. The left hand side of (5) is of degree $2(m-1)$ and, on the right hand side, the first term is of degree $m-1$ while the second and third are of degree m. In particular, as easy computation shows, $k(3) = 7$.

2. Similar result can be obtained for harmonic λ_k-eigenmaps $f : S^m \to S^n$ (omitting the condition that the maps are homothetic immersions [3]). By contrast, note that, for $m = 3$ and $k = 2$, up to equivalence, the only full linearly rigid λ_2-eigenmap is the Hopf map $f : S^3 \to S^2$ (and its 'dual') [5].

3. For further explicit examples of λ_2-eigenmaps consider the case $m = 4$ and $k = 2$. Let S^4 be the unit sphere in $\mathbf{C}^2 \times \mathbf{R}$ with coordinates $z, w \in \mathbf{C}$ and $t \in \mathbf{R}$. Define $f : S^4 \to S^7$ and $g : S^4 \to S^4$ by

$$f(z,w,t) = (\frac{1}{4}(|z|^2 + |w|^2) - t^2, \frac{\sqrt{15}}{4}(|z|^2 - |w|^2), \frac{\sqrt{15}}{2}zw, \sqrt{\frac{5}{2}}zt, \sqrt{\frac{5}{2}}wt)$$

and

$$g(z,w,t) = (\frac{1}{2}|z|^2 - |w|^2 + t^2, \frac{\sqrt{3}}{2}z^2 - 2wt, \sqrt{3}\bar{z}w + zt).$$

Computation shows that f and g are both linearly rigid. (Note that g is the gradient of a cubic isoparametric function on S^4 [3].)

To prove Theorem 1, in §2, we introduce an $SO(m+1)$-saturation on $\mathcal{L}_{\lambda_k}$ with the property that the one point cells correspond to the linearly rigid immersions. In §3 we then use an inductive argument with respect to the dimension of the cells of the saturation of $\mathcal{L}_{\lambda_k}$ to get an upper bound for $\dim \mathcal{L}_{\lambda_k}$ provided that the cardinality of the saturation modulo $SO(m+1)$ is $\leq \aleph_0$. This, for k large, will contradict to the lower estimate of $\dim \mathcal{L}_{\lambda_k}$ given in (4).

2 The fine structure of the parameter space

Let $f : S^m \to S^n$ and $f' : S^m \to S^{n'}$ be full minimal immersions of degree k and k', respectively. f' is said to be *derived* from f, written as $f' \leftarrow f$, if there exists an $(n'+1) \times (n+1)$-matrix A such that $f' = A \cdot f$. In this case $k = k'$.

Given a full minimal immersion $f : S^m \to S^n$ of degree k, we define

$$\mathcal{L}_f = \{C' \in \mathcal{F}_f | C' + I \geq 0\}, \tag{6}$$

where the linear subspace $\mathcal{F}_f \subset S^2(\mathbf{R}^{n+1})$ is the orthogonal complement of the set of all projections

$$\operatorname{proj}[f_*(X)] \in S^2(\mathbf{R}^{n+1}),\ X \in T(S^m).$$

Clearly, we have $\mathcal{L}_{f_{\lambda_k}} = \mathcal{L}_{\lambda_k}$ and $\mathcal{F}_{f_{\lambda_k}} = \mathcal{F}_{\lambda_k}$ and the argument of Do Carmo and Wallach applies yielding that $\mathcal{L}_f \subset \mathcal{F}_f$ is a compact convex body that parametrizes the range equivalence classes of full minimal immersions $f' : S^m \to S^{n'}$ that are derived from f. Let

$$\iota : S^2(\mathbf{R}^{n+1}) \to S^2(\mathbf{R}^{n(\lambda_k)+1})$$

be the affine map defined by

$$\iota(C') = A^\top C' A + \langle f \rangle_{\lambda_k} = A^\top (C' + I)A - I,\ C' \in S^2(\mathbf{R}^{n+1}), \tag{7}$$

where $f = A \cdot f_{\lambda_k}$.

Proposition 1 *ι is injective and maps $\mathcal{F}_f$ into $\mathcal{F}_{\lambda_k}$. Moreover, we have*

$$\iota(\mathcal{L}_f) = \iota(\mathcal{F}_f) \cap \mathcal{L}_{\lambda_k}.$$

Proof. f is full so that A is of maximal rank. Hence $A^\top$ has zero kernel and injectivity of ι follows. Given $C' \in \mathcal{F}_f$, for $X \in T(S^m)$, we have

$$\begin{aligned}
\langle \iota(C'), \operatorname{proj}[(f_{\lambda_k})_*(X)]\rangle &= \langle \iota(C')(f_{\lambda_k})_*(X), (f_{\lambda_k})_*(X)\rangle \\
&= \langle (C' + I) f_*(X), f_*(X)\rangle - \langle (f_{\lambda_k})_*(X), (f_{\lambda_k})_*(X)\rangle \\
&= \langle C', \operatorname{proj}[f_*(X)]\rangle,
\end{aligned}$$

where the last equality is because f and f_{λ_k} have the same homothety constant. The second assertion follows. Finally, the third is obtained by comparing (1) and (6) via (7).

For a full minimal immersion $f : S^m \to S^n$, we define

$$I_f = \iota(\mathcal{L}_f^\circ) \subset \mathcal{L}_{\lambda_k},$$

where the circle stands for the interior. Clearly, I_f is convex and open in $\iota(\mathcal{F}_f)$ and contains $\langle f\rangle_{\lambda_k}$. Then I_f is said to be the *cell* associated to f. Its points correspond to those full minimal immersions $f' : S^m \to S^n$ of degree k that are derived from f, i.e. $f' = A \cdot f$ with A invertible. The cells I_f, for the various f, give rise to a decomposition of $\mathcal{L}_{\lambda_k}$ into mutually disjoint convex sets that comprise the natural saturation $\mathcal{I}_{\lambda_k} = \{I_f | f \leftharpoondown f_{\lambda_k}\}$ of $\mathcal{L}_{\lambda_k}$. One of the most important property of $\mathcal{I}_{\lambda_k}$ is that when passing to the boundary of a cell the dimension of the range of the corresponding minimal immersions strictly decreases. In particular, $I_{f_{\lambda_k}} = \mathcal{L}^{\circ}_{\lambda_k}$ so that the natural saturation is of interest only on the boundary $\partial\mathcal{L}_{\lambda_k}$.

The action of $SO(m+1)$ respects $\mathcal{I}_{\lambda_k}$, in fact, by (3), we have

$$a \cdot I_f = I_{f o a^{-1}}, \quad a \in SO(m+1).$$

We obtain that $\mathcal{I}_{\lambda_k}$ is an $SO(m+1)$-saturation of $\mathcal{L}_{\lambda_k}$. By construction, a full minimal immersion $f : S^m \to S^n$ of degree k is linearly rigid iff I_f is a one point cell.

Two full minimal immersions $f' : S^m \to S^{n'}$ and $f'' : S^m \to S^{n''}$ of degree k are said to be *geometrically distinct* if, for each $U' \in O(n'+1)$, $U'' \in O(n''+1)$ and $a', a'' \in SO(m+1)$, none of the minimal immersions

$$U' \cdot f' \circ a' \text{ and } U'' \cdot f'' \circ a''$$

can be derived from each other. In terms of $\mathcal{I}_{\lambda_k}$, this holds iff none of the orbits

$$SO(m+1) \cdot I_{f'} \text{ and } SO(m+1) \cdot I_{f''}$$

is contained in the other. In the special case when f' and f'' are both linearly rigid, f' and f'' are geometrically distinct iff they are inequivalent. Theorem 1 will therefore be proved if we show the following:

Theorem 2 *Assume that*

$$\dim \mathcal{L}_{\lambda_k} > (n(\lambda_k) - m)(\dim SO(m+1) + 1). \tag{8}$$

Then the cardinality of the orbit space $\mathcal{I}_{\lambda_k}/SO(m+1)$ is $\aleph_1$. In particular, there exist $\aleph_1$ geometrically distinct full minimal immersions $f : S^m \to S^n$ of degree k. Moreover, there exist $\aleph_1$ inequivalent full linearly rigid minimal immersions $f : S^m \to S^n$ of degree k.

3 Proof of Theorem 2

Assume that $\mathcal{I}_{\lambda_k}/SO(m+1)$ is countable. Since the interior of $\mathcal{L}_{\lambda_k}$ is a cell the set of $SO(m+1)$-orbits of cells on $\partial\mathcal{L}_{\lambda_k}$ is also countable. By the Baire Category Theorem, at least one $SO(m+1)$-orbit of a cell has nonempty interior. Let $\langle f_1\rangle_{\lambda_k}$ be an interior point. Then $f_1 : S^m \to S^{n_1}$ is a full minimal immersion of degree k with $n_1 \le n(\lambda_k) - 1$. Moreover, we have

$$\dim \mathcal{L}_{\lambda_k} = \dim(SO(m+1)\cdot I_{f_1}) + 1 \le \dim SO(m+1) + 1 + \dim I_{f_1}. \quad (9)$$

We now take the set of $SO(m+1)$-orbits of cells which lie on the boundary $\partial I_{f_1} = \bar{I}_{f_1} \setminus I_{f_1}$. The intersections of these $SO(m+1)$-orbits with ∂I_{f_1} give a partition of ∂I_{f_1} into countably many subsets. Again by the Baire Category Theorem at least one $SO(m+1)$-orbit of a cell intersects ∂I_{f_1} in a set with nonempty interior in ∂I_{f_1}. Let $\langle f_2\rangle_{\lambda_k}$ be an interior point. Then $f_2 : S^m \to S^{n_2}$ is a full minimal immersion of degree k with $n_2 \le n_1 - 1 \le n(\lambda_k) - 2$. Moreover, I_{f_2} is contained in this intersection so that we have

$$\begin{aligned}\dim I_{f_1} &= \dim(SO(m+1)\cdot I_{f_2} \cap I_{f_1}) + 1\\ &\le \dim(SO(m+1)\cdot I_{f_2}) + 1\\ &\le \dim SO(m+1) + 1 + \dim I_{f_2}.\end{aligned}$$

Combining this with (8), we obtain

$$\dim \mathcal{L}_{\lambda_k} \le 2(\dim SO(m+1) + 1) + \dim I_{f_2}$$

and

$$n_2 \le n(\lambda_k) - 2.$$

Repeating this, in the ℓth step we obtain

$$\dim \mathcal{L}_{\lambda_k} \le \ell(\dim SO(m+1) + 1) + \dim I_{f_\ell}$$

and

$$n_\ell \le n(\lambda_k) - \ell.$$

The procedure clearly stops in $n(\lambda_k) - m$ steps yielding

$$\dim \mathcal{L}_{\lambda_k} \le (n(\lambda_k) - m)(\dim SO(m+1) + 1)$$

which contradicts to (8).

To prove the second statement, choose geometrically distinct full minimal immersions $f_\sigma : S^m \to S^{n_\sigma}$, $\sigma \in \Sigma$, where Σ is of cardinality $\aleph_1$. For $\sigma \in \Sigma$, choose a finite set $\Lambda_\sigma \subset \bar{I}_{f_\sigma}(\subset \mathcal{L}_{\lambda_k})$ consisting of points that correspond to linearly rigid full minimal immersions such that the affine span of Λ_k is equal to that of $\bar{I}_{f_\sigma}$. The existence of Λ_σ follows easily by induction with respect to the dimension of the cells comprising $\bar{I}_{f_\sigma}$.

We now claim that, for $\sigma, \sigma' \in \Sigma$, $\sigma \neq \sigma'$, we have $\Lambda_\sigma \neq \Lambda_{\sigma'}$. In fact, $\Lambda_\sigma = \Lambda_{\sigma'}$ iff $I_{f_\sigma} = I_{f_{\sigma'}}$ so that f_σ and $f_{\sigma'}$ are not geometrically distinct; a contradiction. We obtain that the set $\{\Lambda_\sigma | \sigma \in \Sigma\}$ has cardinality $\aleph_1$. Since the set of all finite subsets of a countable set is countable, $\cup_{\sigma\in\Sigma}\Lambda_\sigma$ has cardinality $\aleph_1$. The proof of Theorem 2 is complete.

Remark. The action of $SO(m+1)$ on $\mathcal{L}_{\lambda_k}$ has further interesting properties that are related to the ones used in the proof above. It can be shown, e.g. that the principal isotropy type is finite provided that $\mathcal{L}_{\lambda_k}$ is nontrivial. Furthermore, it can be proved that the orbit of the center of mass of any cell I_f is always transversal (in the weak sense) to I_f. In particular, equality holds in (9), provided that the orbit of the center of mass is principal. For $m = 3$ this is actually the case for the top (5-)dimensional cell on the boundary of the parameter space for λ_2-eigenmaps [5].

References

1. Calabi E., *Minimal immersions of surfaces in Euclidean spheres*, J. Diff. Geom., 1 (1967) 111-125.

2. Do Carmo M. and Wallach N., *Minimal immersions of spheres into spheres*, Ann. of Math., 93 (1971) 43-62.

3. Eells J. and Lemaire L., *A report on harmonic maps*, Bull. London Math. Soc., 10 (1978) 1-68.

4. Muto Y., *The space W_2 of isometric minimal immersions of the three-dimensional sphere into spheres*, Tokyo J. Math., Vol.7, No.2 (1984) 337-358.

5. Toth G., *Classification of quadratic harmonic maps of S^3 into spheres*, Indiana Univ. Math. J., Vol.36, No.2 (1987) 231-239.

6. Wallach N., *Minimal immersions of symmetric spaces into spheres*, in 'Symmetric Spaces', Dekker, New York (1972) 1-40.